AF326867

228
Topics in Current Chemistry

Springer

Berlin
Heidelberg
New York
Hong Kong
London
Milan
Paris
Tokyo

Dendrimers V

Functional and Hyperbranched Building Blocks, Photophysical Properties, Applications in Materials and Life Sciences

Volume Editors: Christoph A. Schalley
Fritz Vögtle

With contributions by
V. Balzani, P. Ceroni, J. Dennig, A. Harada, M. Kawa,
K. Kim, J. W. Lee, M. Maestri, S. Mashiko, H. Mori,
A. Müller, K. Naka, T. Nakahama, J.-F. Nierengarten,
Y. Okuno, K. Onitsuka, A. Otomo, C. Saudan,
K. Sugiura, S. Takahashi, V. Vicinelli, H. Yamaguchi,
S. Yokoyama

Springer

The series *Topics in Current Chemistry* presents critical reviews of the present and future trends in modern chemical research. The scope of coverage includes all areas of chemical science including the interfaces with related disciplines such as biology, medicine and materials science. The goal of each thematic volume is to give the non-specialist reader, whether at the university or in industry, a comprehensive overview of an area where new insights are emerging that are of interest to a larger scientific audience.

As a rule, contributions are specially commissioned. The editors and publishers will, however, always be pleased to receive suggestions and supplementary information. Papers are accepted for *Topics in Current Chemistry* in English.

In references *Topics in Current Chemistry* is abbreviated Top. Curr. Chem. and is cited as a journal.

Springer WWW home page: http://www.springer.de
Visit the TCC home page at http://www.springerlink.com/series/tcc/

ISSN 0340-1022
ISBN 3-540-00669-9
DOI 10.1007/b11215
Springer-Verlag Berlin Heidelberg New York

Library of Congress Catalog Card Number 74-644622

Springer-Verlag Berlin Heidelberg New York
a member of BertelsmannSpringer Science+Business Media GmbH

http://www.springer.de

The use of general descriptive names, registered names, trademarks, etc. in this publication does not imply, even in the absence of a specific statement, that such names are exempt from the relevant protective laws and regulations and therefore free for general use.

Cover design: KünkelLopka, Heidelberg/design & production GmbH, Heidelberg
Typesetting: Fotosatz-Service Köhler GmbH, 97084 Würzburg

02/3020 ra   –   5 4 3 2 1 0 – Printed on acid-free paper

Topics in Current Chemistry
Also Available Electronically

For all customers with a standing order for Topics in Current Chemistry we offer the electronic form via SpringerLink free of charge. Please contact your librarian who can receive a password for free access to the full articles by registration at:

http://www.springerlink.com/orders/index.htm

If you do not have a standing order you can nevertheless browse through the table of contents of the volumes and the abstracts of each article at:

http://www.springerlink.com/series/tcc/

There you will also find information about the

- Editorial Board
- Aims and Scope
- Instructions for Authors

Preface

The pleasingly favourable reception of the tetralogy on dendrimer chemistry in the Topics in Current Chemistry series prompted us to assemble the present follow-up issue "Dendrimers V". This is also justified by the quickly growing field, in particular, because the focus had not been directed towards the exciting applications that dendrimers have found recently. Dendrimer chemistry is experiencing new challenges coming from supramolecular chemistry, nanotechnology, and biochemistry and the new developments emerging from these interdisciplinary marriages of different fields are invaluable for the future of materials and life sciences. This is also reflected in the large number of recent publications in these areas dealing with dendritic molecules.

In this issue, the reader will find contributions from the whole scope of dendrimer chemistry, some of which are shorter as they describe new developments which are currently emerging. We start with the synthesis of hyperbranched acrylates reviewed by Hideharu Mori and Axel Müller who give profound insight into the properties of hyperbranched polymers in solution, the melt and on surfaces. The exciting new results in the field of metallodendrimers form the focus of the next chapter written by Kiyotaka Onitsuka and Shigetoshi Takahashi, which at the same time introduces a "mini-series" of articles dealing with dendrimers containing particular building blocks implementing exactly defined properties which may lead to function. These are porphyrin-containing dendrimers which are highlighted by Ken-ichi Sugiura, dendrimers that bear fullerenes as expertly discussed by Jean-François Nierengarten, and dendrimers with mechanically bound entities incorporated within their scaffold summarized and beautifully categorized by Kimoon Kim and Jae Wook Lee. Several reviews deal with the applications of dendrimers: Kensuke Naka gives an excellent overview on the potential of dendrimers to stimulate and control the crystallization of calcium carbonate, Vincenzo Balzani, Manabu Kawa, Shiyoshi Yokoyama and their coworkers acted as pioneering authors of three chapters containing a firework of applications of dendrimers in photochemistry and their photochemical analysis. Luminescent dendrimers, antenna effects, and the use of optoelectronics are described in depth in these reviews. Finally, the journey through this volume and the chemistry of dendrimers ends with two intriguing biochemistry- and biology-oriented contributions on gene transfection by Jörg Denning and on antibody dendrimers by Hiroyasu Yamaguchi and Akira Harada which demonstrate the enormous potential of dendritic structures and the broad scope this field has meanwhile developed.

It is with great pleasure and enthusiasm that we present this fifth volume of what was originally intended to be a tetralogy. This volume is not only a heavyweight because of its size and the sheer number of contributions (eleven!), but also due to the variety and quality of its scientific contents. Reflecting the quickly growing field of dendrimer chemistry, we also regard it as a timely update to the two „dendrimer bibles" by Fréchet and Tomalia[1] and Newkome et al.[2] which appeared in print earlier. In this respect, we are looking forward to learning about the further progress made beyond the results discussed here at the 3rd International Dendrimer Symposium, September 17th – 20th, 2003, in Berlin.

Bonn, March 2003   Christoph A. Schalley,
                                                                                Fritz Vögtle

[1] JMJ Fréchet, DA Tomalia (2001) Dendrimers and other Dendritic Polymers. Wiley, Chichester, UK.
[2] GR Newkome, CN Moorefield, F Vögtle (2001) Dendrimers and Dendrons. Wiley-VCH, Weinheim, Germany.

Contents

Contents of Volume 197
Dendrimers

Volume Editor: Fritz Vögtle
ISBN 3-540-64412-1

Contents of Volume 210
Dendrimers II
Architecture, Nanostructure and Supramolecular Chemistry

Volume Editor: Fritz Vögtle
ISBN 3-540-67097-1

Contents of Volume 212

Dendrimers III
Design, Dimension, Function

Volume Editor: Fritz Vögtle
ISBN 3-540-67828-X

Contents of Volume 217

Dendrimers IV
Metal Coordination, Self Assembly, Catalysis

Volume Editor: Fritz Vögtle, Christoph A. Schalley
ISBN 3-540-42095-9

Top Curr Chem (2003) 228: 1–37
DOI 10.1007/b11004

Hyperbranched (Meth)acrylates in Solution, Melt, and Grafted From Surfaces

Hideharu Mori · Axel H. E. Müller

Makromolekulare Chemie II, Bayreuther Institut für Makromolekülforschung
and Bayreuther Zentrum für Kolloide und Grenzflächen, Universität Bayreuth,
95440 Bayreuth, Germany
E-mail: Axel.Mueller@uni-bayreuth.de

This review summarizes recent advances in the synthesis and characterization of hyperbranched (meth)acrylates. We will focus on self-condensing vinyl (co)polymerization as an effective method for the synthesis of hyperbranched polymers. Molecular parameters of hyperbranched polymers obtained by self-condensing vinyl (co)polymerization are discussed from a theoretical point of view. Solution properties and melt properties of the resulting hyperbranched poly(meth)acrylates and poly(acrylic acid)s are reviewed. A novel synthetic concept for preparing hyperbranched polymer brushes on planar surfaces and nanoparticles is also described.

Keywords. Hyperbranched polymers, (Meth)acrylates, Self-condensing vinyl polymerization, Controlled polymerization, Surface-grafted hyperbranched polymers, Polymer brushes, Hybrid nanoparticles

Abbreviations

DB   degree of branching
g, g′   contraction factor
γ   comonomer ratio=$[\text{monomer}]_0/[\text{inimer}]_0$
α   Mark–Houwink exponent
GPC   gel permeation chromatography
MALS   multi-angle light scattering
UNICAL   universal calibration
NMR   nuclear magnetic resonance
MW   molecular weight
MWD   molecular weight distribution
SCVP   self-condensing vinyl polymerization
SCVCP   self-condensing vinyl copolymerization
t-BuA   *tert*-butyl acrylate
Pt-BuA   poly(*tert*-butyl acrylate)
MMA   methyl methacrylate
PMMA   poly(methyl methacrylate)
PAA   poly(acrylic acid)
ATRP   atom transfer radical polymerization
GTP   group transfer polymerization

1
Introduction

In the past decade, the field of arborescent polymers (dendrimers, hyperbranched, and highly branched polymers) has been well established with a large variety of synthetic approaches, fundamental studies on structure and properties of these unique materials, and possible applications [1–4]. Dendrimers are monodisperse molecules with well-defined, perfectly branched architectures, made in a multi-step organic synthesis. In contrast, hyperbranched polymers are made in a one-pot polymerization, making them promising candidates for industrial applications where ultimate perfection in structural uniformity is less needed. However, they are less regular in structure and their degree of branching (DB) typically does not exceed 50% of that of dendrimers.

As early as 1952, Flory [5, 6] pointed out that the polycondensation of AB_x-type monomers will result in soluble "highly branched" polymers and he calculated the molecular weight distribution (MWD) and its averages using a statistical derivation. Ill-defined branched polycondensates were reported even earlier [7, 8]. In 1972, Baker et al. reported the polycondensation of polyhydroxymonocarboxylic acids, $(OH)_nR\text{-}COOH$, where n is an integer from two to six [9]. In 1982, Kricheldorf et al. [10] published the cocondensation of AB and AB_2 monomers to form branched polyesters. However, only after Kim and Webster published the synthesis of pure "hyperbranched" polyarylenes from an AB_2 monomer in 1988 [11–13], this class of polymers became a topic of intensive research by many groups. A multitude of hyperbranched polymers synthesized via polycondensation of AB_2 monomers have been reported, and many reviews have been published [1, 2, 14–16].

The interest in hyperbranched polymers arises from the fact that they combine some features of dendrimers, for example, an increasing number of end groups and a compact structure in solution, with the ease of preparation of linear polymers by means of a one-pot reaction. However, the polydispersities are usually high and their structures are less regular than those of dendrimers. Another important advantage is the extension of the concept of hyperbranched polymers towards vinyl monomers and chain growth processes, which opens unexpected possibilities.

Several strategies for the preparation of hyperbranched polymers are currently employed. The polymerization reactions are classified into three categories [1, 4, 17]: (1) step-growth polycondensation of AB_x monomers; (2) chain-growth self-condensing vinyl polymerization (SCVP) of AB* initiator–monomers ("inimers"); (3) chain-growth self-condensing ring-opening polymerization of

Table 1. Classification of different types of monomers for synthesis of hyperbranched polymers

Step growth AB_2

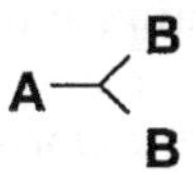

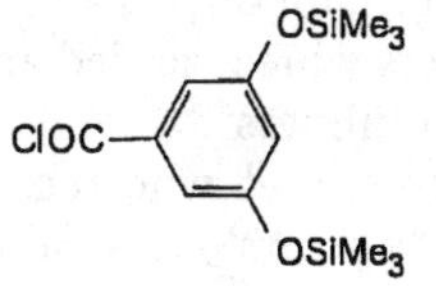

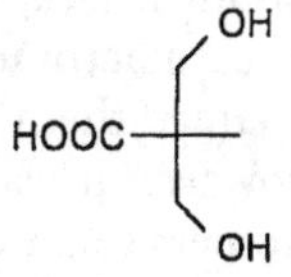

Chain growth AB*

Self-condensing vinyl polymerization

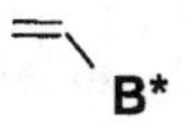

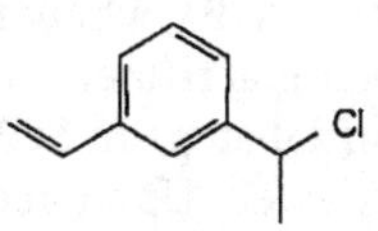

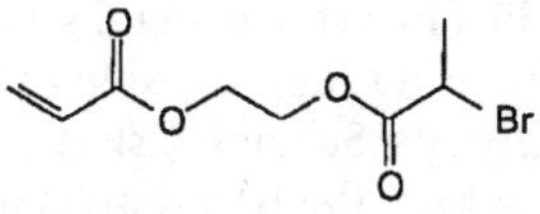

Self-condensing ring-opening polymerization

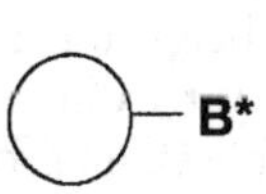

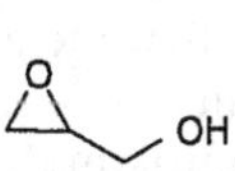

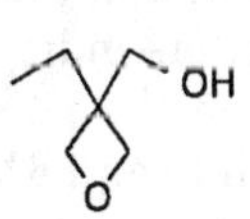

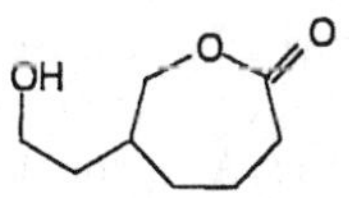

cyclic inimers. Table 1 compares the step-growth and chain-growth methods. The most common method is the polycondensation of AB_x monomers. However, vinyl monomers cannot be polymerized by that approach. The recent discovery of SCVP made it possible to use vinyl monomers for a convenient, one-pot synthesis of hyperbranched vinyl polymers [18–25]. By copolymerizing AB* inimers with conventional monomers, the SCVP technique was extended to self-condensing vinyl copolymerization (SCVCP), leading to highly branched copolymers where the degree of branching (DB) is controlled by the comonomer ratio [26–31]. By using these techniques, a variety of hyperbranched polymers can be synthesized. The scope of this review will cover hyperbranched (meth)acrylates that have been made using SCV(C)P.

2
Synthesis

2.1
Self-Condensing Vinyl Polymerization (SCVP)

In attempts to synthesize ω-styrylpolyisobutylene using m-/p-(chloromethyl)-styrene as an initiator and aluminum alkyls/H_2O as co-initiators, Kennedy and Frisch [32] found significantly less than 100% of vinyl groups in the product. They concluded that copolymerization of the vinyl group of the "initiator" with isobutylene occurred as a "deleterious side reaction". In a similar experiment, Nuyken et al. [33] observed the formation of soluble polymers with much higher than calculated molecular weights (MWs) and broad MWD and attributed this to the formation of branched copolymers. On the other hand, attempted copolymerizations of isobutylene with p-(chloromethyl)styrene or p-(chloromethyl)-α-methylstyrene initiated with boron trifluoride or $EtAlCl_2$ were reported earlier [34, 35]. All these results indicate that p-(chloromethyl)styrene acts both as a cationic initiator and as a monomer. In 1991, Hazer [36, 37] polymerized macromonomers which carried an azo function at the other terminus and named them "macroinimers".

However, it was Fréchet et al. who recognized the importance of initiator–monomers (later called "inimers") to synthesize hyperbranched polymers from vinyl monomers and in 1995 they named the process "self-condensing vinyl polymerization" (SCVP) [18]. Initiator–monomers (later called "inimers" [38, 39]) have the general structure AB*, where the double bond is designated A and B* is a group capable of being activated to initiate the polymerization of vinyl groups. Scheme 1 shows initial steps in SCVP. In order to initiate the polymerization, the B* group is activated. Upon activation of a B* group, the polymerization starts by addition of the B* group to the double bond of another inimer, resulting in the formation of the dimer, Ab–A*B*. The asterisk indicates that a structural group can add monomer; it can be either in its active or dormant form. Lowercase letters indicate that the group has been consumed and can no longer participate in the polymerization. The resulting dimer has two active sites, A* (propagating) and B* (initiating), for possible chain growth besides the vinyl group. Addition of a third monomer unit at either site results in the for-

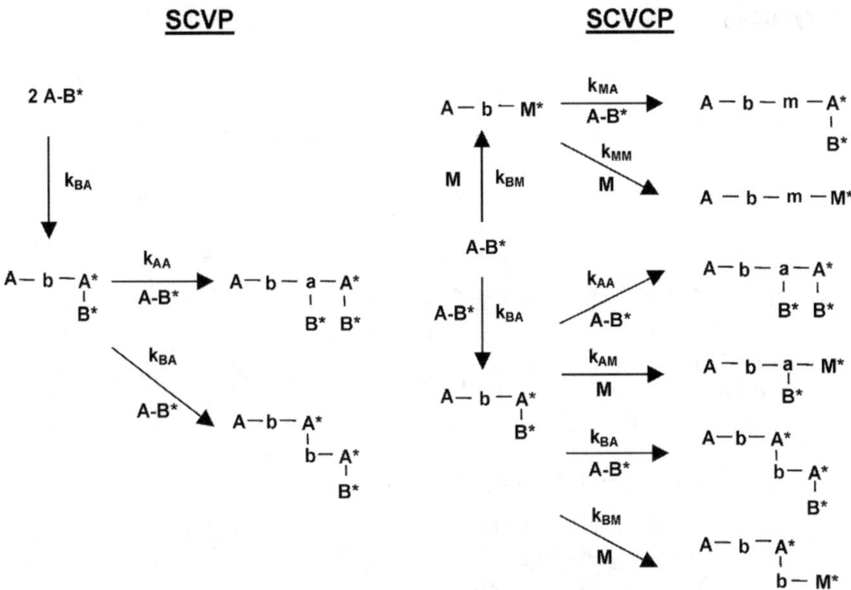

Scheme 1. Initial steps in SCVP and SCVCP. Capital letters indicate vinyl groups (A and M) and active centers (A*, B*, M*), lowercase letters stand for reacted ones (a, b, m)

mation of the trimer which can now grow in three directions. Also, oligomers (i.e., two dimers) or polymers can react with each other, similar (but mechanistically different) to a polycondensation. The polymerization can also be initiated by a mono- of multifunctional initiator, leading to better control of MWD and DB, especially when the inimer is added slowly [40].

AB* inimers used for SCV(C)P are listed in Fig. 1. A variety of acrylate-type inimers have been reported, including **1**, **2** [23–25], **3** [41], **4** [42], **5** [43], **6** [44], **7** [45], and **8** [46]. In general, Cu-based atom transfer radical polymerization (ATRP) was employed for SCVP of these acrylate-type inimers with an acrylate (A) and a bromoester group (B*) capable of initiating ATRP. For example, it was demonstrated that the polymerization of the inimer **1** catalyzed by CuBr/ 4,4′-di-*tert*-butyl-2,2′-bipyridine at 50 °C provided a hyperbranched polymer (DB=0.49) [23, 24]. In addition, kinetics and mechanism of chain growth for SCVP of the inimer **1** and evaluation of the DB of the resulting polymers were investigated. The significant influence of solubility of the deactivator and of the polymerization temperature, which are closely related to the concentration of Cu(II), on the topology of the resulting polymers has been also reported [25]. However, methacrylate-type inimers **9** and **10** as well as the acrylate-type inimer **2** could not be successfully polymerized by such Cu-based ATRP despite variations in ligand and temperature [47]. It was speculated that the tertiary radical sites generated from methacrylate moieties (A) and/or the 2-bromoisobutyryloxy moieties (B*) coupled rapidly, forming an excess amount of deactivating Cu(II) species and prevented polymerization. For the preparation of hyperbranched methacrylates, Cu-based ATRP with addition of zero-

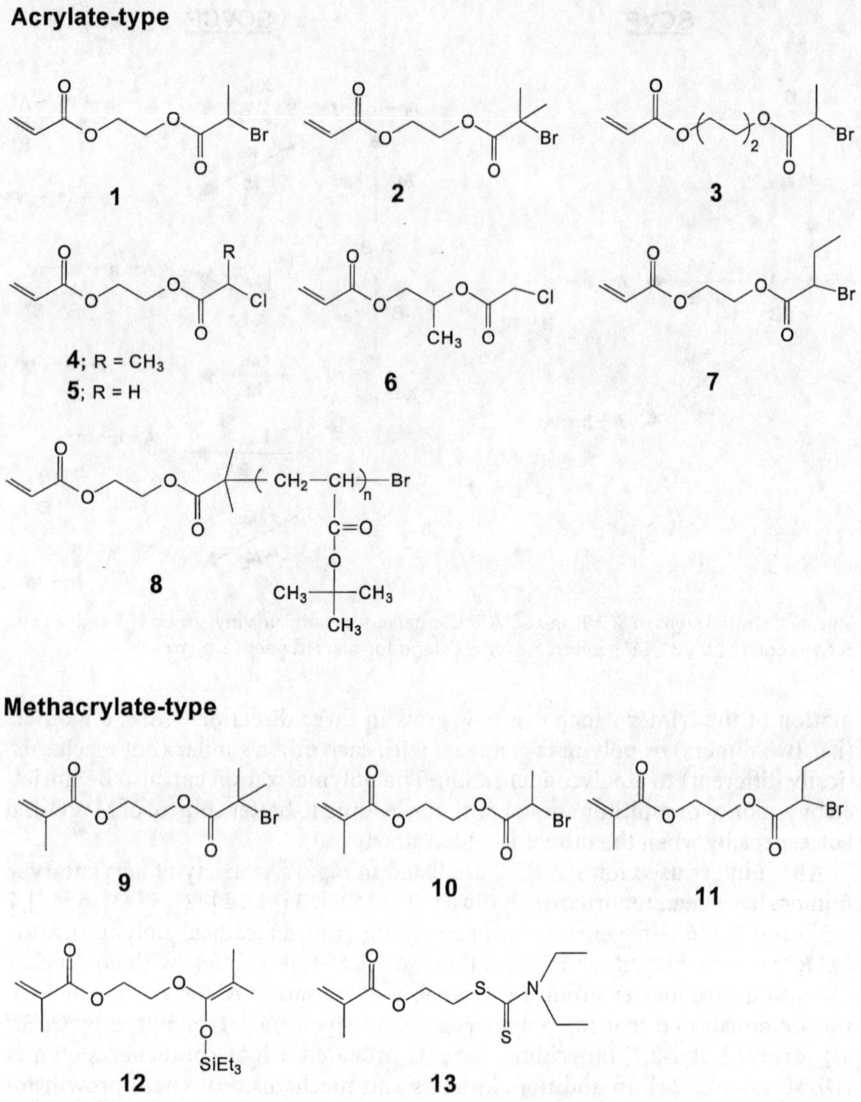
Acrylate-type
1
2
3
4; R = CH3
5; R = H
6
7
8
Methacrylate-type
9
10
11
12
13

Fig. 1. AB* vinyl inimers (initiator–monomers) used for SCVP and SCVCP

valent copper for **9** [47], Ni-based controlled radical polymerization for **9** [48], Cu-based ATRP for **11** [49], group transfer polymerization (GTP) for **12** [20, 50], and photo-initiated radical polymerization for **13** [51] have all been employed.

For SCVP of styrenic inimers, the mechanism includes cationic (**14** [18], **19** [29]), atom transfer radical (**15** [22, 27]), nitroxide-mediated radical (**16** [21]), anionic (**20** [19]), photo-initiated radical (**17** [2], **18** [52–55]), and ruthenium-catalyzed coordinative (**21** [56]) polymerization systems. Another example in-

Styrene-type

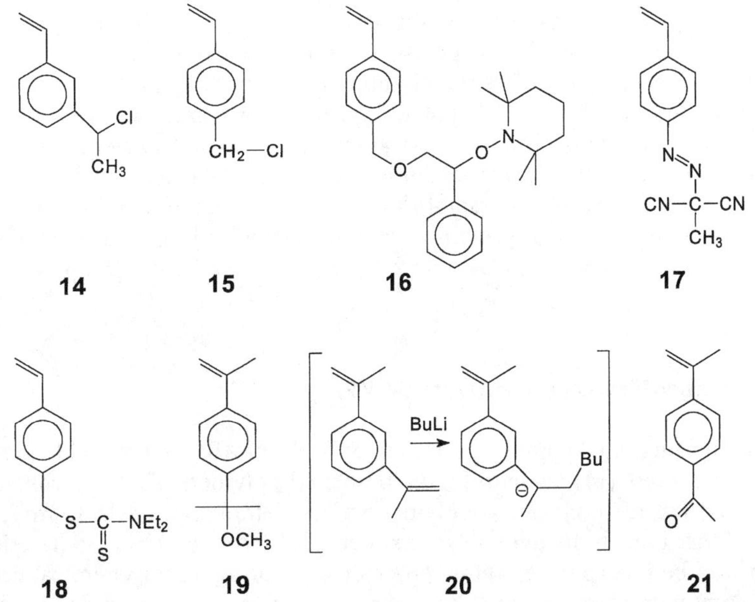

14 **15** **16** **17**

18 **19** **20** **21**

Vinyl ether-type

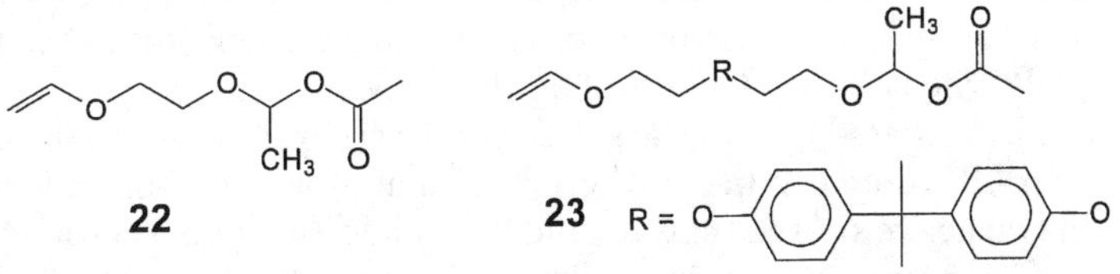

22 **23** R =

Fig. 1 (continued)

volves vinyl ether-type inimers **22** and **23** which undergo cationic polymerization in the presence of a Lewis acid activator, such as zinc chloride [26, 57, 58]. Self-condensing anionic and cationic ring-opening polymerizations have been also reported [4, 59] (see examples in Table 1).

Note that polymerization of compounds **13**, **17**, and **18** is based on a dithiocarbamate radical formed by UV radiation and the methylmalodinitrile radical formed by decomposition of an azo group. These fragments act as reversible terminating/transfer agents, but not as initiators. Hence, these compounds cannot be regarded as initiator–monomers, but the mechanism is analogous to SCVP. There are also several accounts of preparing branched polymers from (meth)acrylate derivatives [60–63], in which polymerization systems can be regarded as SCVP analogs.

In an ideal SCVP process, living polymerization systems are preferred in order to avoid crosslinking reactions and gelation caused by chain transfer or recombination reactions. In addition to cationic, anionic, group-transfer, and transition-metal-catalyzed polymerizations, controlled radical polymerization processes are most frequently employed. These include atom transfer radical polymerization (ATRP) [64–67], nitroxide-mediated radical polymerization [68], and reversible addition–fragmentation chain transfer polymerization (RAFT) [69]. Similar to cationic and group transfer polymerization, these systems are based on establishing a rapid dynamic equilibration between a minute amount of growing free radicals and a large majority of dormant species; however, they are more tolerant of functional groups and impurities.

2.2
Self-Condensing Vinyl Copolymerization (SCVCP)

Self-condensing vinyl copolymerization (SCVCP) of AB* inimers with conventional monomers (M) leads to highly branched polymers, allowing control of MWD and DB [26–31]. The copolymerization method is a facile approach to obtain functional branched polymers, because different types of functional groups can be incorporated into a polymer, depending on the chemical nature of the comonomer. In addition, the chain architecture can be modified easily by a suitable choice of the comonomer ratio in the feed. Because the number of linear units is higher, the DB of the copolymers is lower than that of SCVP homopolymers. However, the effect on solution properties like intrinsic viscosity and radius of gyration should be less than proportional, because branched polymers above a limiting molecular weight are self-similar objects. Therefore, the copolymerization is an economic approach to obtain highly branched polymers, especially when aiming at controlling rheology.

SCVCP can be initiated by two ways (Scheme 1): (i) by addition of the active B* group in an AB* inimer to the vinyl group A of another AB* inimer forming a dimer with two active sites, A* and B*, and (ii) by addition of a B* group to the vinyl group of monomer M forming a dimer with one active site, M*. Both the initiating B* group and the newly created propagating centers A* and M* can react with any vinyl group in the system. Thus, we have three different types of active centers, A*, B*, and M* in the dimers, which can react with double bonds A (inimer and macromolecules; each macromolecule contains strictly one double bond) and M (monomer).

Several approaches have been reported for the synthesis of hyperbranched (meth)acrylates via SCVCP. Highly branched PMMA was synthesized by SCVCP of MMA with the inimer 12 having a methacrylate group (A) and a silylketene acetal group (B*), capable of initiating GTP [28]. Highly branched poly(*tert*-butyl methacrylate) was also obtained by SCVCP of the inimer 12 with *tert*-butyl methacrylate, which is a precursor of branched poly(methacrylic acids). A series of hyperbranched acrylates with different DBs and MWs have been synthesized by SCVCP of the acrylate-type inimer 1 with *t*-BuA via ATRP [31]. SCVCP of the inimer 1 with methyl acrylate was also reported [30].

In an alternate way, highly branched Pt-BuA has been obtained by SCVP via ATRP of macroinimer **8** which is a heterotelechelic Pt-BuA possessing both an initiating and a polymerizable moiety [46].

3
Molecular Parameters of Hyperbranched Polymers Obtained by SCVP and SCVCP

A series of theoretical studies of the SCV(C)P have been reported [38, 40, 70–74], which give valuable information on the kinetics, the molecular weights, the MWD, and the DB of the polymers obtained. Table 2 summarizes the calculated MWD and DB of hyperbranched polymers obtained by SCVP and SCVCP under various conditions. All calculations were conducted, assuming an ideal case, no cyclization (i.e., intramolecular reaction of the vinyl group with an active center), no excluded volume effects (i.e., rate constants are independent of the location of the active center or vinyl group in the macromolecule), and no side reactions (e.g., transfer or termination).

3.1
Molecular Weight Distribution (MWD)

Our theoretical studies [38] showed that the hyperbranched polymers generated from an SCVP possess a very wide MWD which depends on the reactivity ratio of propagating and initiating groups, $r=k_A/k_B$. For $r=1$, the polydispersity index $M_w/M_n \approx P_n$, where P_n is the number-average degree of polymerization. This value is similar to that obtained in the polycondensation of AB_2 monomers ($M_w/M_n \approx P_n/2$) [5]. This can be explained by the fact that large molecules have a higher probability of reacting with vinyl groups than smaller ones since they have more active centers. In fact, even at high conversion there is still a consid-

Table 2. Molecular parameters of polymers obtained by self-condensing vinyl polymerization and copolymerization

	SCVP of AB*	Copolymerization of AB*+M
Polymerization without initiator	$DB \approx 1/2$ $\overline{M}_w/\overline{M}_n \approx 1 + \overline{DP}_n$ [38, 70]	$DB \approx 2/(\gamma + 1)\ (\gamma \gg 1)$ $\overline{M}_w/\overline{M}_n = 1 + \overline{DP}_n/(\gamma + 1)$ [73, 74]
Polymerization with multifunctional initiator (batch)	$DB \approx 1/2$ $\overline{M}_w/\overline{M}_n = 1 + \overline{DP}_n/f^2$ [40]	$DB \approx 2/(\gamma + 1)\ (\gamma \gg 1)$ $\overline{M}_w/\overline{M}_n = 1 + \overline{DP}_n/(\gamma + 1)\,f^2$ [72]
Polymerization with multifunctional initiator (semi-batch)	$DB \approx 2/3$ $\overline{M}_w/\overline{M}_n = 1 + 1/f$ [71]	$DB \approx 2/(\gamma + 1)\ (\gamma \gg 1)$ $\overline{M}_w/\overline{M}_n = 1 + 1/f$ [72]

DB degree of branching; f initiator functionality; γ $[M_0]/[AB^*]_0$; Semi-batch = slow monomer addition.

erable amount of inimer left (in reality this fraction is smaller because of the higher accessibility of active centers compared to larger molecules). The polydispersity index increases for $r>1$ and decreases for $r<1$. SCVP follows the Carothers equation, that is, the degree of polymerization is inversely proportional to the fraction of unreacted vinyl groups, $P_n=1/(1-x)$. However, in reality the degree of polymerization is limited by cyclization reactions which transform vinyl groups into active centers.

A way to narrow the MWD and to approach the structure of dendrimers is the addition of a small fraction of a f-functional initiator, G_f, to inimers [40, 71]. In this process the obtainable degree of polymerization is limited by the ratio of inimer to initiator. It can be conducted in two ways: (i) inimer molecules can be added so slowly to the initiator solution that they can only react with the initiator molecules or with the already formed macromolecules, but not with each other (semi-batch process). Thus, each macromolecule generated in such a process will contain one initiator core but no vinyl group. Then, the polydispersity index is quite low and decreases with f: $M_w/M_n \approx 1+1/f$. (ii) Alternatively, initiator and monomer molecules can be mixed instantaneously (batch process). Here, the normal SCVP process and the process shown above compete and both kinds of macromolecules will be formed. For this process the polydispersity index also decreases with f, but is higher than for the semi-batch process, $M_w/M_n \approx P_n/f^2$.

There have been a few approaches to calculate the effect of cyclization on the MWD [75–77]. Since each cyclization event generates a new multifunctional initiator, it becomes clear that cyclization is expected to narrow the MWD.

For SCVCP, the PDI is decreased in proportion to the comonomer ratio, $\gamma=[M]_0/[I]_0$: $M_w/M_n=1+P_n/(\gamma+1)$ for $\gamma>>1$ [73]. The addition of a multifunctional initiator again affects the polydispersity index [72]. In the batch process it decreases with initiator functionality as $M_w/M_n \approx P_n/(\gamma+1)f^2$, similar to homo-SCVP. The effect is even more pronounced for the semi-batch process where the concentration of the inimer and the comonomer is kept infinitesimally low and $M_w/M_n \approx 1+1/f$. This result is identical to the value obtained in homo-SCVP, that is, addition of comonomer does not decrease polydispersity any further.

3.2
Degree of Branching (DB)

The degree of branching can be regarded as the ratio of branched units in the polymer to those in a perfect dendrimer. Thus, the limiting values are DB=0 for linear polymers and DB=1 for a perfect dendrimer. Various definitions of DB have been given. If we do not take into account the vinyl group or initiator unit (the "core unit"), the DB is defined as

$$DB = \frac{(\text{number of branched units}) + (\text{number of terminal units}) -1}{(\text{total number of units}) -1} \tag{1}$$

Here, one unit has been subtracted from the numerator and the denominator to take into account that even a linear polymer has one initiating and one terminal unit.

From the topology of branched systems with trifunctional branchpoints, for any given molecule the number of branched units is equal to the number of terminal unit minus one. Thus, Eq. 1 can be further simplified to

$$DB = \frac{2 \times (\text{number of branched units})}{(\text{total number of units}) - 1} \qquad (2)$$

Alternatively, the fraction branchpoints can be defined as

$$\overline{FB} = \frac{\text{number of branched units}}{(\text{total number of units}) - (\text{number of monomers})} \qquad (3)$$

The DB obtainable in SCVP is DB=0.465 for $r=k_A/k_B=1$ and reaches its maximum, DB=0.5, for $r=2.6$ [70, 78]. This value is identical to that obtained in AB_2 polycondensation when both B functions have the same reactivity [70, 78]. Thus, hyperbranched polymers prepared by bulk polycondensation or polymerization contain at least 50% linear units, making this approach less efficient than the synthesis of dendrimers.

The presence of a polyinitiator has a small effect only in batch mode, whereas the polymer obtained in a semi-batch process is more strongly branched (DB=2/3 as compared to DB=0.465 without initiator) [40]. A similar result was found by Hanselmann et al. for AB_2 monomers in the presence of a core-forming molecule [79].

For SCVCP in general, DB strongly depends on the comonomer ratio ($\gamma=[\text{monomer}]_0/[\text{inimer}]_0$) [73, 74]. In the ideal case, when all rate constants are equal, for $\gamma>>1$, the final value of DB decreases with γ as DB~2/(γ+1) which is four times higher than the value expected from dilution of inimer molecules by monomers. For low values of ($\gamma\leq1$), DB even exceeds the values for a homo-SCVP; a maximum of DB=0.5 is reached at $\gamma\approx0.6$. Depending on the reactivity ratios, the structure of polymer obtained can change from "macroinimers" when the monomer M is much more reactive than the vinyl groups of inimer or polymer molecules to "hyperstars" in the opposite limiting case.

Theoretical calculations were also conducted on the influence of f-functional initiators on DB in SCVCP [72]. In the semi-batch system, DB is only slightly affected by the presence of polyinitiator and is mostly governed by the comonomer content. The calculations are also applied to polymerizations from surface-bound initiators (see later).

3.3
Experimental Evaluation of DB

Hyperbranched polymers are generally composed of branched (dendritic), linear, and terminal units. In contrast to AB_2 systems, there are two different types of linear units in SCVP: one resembles a repeat unit of a polycondensate ($\sim\sim\sim A^x\text{-}b\sim\sim\sim$) and one a monomer unit of a vinyl polymer ($\sim\sim\sim a(B^*)\sim\sim\sim$).

The DB of the hyperbranched acrylates obtained by SCV(C)P of the acrylate-type inimer 1 can be determined by ^{1}H NMR spectroscopy [23, 31]. Figure 2a shows the respective ^{1}H NMR spectrum of a hyperbranched polymer obtained

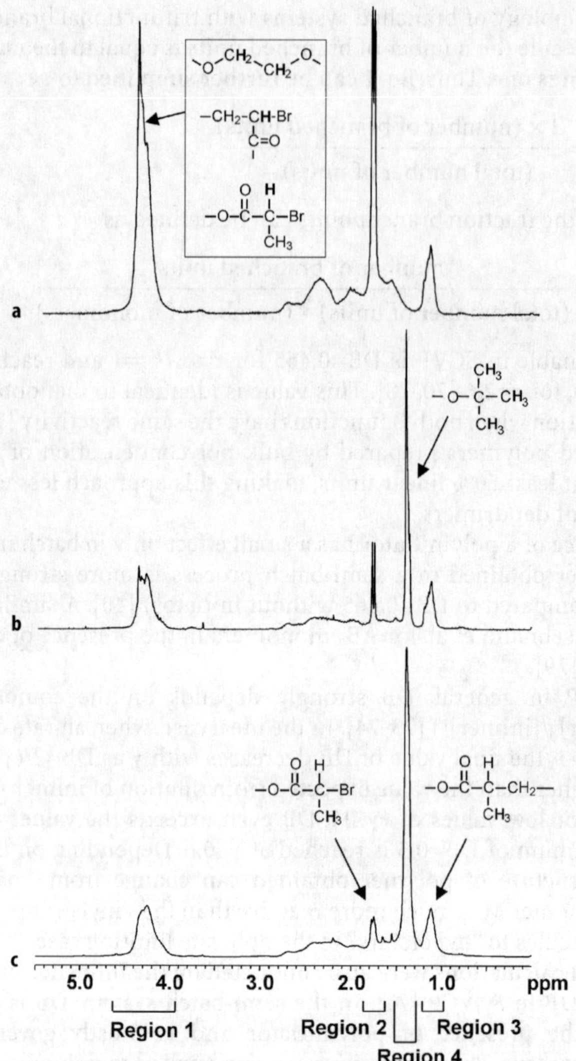

Fig. 2. ^{1}H NMR spectra of the polymers obtained by SCVP of the acrylate-type inimer **1** (**a**), and by copolymerizations of the inimer **1** and *t*-BuA: $\gamma=[t\text{-BuA}]_0/[\mathbf{1}]_0=0.5$ (**b**), $\gamma=1.1$ (**c**). (Reproduced with permission from [31]. Copyright 2002 American Chemical Society.)

by a homo-SCVP of the inimer **1**. The large doublet at 1.85 ppm, region 2, is assigned to CH_3 of the 2-bromopropionyloxy group, B* (corresponding to an end group), while the broad peak at 1.0–1.3 ppm, region 3, is assigned to b, which is formed by activation of the B* and subsequent addition of monomer. The determinations of the proportion of B*, the reactivity ratio of A* and B* groups, $r=k_A/k_B$, and the DB can be performed by evaluation of these peaks [23, 24, 70]. The proportions of b and B* can be calculated from these peaks: $b=$(signal at

1.0–1.3 ppm)/(sum of signals at 1.0–1.3 ppm and signals at 1.75 ppm); $B^*=1-b$. The reactivity ratio of A* and B* groups, $r=k_A/k_B$, was determined from Eq. (4) [24]:

$$r = k_A/k_B = (x + B^* - 1)/[-\ln(B^*) + B^* - 1] \tag{4}$$

where x is the conversion of double bonds.

Figure 2 (b, c) shows ^{1}H NMR spectra of the branched Pt-BuA obtained by SCVCP of the inimer 1 with t-BuA. The broad peak of region 1 corresponds to the protons of the ethylene linkage and the protons which are geminal to bromine in either A*, B*, or M*, all of which are derived from the inimer 1. The later protons correspond to the end groups. Although B* in the inimer 1 is consumed during the copolymerization, for every B* consumed one A* or M* is formed and consequently, original $B^*=B^*_{\text{left}}+A^*+M^*$. Hence, the sum of protons of the ethylene linkage and geminal to bromine is proportional to the fraction of the inimer 1 in the copolymer but independent of the DB. The peak at 1.3–1.4 ppm, region 4, is assigned to the $tert$-butyl group of Pt-BuA segment. The comonomer composition calculated from the ratio of these peaks is in good agreement with the comonomer composition in the feed which corresponds to the $y=[t\text{-BuA}]_0/[\text{inimer 1}]_0$ value. For the copolymer obtained by SCVCP, these peaks at 1.85 ppm (region 2, which is assigned to B*) and at 1.0–1.3 ppm (region 3, which is assigned to b) should be related to the DB and the comonomer composition. For example, the proportion of b calculated by the equation, $b=(\text{region 3})/(\text{region 2}+\text{region 3})$, was 0.62 in the case of copolymerization at $y=1.1$. For equal reactivity of active sites, the DB determined by NMR, DB_{NMR}, at full conversion is given as [28]

$$DB_{\text{NMR}} = 2\left(\frac{b}{y+1}\right)\left[1 - \left(\frac{b}{y+1}\right)\right] \tag{5}$$

According to the theory of SCVCP, the comonomer ratio, y, can be directly related to the DB [73]. Assuming equal reactivity of all active sites, the DB obtained from the theory, DB_{theo}, at full conversion can be represented as

$$DB_{\text{theo}} = \frac{2(1 - e^{-(y+1)})(y + e^{-(y+1)})}{(y+1)^2} \tag{6}$$

From these approaches, $DB_{\text{NMR}}=0.42$ and $DB_{\text{theo}}=0.49$ can be obtained at $y=1.1$ ($b=0.62$), respectively. Note that these values represent a rough estimate, as they are calculated based on the assumption of equal rate constants for copolymerization. For low y values ($y=0.5$), the DB ($DB_{\text{NMR}}=0.48$) even exceeds the value for poly(inimer 1) ($DB_{\text{NMR}}=0.43$) obtained by a homo-SCVP. This is an accordance with theoretical prediction that a maximum of DB=0.5 is reached at $y=0.6$ [73]. The effect can be explained by the addition of monomer molecules to in-chain active centers (i.e., in linear segments), leading to very short branches. For $2.5 \geq y \geq 0.5$, DB_{NMR} decreases with y, as predicted by calculations.

Although NMR experiments afford a conclusive measurement of the degree of branching for lower y values, the low concentration of branchpoints in the copolymer does not permit the determination of the DB directly by the spectroscopic method, because of low intensities of the peaks. However, for the case of high comonomer ratios, $\gamma \gg 1$, the relation between DB_{theo} and γ becomes very simple and does not depend on the reactivity ratios of the various active centers. It is represented as $DB_{theo} \approx 2/(\gamma+1)$. In the case of $\gamma=25$, theory predicts $DB_{theo}=0.077$. We calculated a fraction of branched units $f_B=DB_{theo}/2=0.038$, that is, which corresponds to 38 branching points in 1,000 monomer units or an average of 25 monomer units between branch points. Note that the dependencies are more complex and the DB may be higher or lower, depending upon the systems, when the reactivities of the various active centers are not equal and comonomer ratios are low.

The direct determination of DB for hyperbranched methacrylates obtained by SCVP of **9** and **10** via ATRP was reported to be impossible due to overlapping signals in the ^{1}H NMR spectra [47]. However, DB of some hyperbranched methacrylates could be determined by NMR. In order to experimentally determine DB, copolymers of **12** and MMA-d$_8$ made by GTP and terminated with protons were analyzed by ^{1}H NMR spectroscopy [28]. In the corresponding copolymers, linear monomer units, m, are "invisible" and therefore the protons in A*, B*, and M* centers could be detected even at high comonomer ratios. In this system, the chemical structure of B* centers is different from that of A* and M* centers. To estimate the proton chemical shifts, model compounds were analyzed by ^{1}H NMR spectroscopy. The ^{1}H NMR spectra of the copolymers show two distinct signals at 2.39 and 2.26 ppm. According to the reference spectra, the former signal group can be attributed to the sum b=A*+M* of protonated A* and M* centers, whereas the later signal is due to the protonated B* center. The DB determined by this approach, DB_{NMR}, agrees qualitatively with the theoretical predictions.

Often in hyperbranched polymers obtained via SCVP, it is not possible to determine the DB directly via NMR analysis. Therefore, other methods, for example, viscosity measurements and light-scattering methods have to be used to confirm the compact structure of a hyperbranched polymer. Such characterizations of hyperbranched (meth)acrylates will be discussed in the next section.

4
Solution Properties

4.1
In Organic Solvent

The hyperbranched structures lead to characteristic properties, such as a relatively compact shape, and absence of entanglements, in pronounced contrast to linear polymer chains. Furthermore, low viscosity in bulk and solution is generally observed. Such phenomena strongly depend on MW and DB. The determination of the MW of branched polymers is complicated by fact that the hydrodynamic volume for a given MW differs significantly from that of a linear sample. Therefore, the use of a calibration curve generated by the use of linear standards in SEC leads to erroneous results. This problem can be overcome by the use of mass-sensitive on-line detectors such as a multi-angle light-scattering photometer (MALS) [80, 81] or a viscosity detector using the universal calibration (UNICAL) principle [82, 83].

Relationships between dilute solution viscosity and MW have been determined for many hyperbranched systems and the Mark–Houwink constant typically varies between 0.5 and 0.2, depending on the DB. In contrast, the exponent is typically in the region of 0.6–0.8 for linear homopolymers in a good solvent with a random coil conformation. The contraction factors [84], $g=<R_g^2>_{branched}/<R_g^2>_{linear}$, $g'=[\eta]_{branched}/[\eta]_{linear}$, are another way of expressing the compact structure of branched polymers. Experimentally, g' is computed from the intrinsic viscosity ratio at constant MW. The contraction factor can be expressed as the averaged value over the MWD or as a continuous fraction of MW.

Highly branched PMMA synthesized by SCVCP of MMA with the inimer **12** via GTP was characterized by GPC using universal calibration and MALS [28]. The corresponding Mark–Houwink plots, $\log[\eta]$ versus $\log M$, and contraction factor, $g'=[\eta]_{branched}/[\eta]_{linear}$, as a function of $\log M$ are presented in Fig. 3 for different comonomer ratio $\gamma=[MMA]_0/[\mathbf{12}]_0$. For $M>10^4$, the viscosity of the

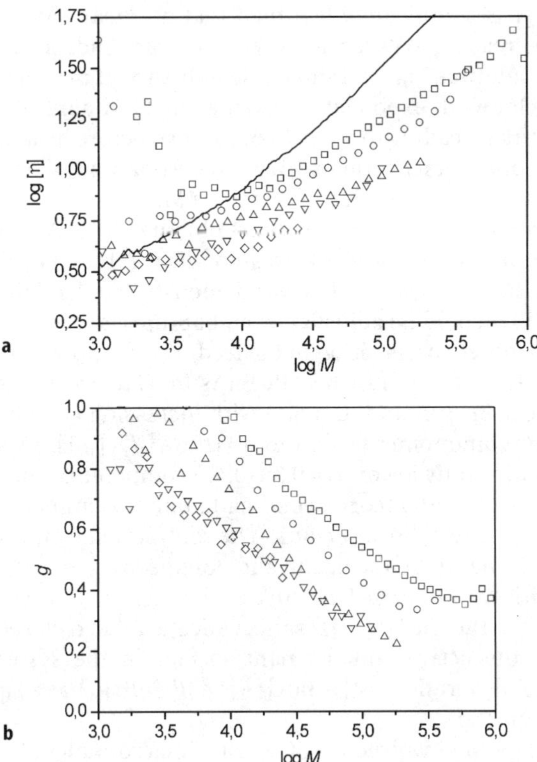

Fig. 3. Mark-Houwink plot (**a**) and contraction factors (**b**), $g'=[\eta]_{branched}/[\eta]_{linear}$, as a function of the molecular weight for the copolymerization of the methacrylate-type inimer **12** with MMA under different comonomer ratios, $\gamma=[MMA]_0/[\mathbf{12}]_0=1.2$ (+), 5.2 (◊), 9.8 (▽), 26 (△), 46.8 (○), 86.5 (□), respectively. The intrinsic viscosities of PMMA (—) are given for comparison. (Reproduced with permission from [28]. Copyright 2001 American Chemical Society.)

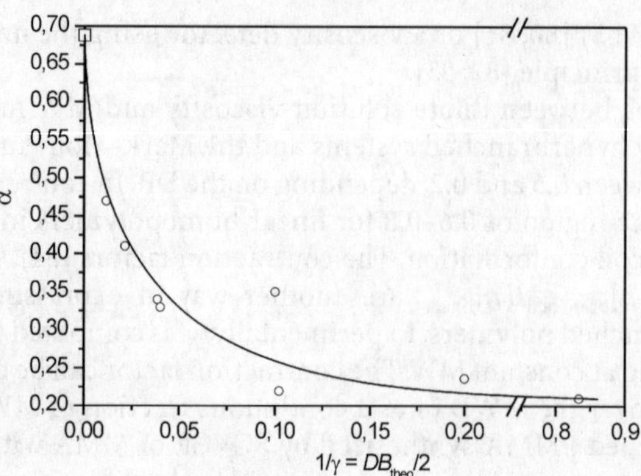

Fig. 4. Dependence of the Mark-Houwink exponent, α, on comonomer ratio, $\gamma=[\text{MMA}]_0/[\mathbf{12}]_0$, for the copolymerization of the inimer **12** with MMA. Linear PMMA ($\square$): $\alpha=0.68$. (Reproduced with permission from [28]. Copyright 2001 American Chemical Society.)

branched polymer is significantly less than that of linear PMMA. The g' value decreases with increasing MW for all polymers, which indicates a highly compact structure in solution. Figure 4 shows the influence of the comonomer ratio, γ, on the Mark–Houwink exponent, α. Even a small amount of the inimer **12** considerably lowers α, leading to a more compact structure in solution than linear PMMA. At a comonomer ratio of $\gamma=26$ (corresponding to only 4% inimer), α is approximately 50% of the value of linear PMMA.

Characterization of the branched Pt-BuAs obtained by SCVCP of t-BuA with the acrylate-type inimer **1** via ATRP was conducted by GPC, GPC/viscosity, GPC/MALS, and NMR analysis [31]. It was demonstrated that DB, the composition, MW, and MWD could be adjusted by an appropriate choice of the catalyst system, the comonomer composition in the feed, and the polymerization conditions. The viscosities of the branched Pt-BuAs in THF are significantly lower than those of linear Pt-BuA and decrease with increasing DB which, in turn, is determined by the comonomer feed ratio, $\gamma=[t\text{BuA}]_0/[\mathbf{1}]_0$. The Mark–Houwink exponents are significantly lower ($\alpha= 0.38$–0.47) compared to that for linear Pt-BuA ($\alpha=0.80$). Even at $\gamma=100$ (corresponding to only 1% inimer), the α value is only 50–60% of the value of linear Pt-BuA. The contraction factors decrease with increasing MW. The nature of the ligand and polymerization temperature affect the MWs, viscosities, and Mark–Houwink exponents (Fig. 5). The phenomena may be attributed to the Cu(I)/Cu(II) ratio, which is related closely to the equilibrium between the active and dormant species in the system. While the comonomer-to-catalyst ratio, $\mu=([t\text{-BuA}]_0+[\mathbf{1}]_0)/[\text{CuBr}]_0$, has a slight influence only on these parameters.

Another investigation involved the SCVP of a "macroinimer **8**" via ATRP [46]. GPC/viscosity measurements indicated that the intrinsic viscosity of the branched polymer is less than 40% of that of the linear one at highest MW area (Fig. 6). A significantly lower value for the Mark–Houwink exponent ($\alpha=0.47$ compared to $\alpha=0.80$ for linear Pt-BuA) was also observed, indicating the compact nature of the branched macromolecules.

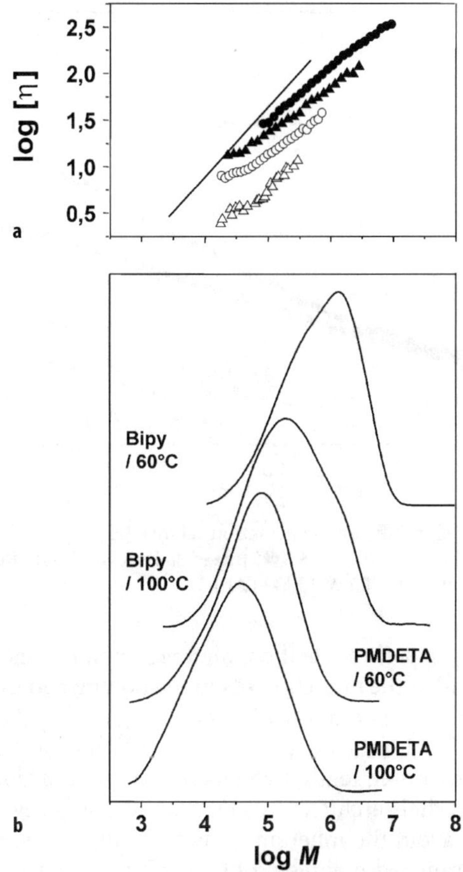

Fig. 5. Mark-Houwink plots (**a**) and RI signals (**b**) for polymers obtained by SCVCP of t-BuA with the inimer **1**; CuBr/PMDETA at 60 °C (○) and 100 °C (△); CuBr/Bipy at 60 °C (●) and 100 °C (▲). The viscosity result for a linear Pt-BuA (—) is given for comparison. (Reproduced with permission from [31]. Copyright 2001 American Chemical Society.)

Hyperbranched poly(ethyl methacrylate)s prepared by the photo-initiated radical polymerization of the inimer **13** were characterized by GPC with a light-scattering detector [51]. The hydrodynamic volume (R_h) and radius of gyration (R_g) of the resulting hyperbranched polymers were determined by DLS and SAXS, respectively. The ratios of R_g/R_h are in the range of 0.75–0.84, which are comparable to the value of hard spheres (0.775) and significantly lower than that of the linear unperturbed polymer coils (1.25–1.37). The compact nature of the hyperbranched poly(ethyl methacrylate)s is demonstrated by solution properties which are different from those of the linear analogs.

Experimental data on the solution properties and melt rheology of highly branched structures are scarcely found in the literature. This might be because of the structural nonuniformity of hyperbranched polymers, which makes it difficult to obtain reliable data. Because of the purely statistical nature of the poly-

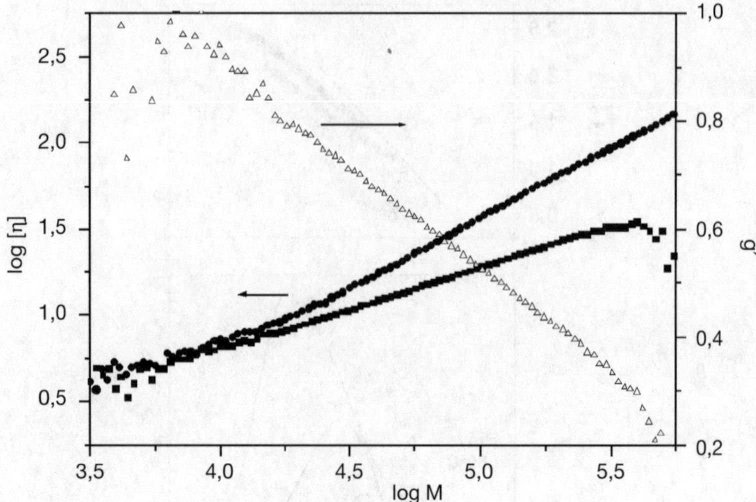

Fig. 6. Mark-Houwink plots (■) and contraction factors (△) of branched Pt-BuA (α=0.47) obtained by SCVP of the macroinimer **8**. (●) linear Pt-BuA (α=0.80). (Reproduced with permission from [46]. Copyright 2000 WILEY-VCH.)

merization process, hyperbranched copolymers exhibit rather broad distributions of the MWs and of the branchpoints in the polymer. At this point, it should be stressed that the latter nonuniformity is not reflected by the DB, because this parameter deals only with the number but not with the location of the branchpoints in a polymer. Consequently, two macromolecules exhibiting an equal DB value could differ in their architectures [85, 86]. Recently, several attempts have been conducted to avoid the inherent nonuniformities. A promising way is to fractionate hyperbranched polymers obtained by polymerization and characterize the fractionated products, respectively, which may give more useful information on the architectures.

Jackson et al. [87] investigated the properties of dilute solutions by comparing the intrinsic viscosity and radius of gyration of fractionated and unfractionated moderately branched PMMAs prepared by free-radical copolymerization of MMA with ethylene glycol dimethacrylate using multi-detector GPC. The authors found the z-average radius of gyration of these polymers to be insensitive to branching and detected some evidence for increased polydispersity after GPC separation at higher MWs.

The solution properties of a highly branched PMMA prepared by the SCVCP of the inimer **12** with MMA (γ=25) via GTP were investigated in detail [88]. To overcome the drawback of a broad MWD, preparative size-exclusion chromatography was employed in order to fractionate the branched polymer into a number of samples with low MWD ($M_w/M_n \leq 1.4$, Fig. 7b). These fractions were characterized using multi-detector SEC with on-line viscosity and light-scattering detection. Using the fractionated samples, the UNICAL and MALS calibration curves were established, which are in good agreement with the results of the unfractionated feed sample. As can be seen in Fig. 8a, for all of the fractionated

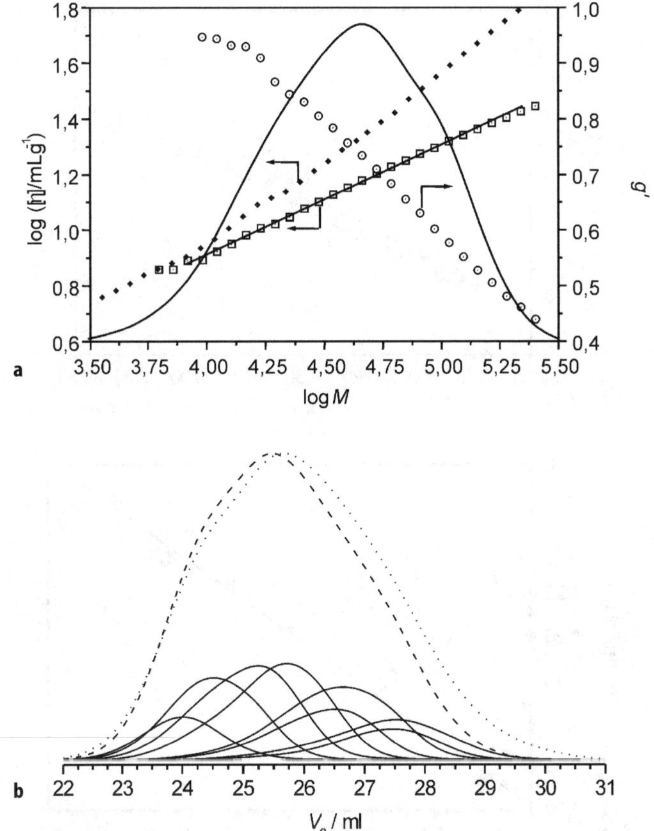

Fig. 7. a Mark-Houwink plot of highly branched PMMA obtained by SCVCP of MMA with the inimer **12**. (–)RI signal; (□) intrinsic viscosity of feed, (♦) intrinsic viscosity of linear PMMA; (○) contraction factor, g'. **b** Separation of feed polymer into fractions by preparative SEC. (–) RI signal of fractions; (– – –) accumulated RI signals; (······) RI signal of feed polymer. (Reproduced with permission from [88]. Copyright 2001 American Chemical Society.)

samples, the Mark–Houwink curves are nearly identical. To crosscheck the results, a Mark–Houwink plot was constructed from the average intrinsic viscosities, $[\eta]$, of each fraction and the viscosity-average molecular weight, M_η (Fig. 8b). A linear fit of the data in Fig. 8b results in a slope of $\alpha=0.402\pm0.017$, which is in excellent agreement with the value determined in Fig. 7a. Hence, the hyperbranched polymer was fractionated using preparative SEC. No differences in solution properties could be detected between the SEC slices of the feed polymer and the average values of the different fractions, indicating that the separation is attributable only to differences in hydrodynamic volume and not to polymer structure or MW alone.

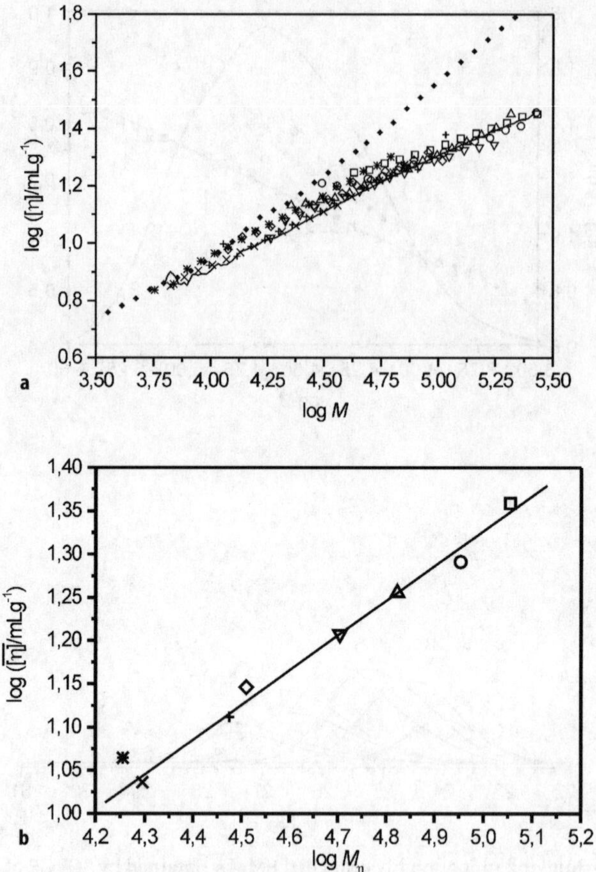

Fig. 8. a Mark-Houwink plot for various fractions of highly branched PMMA obtained by SCVCP of MMA with the inimer **12**. $M_{w,\text{UNICAL}}$ ($\times 10^{-4}$) = 12.06 ($\square$), 9.64 ($\bigcirc$), 7.2 ($\triangle$), 5.58 ($\triangledown$), 3.52 ($\lozenge$), 3.26 (+), 2.18 ($\times$), 2.00 (*); (—) unfractionated feed polymer; ($\blacklozenge$) linear PMMA. **b** Mark-Houwink plot obtained from intrinsic viscosities of each fraction of branched PMMA. (Reproduced with permission from [88]. Copyright 2001 American Chemical Society.)

4.2
In Aqueous Solution

Branched polyelectrolytes have become of special interest because of their industrial importance and scientifically interesting properties. Poly(ethyleneimine), which is important in various industrial applications, can provide an excellent example: branched and linear polyelectrolytes have quite different properties due to both their different topographies and structures [89–91]. As another practical point, branched polyelectrolytes can act as precursor or fragments of polyelectrolyte gels. A variety of theoretical approaches have been reported on the investigations of branched polyelectrolytes [92–97]. However,

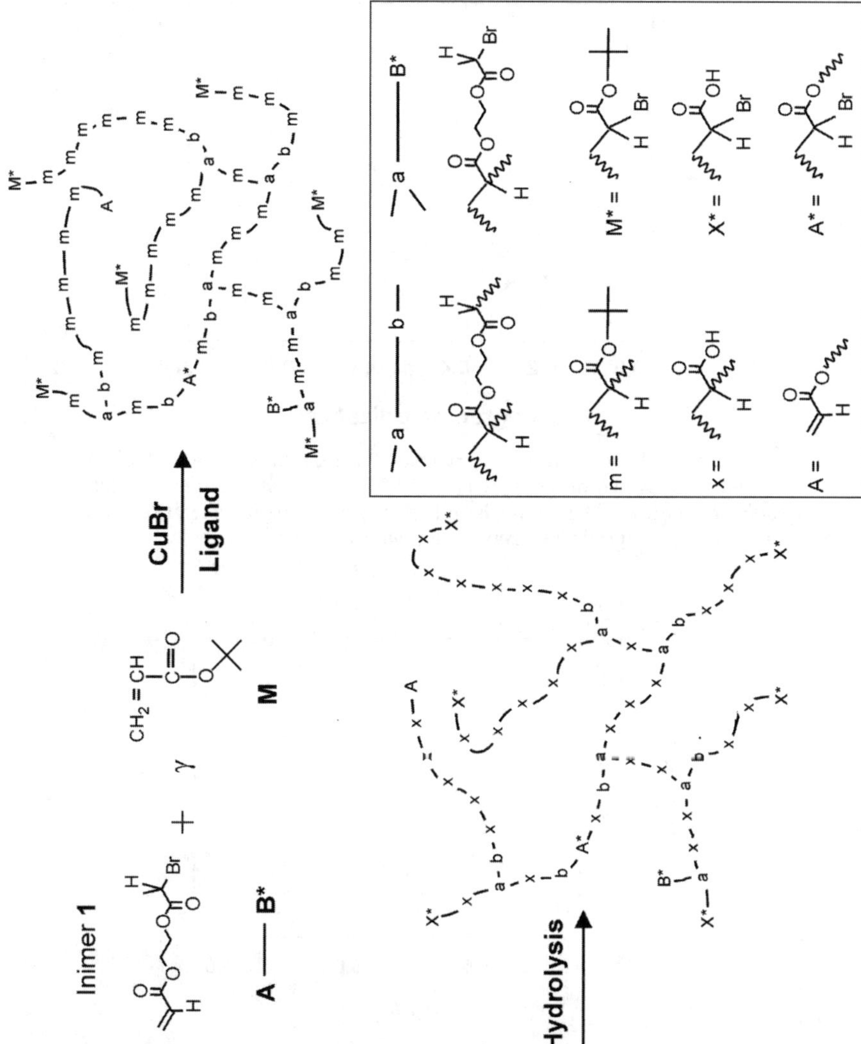

Scheme 2. General route to branched poly(acrylic acid) via SCVCP, followed by hydrolysis

the correlation of the topology and the properties of branched polyelectrolytes have not been studied very much experimentally, because of difficulties in the synthesis of well-defined branched polymers with ionic or ionizable groups. One challenge in this field is, therefore, to produce randomly or regularly branched polyelectrolytes which are suitable for various applications as well as for quantitative analysis of the relation between the properties and the architectures.

The synthesis of randomly branched PAA was conducted by SCVCP of *t*-BuA with the acrylate-type inimer 1 via ATRP, followed by hydrolysis of the *tert*-butyl

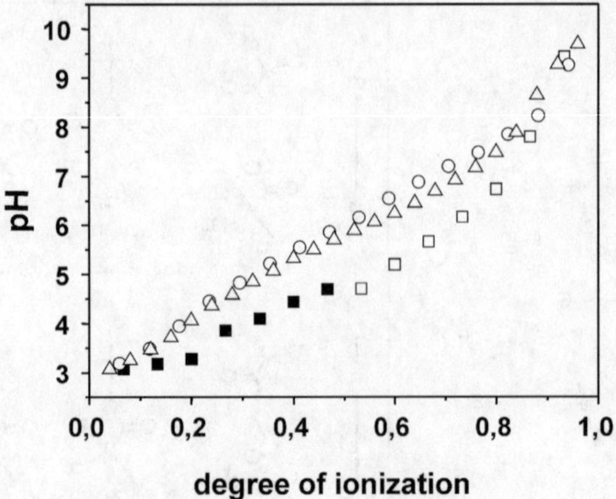

Fig. 9. Potentiometric titration curves for branched PAAs obtained by SCVCP of *t*-BuA with the inimer **1**, followed by hydrolysis: γ=100 (○), 10 (△), 2.5 (□, ■) in aqueous solutions. The *filled symbols* (■) indicate the region where PAA was insoluble in water. (Reproduced with permission from [31]. Copyright 2001 American Chemical Society.)

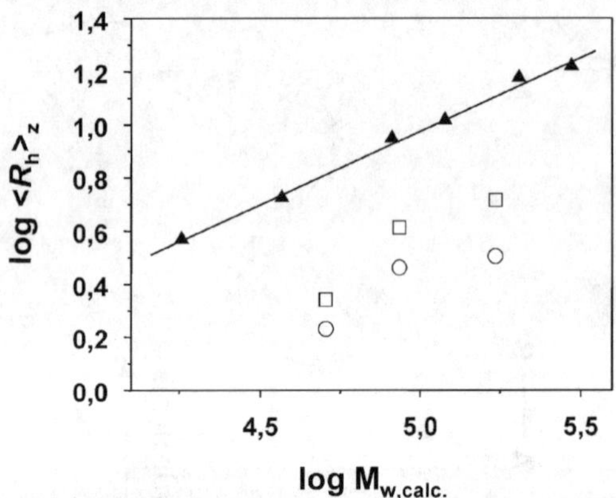

Fig. 10. Dependence of z-average hydrodynamic radius $<R_\mathrm{h}>_\mathrm{z}$ on weight-average molecular weight ($M_\mathrm{w,calc.}$) of the branched PAAs obtained by SCVCP of *t*-BuA with the inimer **1**: pH=3 (○) and pH=10 (□). Linear PAAs (-▲-) at pH=6–8 used as a reference are derived from [137]. (Reproduced with permission from [31]. Copyright 2001 American Chemical Society.)

groups (Scheme 2) [31]. The water solubility of the branched PAAs decreases with increasing DB and decreasing pH. Potentiometric titration curves (Fig. 9) suggest that the apparent pK_a values (taken as the pH at 50% ionization) of the branched PAAs with a comonomer ratio γ=10 and 100 are comparable to the corresponding value for linear PAA homopolymer of $pK_\mathrm{a,app}\cong5.8$ [98]. The branched PAA at γ=2.5 become insoluble at pH≤4.7. For the PAAs with higher DB (γ≤1.5), solubility is only obtained at pH>8, suggesting that these polymers are soluble in water only at a high degree of ionization. However, for low comonomer ratios it has to be taken into account that a large fraction of the polymer (i.e., the linking inimer groups) is in fact non-ionic in nature. Aqueous-phase GPC and dynamic light scattering confirm the compact structure of the randomly branched PAAs. The comparison of the hydrodynamic radius of the branched PAAs and linear ones (obtained from dynamic light scattering, Fig. 10) as a function of M_w suggested that highly compact structures of the polyelectrolytes can be obtained due to their branched architectures. Studies at different pH indicate that a marked stretching of the branched chains takes place when going from a virtually uncharged to a highly charged stage.

Highly branched poly(methacrylic acid) was synthesized by SCVCP of *tert*-butyl methacrylate with the inimer **12** via GTP, followed by hydrolysis [28]. Acid-catalyzed hydrolysis of the *tert*-butyl groups and neutralization with NaOH produced a water-soluble, highly branched poly(sodium methacrylate).

5
Melt Properties

The viscoelastic or rheological behavior of polymer melts, which is governed by the size of the macromolecules and their topology, is strongly related to their industrial applications. For melts of linear monodisperse chains, the viscoelastic spectra provide information about both segmental and chain relaxation times, which, through known scaling dependences, allow for a determination of the macromolecular sizes of test samples. The situation is more complicated for macromolecules with non-linear topologies for which the scaling dependences are different and in many cases not well known or tested. Systems in which both the sizes and the topologies of molecules are not precisely known and can vary in an undefined way seem to be especially difficult. This seems to be the case for the broad distribution of sizes of highly branched, but poorly defined, molecular topologies.

Eight fractions of the highly branched PMMA prepared by the SCVCP of the inimer **12** with MMA ($\gamma=25$) via GTP were characterized by viscoelastic spectroscopy [88]. Figure 11a shows the characteristics of the viscoelastic properties determined for a single fraction of the branched polymer. As for the linear chains, two relaxations can be distinguished here, but they are much less distinctly separated from each other and both have a considerably broader distribution of relaxation times. Nevertheless, four frequency ranges can also be recognized here in the behavior of both the moduli and the viscosity. The flow regime at low frequencies, $\omega \lesssim 1/\tau_c$, is clearly seen and allows for a determination of η_0. The segmental and polymer relaxation times were determined as the intersection point of G' and G'' at high frequencies and as the intersection point of linear extrapolations of logG' and logG'' versus log ω in the distinctly seen terminal-flow range at low frequencies, respectively. The results for the eight fractions are shown in Fig. 11b as a function of M_w determined by means of SEC–UNICAL. This allows one to notice that there is a considerable difference in the segmental mobility between linear and branched polymers. It is also reflected in the glass transition temperatures which are lower by 30–40 K for the

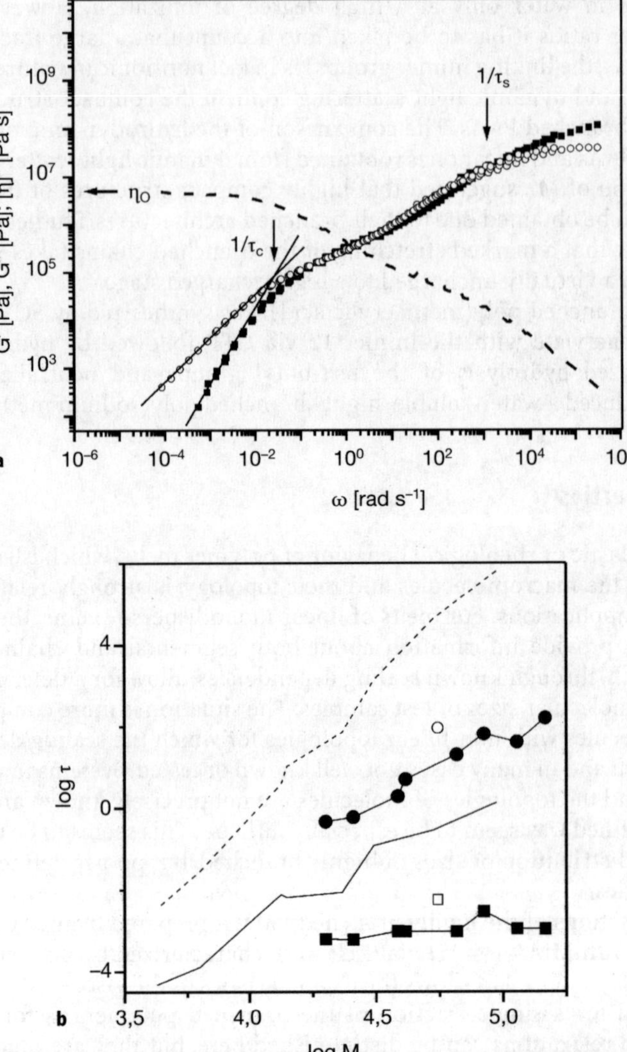

Fig. 11. a Storage and loss moduli, G' (■) and G'' (○), respectively, and melt viscosity (– – –) of a fraction of highly branched PMMA (M_w=96,000) as a function of shear frequency at a reference temperature of 130°C. **b** Dependences of the chain, τ_c (●), and segmental, τ_s (■), relaxation times on the MW for all fractions; the corresponding relaxation times of the feed polymer, τ_c (○) and τ_s (□), are included for reference. Corresponding dependences for linear PMMA are shown by *dashed* and *solid lines*. (Reproduced with permission from [88]. Copyright 2001 American Chemical Society.)

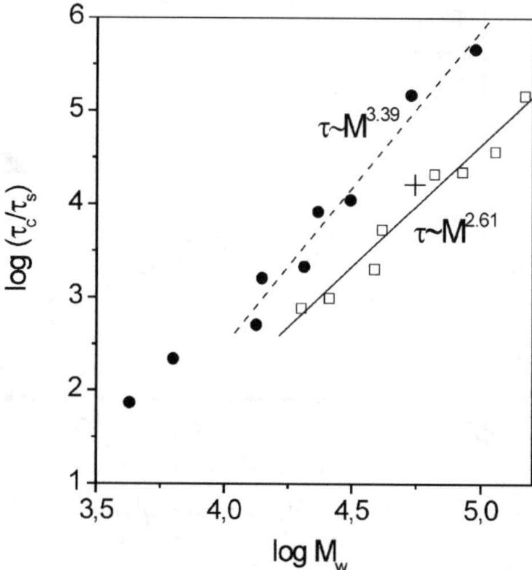

Fig. 12. Molecular weight dependences of the normalized chain relaxation time, τ_c/τ_s, for linear polymers ($\square$), branched fractions ($\bullet$), and branched feed polymer (+). (Reproduced with permission from [88]. Copyright 2001 American Chemical Society.)

branched polymers as compared to the linear ones. This effect can be attributed to a much higher concentration of chain ends in the melts of the branched polymers with respect to the melts of linear chains having the same MWs. To separate this effect from the macromolecular relaxation rates, the ratio τ_c/τ_s is considered further.

The MW dependences of the normalized chain relaxation times in melts of linear and branched samples are compared in Fig. 12. Both can be represented by scaling power laws, but with remarkably different scaling exponents. For the melts of linear chains, the exponent 3.39 is observed close to the typical value of 3.4 for such systems. In contrast, for the fractions of the branched polymer, the exponent is considerably lower (2.61). It is interesting to note that the value of the normalized chain relaxation time for the feed polymer with the broad MWD fits nicely into the data for the fractions with narrow MWDs. This seems to indicate that conclusions can also be drawn from a series of hyperbranched polymers with broad MWDs.

Some information concerning the intramolecular relaxation of the hyperbranched polymers can be obtained from an analysis of the viscoelastic characteristics within the range between the segmental and the terminal relaxation times. In contrast to the behavior of melts with linear chains, in the case of hyperbranched polymers, the range between the distinguished local and terminal relaxations can be characterized by the values of G′ and G″ changing nearly in parallel and by the viscosity variation having a frequency with a considerably different exponent α_{vs}. This can be considered as an indication of the extremely broad spectrum of internal relaxations in these macromolecules. To illustrate this effect, the frequency dependences of the complex viscosities for both linear

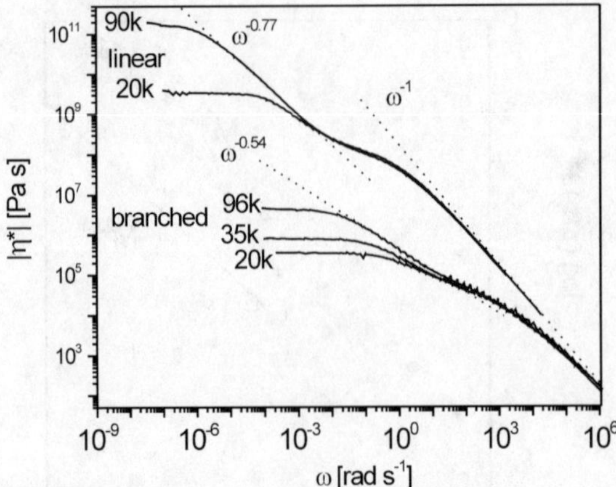

Fig. 13. Frequency dependences of the complex viscosity, η^*, for melts of linear and branched PMMA with different M_w values. (Reproduced with permission from [88]. Copyright 2001 American Chemical Society.)

and hyperbranched polymer melts are compared in Fig. 13. The exponents determined for the longest linear chains and for the fraction of hyperbranched polymers with the highest molecular mass are indicated in the figure. The value of $\alpha_{vs} \approx -0.54$ obtained for the hyperbranched melt strongly resembles the behavior of the microgels [99] and near-critical gels [100, 101] with a scaling exponent of $\alpha_{vs} = -0.5$.

Owing to the lack of chain entanglement that act as physical cross-links in linear chains, hyperbranched polymers are usually not very tough materials. When compared with a linear polymer of the same MW, the additional chain ends in the branched polymers act to decrease the glass transition temperature (T_g), while the restriction in mobility caused by the branch points act to decrease T_g. The influence of the branched architectures on T_g has been investigated using both theoretical and practical approaches. Recently, chain-end free volume theory was extended to study T_g as a function of conversion in hyperbranched polymers [102]. T_g is found to have a nonlinear inverse relationship to the MW for polymers obtained by SCVP. During the monomer conversion process, T_g decreases with the increase in MW in the low conversion range, then levels off in the high conversion range. Experimental T_g values of hyperbranched acrylates obtained by SCVP of the inimer **1** [23], **3** [41], and highly branched PMMA obtained by SCVCP of the inimer **12** with MMA [88] have been reported.

6
Surface-Grafted Hyperbranched Polymers

Highly branched polymers play an increasingly important role in interface and surface sciences, since their distinctive chemical and physical properties can be used advantageously as functional surfaces and as interfacial materials. Due to their highly compact and globular shape, as well as their monodispersity, for

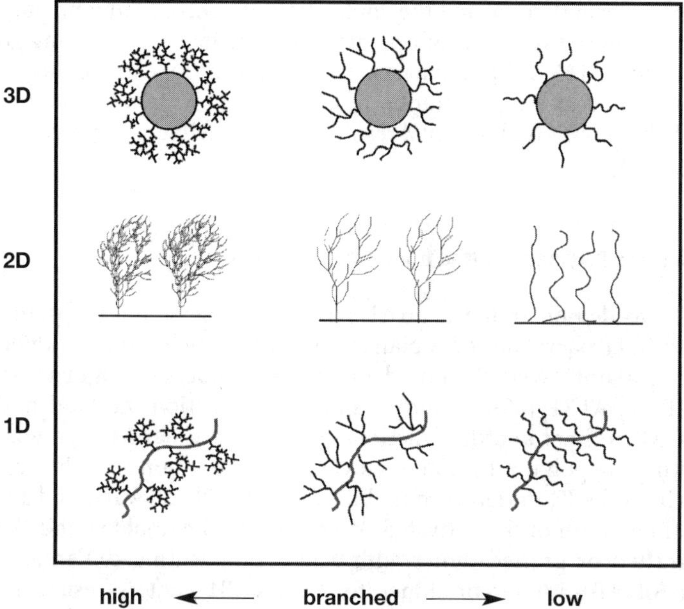

Fig. 14. Surface-grafted hyperbranched, branched, and linear polymers: from 1D to 3D

example, dendrimers attached to flat surfaces [103–106] are useful for many applications, such as data storage or nanolithography systems. A highly branched poly(acrylic acid) film attached to a flat gold surface has been successfully applied to a number of technical applications including corrosion inhibition, chemical sensing, cellular engineering, and micrometer-scale patterning, due to an extremely high density of functional groups at the surface [107]. The surface chemistry and interfacial properties of hyperbranched polymers have also become fields of growing interest [2, 108–115].

Figure 14 summarizes hyperbranched, branched, and linear polymers grafted on surfaces. Depending upon the substrates, it can be divided into 3D, 2D, and 1D hybrids, which correspond to products grafted on spherical particles, planar surfaces, and linear polymers, respectively. Previously, synthesis of dendrimers and hyperbranched polymers grafted onto surfaces has been mainly conducted by "grafting to" techniques [103–106, 115]. A series of repeated "grafting from" steps have also been employed [107, 116, 117]. However, both approaches have the disadvantage that many tedious synthetic steps are necessary to reach the defined surface structures. Dendrimers grafted on a linear polymer chain (1D) have also been synthesized [118–120].

In recent years, much attention has been paid to the use of controlled/ living polymerizations from flat and spherical surfaces [121, 122], because this allows better control over the MW and MWD of the target polymer. By using these techniques, a high grafting density and a controlled film thickness can be obtained, as such brushes consist of end-grafted, strictly linear chains of the same length and the chains are forced to stretch away from the flat surface. Several research groups have recently reported the application of controlled/living polymerization systems to the synthesis of organic/inorganic hybrids involving gold [123, 124] and silica [125–129] nanoparticles. These techniques have also been successfully applied to the preparation of the linear polymers grafted on a linear polymer chain with high density, resulting in the cylindrical polymer brushes [118, 130–133].

6.1
Hyperbranched Polymers Grafted from Planar Surfaces

Recently, we demonstrated a novel synthetic concept for preparing hyperbranched (meth)acrylates on a planar surface in which a silicon wafer grafted with an initiator layer composed of an α-bromoester fragment was used for SCVP via ATRP [48]. Scheme 3 shows the reaction mechanism. Because both the AB* inimer and the functionalized silicon wafer have groups capable of initiating the polymerization of vinyl groups, the chain growth can be started from both the B* initiators immobilized on the silicon wafer and a B* group in the inimer. Both of the activated B* can add to the double bond, A, to form the ungrafted or grafted dimer with a new propagating center, A*. Further addition of AB* inimer or dimer to A* and B* centers results in hyperbranched polymers. The one-step self-condensing ATRP from the surface can be regarded as a novel and convenient approach towards the preparation of smart interfaces.

Self-condensing ATRP of the acrylate-type inimer 1 was found to yield polymer films with a high DB and with a characteristic surface topography. Tapping mode scanning force microscopy (SFM) and X-ray photoelectron spectroscopy (XPS) were used to investigate the surface topography and chemical composition of the grafted hyperbranched polymers. Typical results are shown in Fig. 15. The size and density of the nanoscale protrusions obtained on the surface and the film thickness were observed to depend on the polymerization conditions, such as the ratio $[1]_0:[catalyst]_0$. Similar results were obtained by SCVP of the methacrylate-type inimer 9. In this way, we have been able to create novel surface architectures in which the characteristic nanoprotrusions with different densities and sizes are composed of hyperbranched polymers tethered directly to the surface.

The chain architecture and chemical structure could be modified by SCVCP leading to a facile, one-pot synthesis of surface-grafted branched polymers. The copolymerization gave an intermediate surface topography and film thickness between the polymer protrusions obtained from SCVP of an AB* inimer and the polymer brushes obtained by ATRP of a conventional monomer. The difference in the Br content at the surface between hyperbranched, branched, and linear polymers was confirmed by XPS, suggesting the feasibility to control the surface chemical functionality. The principal result of the works is a demonstration of utility of the surface-initiated SCVP via ATRP to prepare surface-grafted hyperbranched and branched polymers with characteristic architecture and topography.

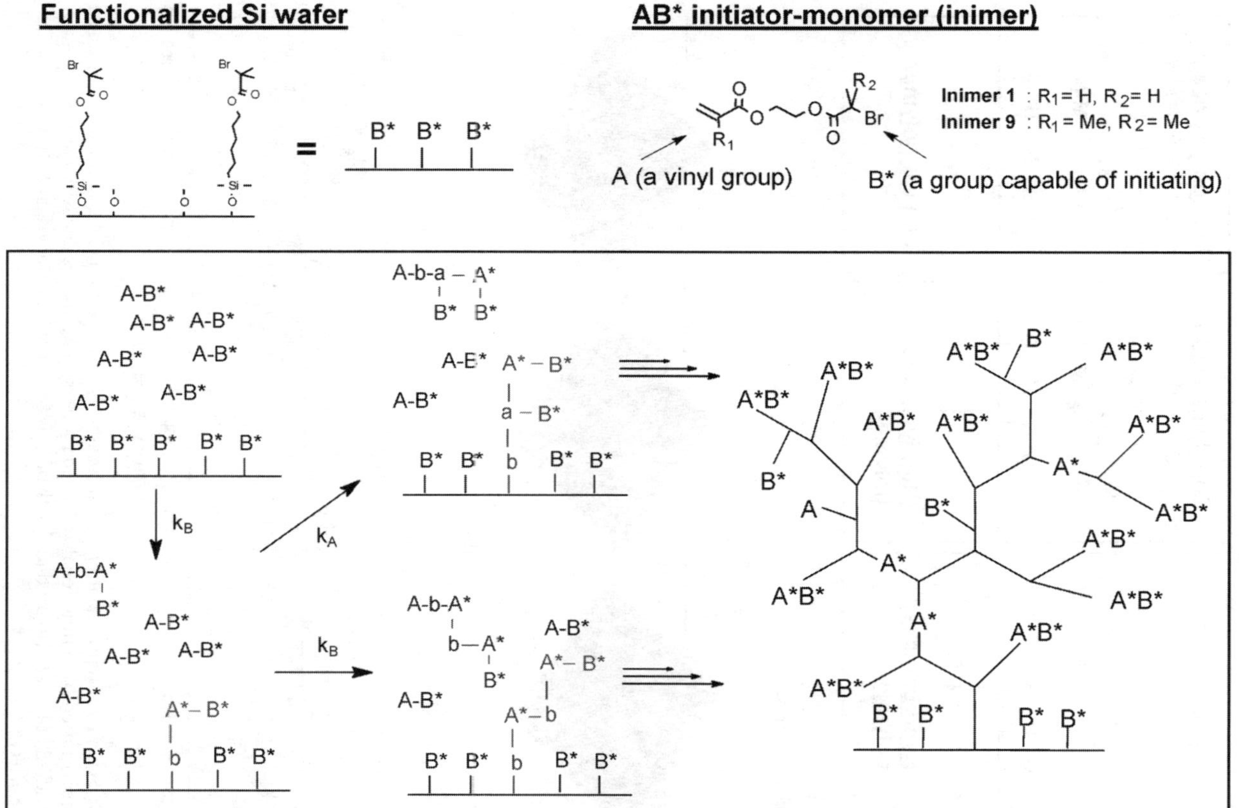

Scheme 3. Schematic illustration of SCVP of an AB* inimer from a functionalized Si surface

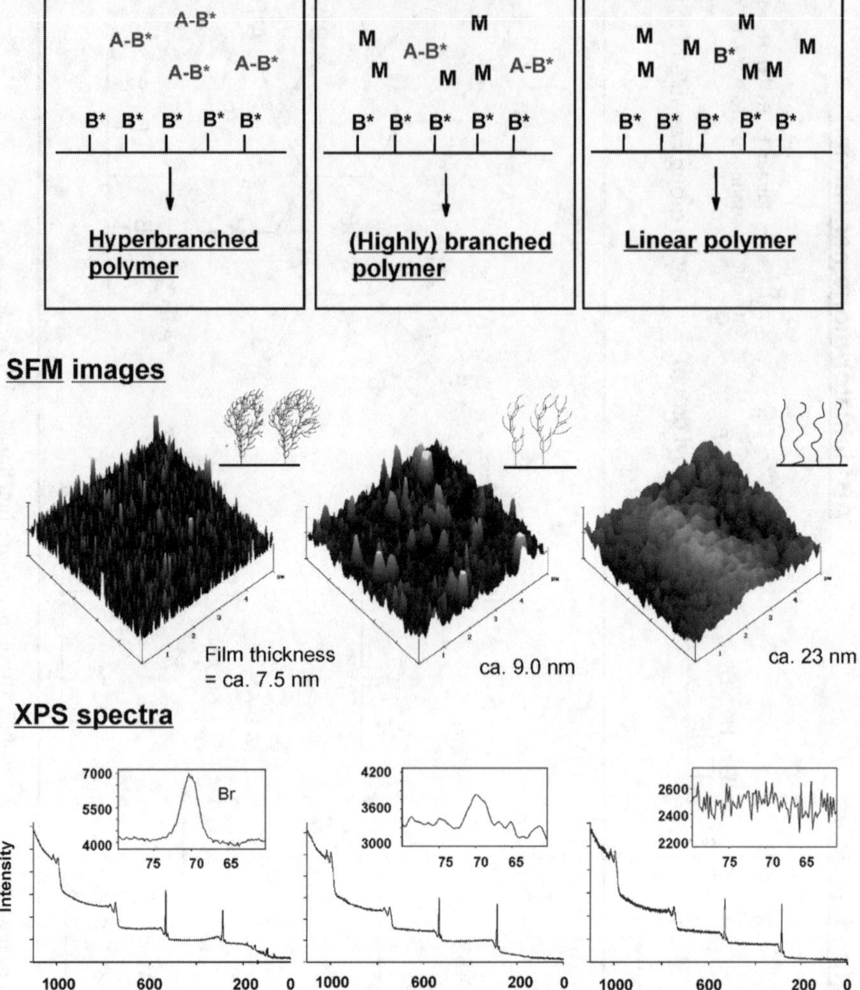

Fig. 15. Schematic representation of the synthesis of hyperbranched, branched, and linear polymers grafted from functionalized silicon wafers; SFM images and XPS spectra of the surface-grafted polymers. (Reproduced with permission from [48]. Copyright 2001 American Chemical Society.)

6.2
Hyperbranched Polymers Grafted from Spherical Particles

Synthesis of hyperbranched polymer-silica hybrid nanoparticles was conducted by SCVP via ATRP from silica surfaces [134]. ATRP initiators were covalently linked to the surface of silica particles, followed by SCVP of the inimer 1 (Scheme 4). Well-defined polymer chains were grown from the surface to yield hybrid nanoparticles comprising silica cores and hyperbranched polymer shells with multifunctional bromoester end groups. Such surface multifunctionality is ideally independent of the surface curvature of the core particle and the layer thickness of the polymer shell, which could not be achieved by linear polymers. SCVCP of the inimer 1 and *t*-BuA from the functionalized silica nanoparticles created branched P*t*-BuA–silica nanoparticles. The functionality of the end groups on the surface, and the chemical composition as well as the structure of the branched polymers grafted on the silica nanoparticles could be controlled by composition in the feed during the SCVCP, as confirmed by elemental analysis and FTIR measurement. Field-emission scanning electron microscopy (FE-SEM, Fig. 16), transmission electron microscopy (TEM), scanning force microscopy (SFM), and dynamic light scattering (DLS) measurements indicate that the hybrid nanoparticles comprising the silica core and the hyperbranched polymer shell exist as isolated and aggregated forms. Novel hybrid nanoparticles with branched polyelectrolytes, poly(acrylic acid) (PAA)–silica, were obtained after hydrolysis of linear segments of the branched P*t*-BuA.

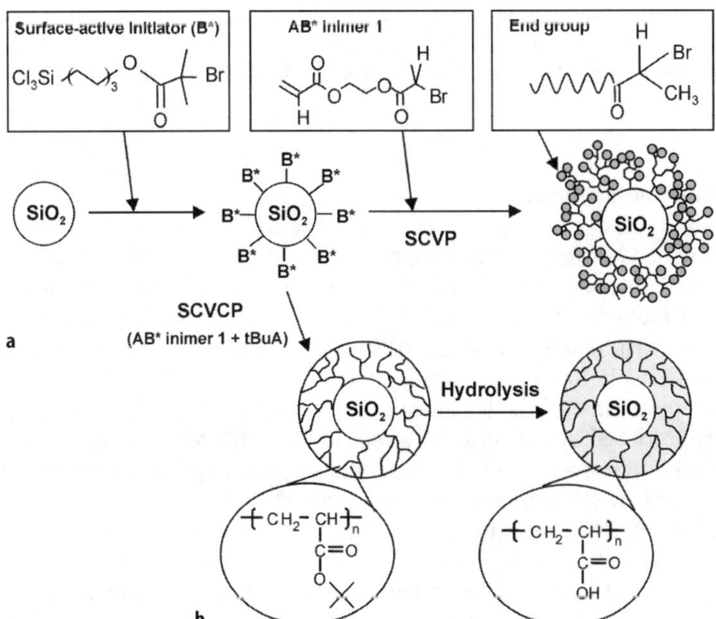

Scheme 4. Synthetic routes for a silica particle with hyperbranched polymer shell (**a**) and branched polyelectrolyte shell (**b**)

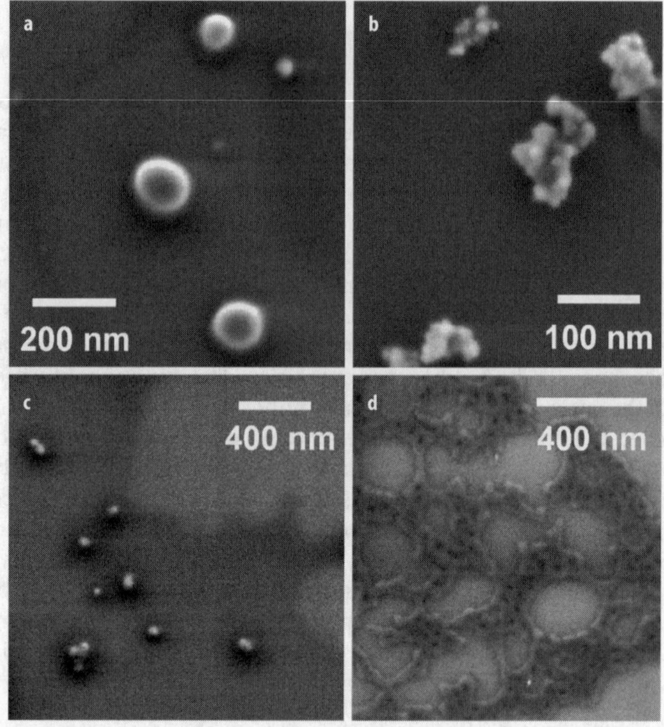

Fig. 16. Representative FE-SEM images of the branched P*t*-BuA-silica hybrid nanoparticles obtained by SCVCP of *t*-BuA with the inimer **1** at γ=6.1 (**a, b**) and γ=1.1(**c, d**). (Reproduced with permission from [134]. Copyright 2001 American Chemical Society.)

6.3
Theoretical Considerations

The calculated MW of polymers formed in SCVP without initiators (conventional SCVP in bulk or solution) is broader than that obtained from SCVP in the presence of *f*-functional initiators [40, 71]. The presence of multifunctional initiators leads to a considerable narrowing of the polydispersity index, which decreases with increasing initiator functionality, *f*. Thus, the MW and MWD of the ungrafted polymer obtained in solution might be different from those of the grafted polymer produced by a surface-initiated SCVP. On the other hand, the effect of *f*-functional initiators on the DB was calculated to be negligible under batch conditions used here (inimers and initiators grafted on the surface are mixed instantaneously) [40]. This indicates that the DB does not depend on whether polymer is formed in solution or on a surface. Therefore, it is reasonable to suppose that SCVP of inimer **1** with functionalized silica particles (or silicon wafers) provides surface-grafted poly(acrylate) with a highly branched structure, even if the correlation of the molecular parameters of the soluble polymers with the polymers grafted on the surface is not confirmed experimentally.

Major drawbacks of hyperbranched polymers are the lack of controlling MW and broad MWD. Hence, a challenging goal in this field is the development of general AB* polymerization methods that achieve control over DB and narrow MWD. Very few examples of the controlled approach to hyperbranched polymers have been reported. Frey et al. reported the controlled chain-growth-type of approach to hyperbranched polymers based on the anionic ring-opening self-condensing polymerization of glycidol, which is another AB* monomer [4, 59]. The MWD of the resulting polyglycerols were reported to be in the range of $M_w/M_n=1.1–1.3$. On the other hand, Bharathi and Moore recently reported a new hyperbranched procedure (one-step AB_2-type polycondensation) which takes place on an insoluble solid support, providing polymers with low polydispersity and controlled MW [135, 136]. Recently, a series of hyperbranched (meth)acrylates were synthesized by SCVP of the inimers 7 [45] and 11 [49] in the presence of a tetrafunctional ATRP initiator capped with bromoester fragment, which was used as the core-forming molecule. A decrease of the polydispersity was observed by the addition of the tetrafunctional initiator.

7
Summary and Perspective

In this review, we have described the synthesis of hyperbranched (meth)acrylates. We have shown that the solution and melt properties are considerably different from their linear analogs, due to their compact, nonentangled structure. SCV(C)P has become a valuable tool in synthesis of hyperbranched polymers from vinyl monomers. Theoretical investigations help to obtain information on the molecular parameters of the resulting hyperbranched polymers which often could not be obtained experimentally. Studies on the solution and melt properties help one to understand the relationship between the properties and molecular parameters (DB, MW, distribution of branching points), which are extremely valuable from both industrial and scientific viewpoints.

A challenging goal in this field, particularly from the synthetic point of view, is the development of general AB* polymerization methods that achieve control over DB and narrow MWDs. Experimental results and theoretical studies mentioned above suggest that the SCV(C)P from surfaces, which are functionalized with monolayers of initiators, permit a controlled polymerization, resulting structural characteristics (molecular weight averages, DB) of hyperbranched polymers. In particular, it is expected that the use of polyfunctional initiators with a different number of initiator functionality, copolymerization, and slow monomer addition techniques lead to control the molecular parameters.

Owing to multi-functionality, physical properties such as solubility and the glass transition temperature and chemical functionality the hyperbranched (meth)acrylates can be controlled by the chemical modification of the functional groups. The modifications of the chain architecture and chemical structure by SCV(C)P of inimers and functional monomers, which may lead to a facile, one-pot synthesis of novel functionalized hyperbranched polymers, is another attractive feature of the process. The procedure can be regarded as a convenient approach toward the preparation of the chemically sensitive interfaces.

The one-step SCV(C)P from the surfaces was used for the development of new hyperbranched polymer-inorganic 2D and 3D hybrid materials. Because these hyperbranched polymers contain a high density of functional groups, they are suitable for a number of technological applications including corrosion inhibition, chemical sensing, cellular engineering, or catalysis. Furthermore, the 3D hybrids with various functional groups, such as carboxylic acid groups, can be used as fundamental buildings blocks for the synthesis of macromolecular clusters with a higher order of complexity. Thus, a well-controlled synthesis for these materials is considered to lead the creation of an entirely new category of materials that are controllable on nanoscopic scale and have chemically sensitive interfaces.

Acknowledgement. The authors are grateful to Wolfgang Radke, Peter F.W. Simon, Universität Mainz; Alexander Böker, Delphine Chan Seng, Guanglou Cheng, Georg Krausch, Hans Lechner, Mingfu Zhang, Universität Bayreuth; Tadeusz Pakula, Max-Planck Institute for Polymer Research, Mainz; Deyue Yan, Shanghai Jiao Tong University; Galina I. Litvinenko, Karpov Institute of Physical Chemistry, Moscow, for providing the theoretical and experimental basis for this review article. The authors wish to thank the Deutsche Forschungsgemeinschaft (DFG) for financial support.

8
References

1. Jikei M, Kakimoto M (2001) Prog Polym Sci 26:1233
2. Voit B (2000) J Polym Sci Part A:Polym Chem 38:2505
3. Ambade AV, Kumar A (2000) Prog Polym Sci 25:1141
4. Sunder A, Heinemann J, Frey H (2000) Chem –Eur J 6:2499
5. Flory PJ (1952) J Am Chem Soc 74:2718
6. Flory PJ (1953) Principles of polymer chemistry. Cornell University Press, Ithaca, NY
7. Hunter WH, Woollett GH (1921) J Am Chem Soc 43:135
8. Jacobson RA (1932) J Am Chem Soc 54:1513
9. Baker AS, Walbridge DJ (1972) US Patent 3 669 939
10. Kricheldorf HR, Zang QZ, Schwarz G (1982) Polymer 23:1821
11. Kim YH, Webster OW (1992) Macromolecules 25:5561
12. Kim YH, Webster OW (1990) J Am Chem Soc 112:4592
13. Kim YH (1992) J Am Chem Soc 114:4947
14. Brenner AR, Schmaljohann D, Voit BI, Wolf D (1997) Macromol Symp 122:217
15. Hult A, Johansson M, Malmstrom E (1999) Adv Poly Sci143:1
16. Malmstroem E, Hult A (1997) J Macromol Sci Rev Macromol Chem Phys C37:555
17. Galina H, Lechowicz JB (2002) e-Polymers:No 12
18. Fréchet JMJ, Henmi M, Gitsov I, Aoshima S, Leduc MR, Grubbs RB (1995) Science (Washington DC) 269:1080
19. Baskaran D (2001) Macromol Chem Phys 202:1569
20. Simon PFW, Radke W, Müller AHE (1997) Macromol Rapid Commun 18:865
21. Hawker CJ, Fréchet JMJ, Grubbs RB, Dao J (1995) J Am Chem Soc 117:10763
22. Weimer MW, Fréchet JMJ, Gitsov I (1998) J Polym Sci Part A Polym Chem 36:955
23. Matyjaszewski K, Gaynor SG, Kulfan A, Podwika M (1997) Macromolecules 30:5192
24. Matyjaszewski K, Gaynor SG, Müller AHE (1997) Macromolecules 30:7034
25. Matyjaszewski K, Gaynor SG (1997) Macromolecules 30:7042
26. Fréchet JMJ, Aoshima S (1996) WO9614345; WO9614346
27. Gaynor SG, Edelman S, Matyjaszewski K (1996) Macromolecules 29:1079
28. Simon PFW, Müller AHE (2001) Macromolecules 34:6206

29. Paulo C, Puskas JE (2001) Macromolecules 34:734
30. Hong C-Y, Pan C-Y (2002) Polymer Int 51:785
31. Mori H, Chan Seng D, Lechner H, Zhang M, Müller AHE (2002) Macromolecules 35:9270
32. Kennedy JP, Frisch KC Jr (1982) US4327201; US4442261
33. Nuyken O, Gruber F, Pask SD, Riederer A, Walter M (1993) Makromolekulare Chemie 194:3415
34. Jones GD, Runyon JR, Ong J (1961) J Appl Polymer Sci 5:452
35. Powers KW, Kuntz I (1978) US4074035
36. Hazer B (1992) Makromol Chem 193:1081
37. Hazer B (1991) Macromol Rep A28 (Supp 1):47
38. Müller AHE, Yan D, Wulkow M (1997) Macromolecules 30:7015
39. Puskas JE, Grasmueller M (1998) Macromol Symp 132:117
40. Radke W, Litvinenko GI, Müller AHE (1998) Macromolecules 31:239
41. Yoo SH, Yoon TH, Jho JY (2001) Macromol Rapid Commun 22:1319
42. Yoo SH, Lee JH, Lee J-C, Jho JY (2002) Macromolecules 35:1146
43. Bibiao J, Yang Y, Xiang J, Rongqi Z, Jianjun H, Wenyun W (2001) Eur Poly J 37:1975
44. Bibiao J, Yang Y, Jian D, Shiyang F, Rongqi Z, Jianjun H, Wenyun W (2002) J Appl Poly Sci 83:2114
45. Hong CY, Pan CY (2001) Polymer 42:9385
46. Cheng G, Simon PFW, Hartenstein M, Müller AHE (2000) Macromol Rapid Commun 21:846
47. Matyjaszewski K, Pyun J, Gaynor SG (1998) Macromol Rapid Commun 19:665
48. Mori H, Böker A, Krausch G, Müller AHE (2001) Macromolecules 34:6871
49. Hong Cy, Pan Cy, Huang Y, Xu Zd (2001) Polymer 42:6733
50. Sakamoto K, Aimiya T, Kira M (1997) Chem Lett 1245
51. Ishizu K, Shibuya T, Mori A (2002) Polym Int 51:424
52. Ishizu K, Mori A, Shibuya T (2002) Des Monomers Polym 5:1
53. Ishizu K, Mori A (2000) Macromol Rapid Commun 21:665
54. Ishizu K, Mori A (2001) Polym Int 51:50
55. Ishizu K, Ohta Y, Kawauchi S (2002) Macromolecules 35:3781
56. Lu P, Paulasaari J, Weber WP (1996) Macromolecules 29:8583
57. Zhang H, Ruckenstein E (1997) Polym Bull (Berlin) 39:399
58. Aoshima S, Fréchet JMJ, Grubbs RB, Henmi M, Leduc L (1995) Polym Prepr (Am Chem Soc Div Polym Chem) 36:531
59. Sunder A, Hanselmann R, Frey H, Mülhaupt R (1999) Macromolecules 32:4240
60. Hobson LJ, Feast WJ (2000) J Mater Chem 10:609
61. Zhang X, Liu WH, Chen YM, Gong AJ, Chen CF, Xi F (1999) Polym Bull (Berlin) 43:29
62. Guan Z (2002) J Am Chem Soc 124:5616
63. Kadokawa J, Ikuma K, Tagaya H (2002) J Macromol Sci Pure Appl Chem A39:879
64. Matyjaszewski K, Editor (2000) Controlled/living radical polymerization. Progress in ATRP, NMP, and RAFT. Proceedings of a Symposium on Controlled Radical Polymerization held on 22-24 August 1999, in New Orleans. ACS Symp Ser 2000:768
65. Qiu J, Charleux B, Matyjaszewski K (2001) Progr Poly Sci 26:2083
66. Matyjaszewski K, Xia J (2001) Chem Rev (Washington DC) 101:2921
67. Kamigaito M, Ando T, Sawamoto M (2001) Chem Rev (Washington DC) 101:3689
68. Hawker CJ, Bosman AW, Harth E (2001) Chem Rev (Washington DC) 101:3661
69. Rizzardo E, Chiefari J, Mayadunne R, Moad G, Thang S (2001) Macromol Symp 174:209
70. Yan D, Müller AHE, Matyjaszewski K (1997) Macromolecules 30:7024
71. Yan D, Zhou Z, Müller AHE (1999) Macromolecules 32:245
72. Litvinenko GI, Müller AHE (2002) Macromolecules 35:4577
73. Litvinenko GI, Simon PFW, Müller AHE (1999) Macromolecules 32:2410
74. Litvinenko GI, Simon PWF, Müller AHE (2001) Macromolecules 34:2418
75. Dušek K, Šomvársky J, Smrcková M, Simonsick WJ, Wilczek L (1999) Polym Bull 42:489
76. Burgath A, Sunder A, Frey H (2000) Macromol Chem Phys 201:782
77. Galina H, Lechowicz JB, Kaczmarski K (2001) Macromol Theory Simul 10:174

78. Hölter D, Frey H (1997) Acta Polym 48:298
79. Hanselmann R, Hölter D, Frey H (1998) Macromolecules 31:3790
80. Wyatt PJ (1993) Anal Chim Acta 272:1
81. Wyatt PJ (1993) J Chromatography 648:27
82. Benoît H, Grubisic Z, Rempp P, Decker D, Zilliox JG (1966) J Chem Phys 63:1507
83. Sanayei RA, Suddaby KG, Rudin A (1993) Makromolekulare Chemie 194:1953
84. Burchard W (1999) Adv Polym Sci 143:113
85. Hobson LJ, Feast WJ (1997) Chem Commun 2067
86. Maier G, Zech C, Voit B, Komber H (1998) Macromol Chem Phys 199:2655
87. Jackson C, Chen Y-J, Mays JW (1996) J Appl Poly Sci 59:179
88. Simon PFW, Müller AHE, Pakula T (2001) Macromolecules 34:1677
89. Dautzenberg H, Jaeger W, Kotz J, Philipp B, Stscherbina D (1994) Polyelectrolytes: Form Charact Appl
90. Smits RG, Koper GJM, Mandel M (1993) J Phys Chem 97:5745
91. Borkovec M, Koper GJM (1998) Prog Colloid Polym Sci 109:142
92. Muthukumar M (1993) Macromolecules 26:3904
93. Wolterink JK, Leermakers FAM, Fleer GJ, Koopal LK, Zhulina EB, Borisov OV (1999) Macromolecules 32:2365
94. Borkovec M, Koper GJM (1997) Macromolecules 30:2151
95. Borisov OV, Birshtein TM, Zhulina EB (1992) Prog Colloid Polym Sci 90:177
96. Borisov OV, Zhulina EB (1998) Eur Phys J B 4:205
97. Misra S, Mattice WL, Napper DH (1994) Macromolecules 27:7090
98. Sato Y, Hashidzume A, Morishima Y (2001) Macromolecules 34:6121
99. Antonietti M, Rosenauer C (1991) Macromolecules 24:3434
100. Chambon F, Petrovic ZS, MacKnight WJ, Winter HH (1986) Macromolecules 19:2146
101. Muthukumar M, Winter HH (1986) Macromolecules 19:1284
102. Liu H, Wilen C-E, Shi W (2002) Macromol Theory Simul 11:459
103. Hierlemann A, Campbell JK, Baker LA, Crooks RM, Ricco AJ (1998) J Am Chem Soc 120:5323
104. Li J, Piehler LT, Qin D, Baker JR, Jr., Tomalia DA, Meier DJ (2000) Langmuir 16:5613
105. Tully DC, Trimble AR, Fréchet JMJ, Wilder K, Quate CF (1999) Chem Mater 11:2892
106. Tully DC, Fréchet JMJ (2001) Chem Commun 1229
107. Zhou Y, Bruening ML, Bergbreiter DE, Crooks RM, Wells M (1996) J Am Chem Soc 118:3773
108. Bergbreiter DE, Tao G, Franchina JG, Sussman L (2001) Macromolecules 34:3018
109. Bergbreiter DE, Tao G (2000) J Polym Sci Part A Polym Chem 38:3944
110. Fujiki K, Sakamoto M, Sato T, Tsubokawa N (2000) J Macromol Sci, Pure Appl Chem A37:357
111. Nakayama Y, Sudo M, Uchida K, Matsuda T (2002) Langmuir 18:2601
112. Hayashi S, Fujiki K, Tsubokawa N (2000) React Funct Polym 46:193
113. Mackay ME, Carmezini G, Sauer BB, Kampert W (2001) Langmuir 17:1708
114. Beyerlein D, Belge G, Eichhorn K-J, Gauglitz G, Grundke K, Voit B (2001) Macromol Symp 164:117
115. Sidorenko A, Zhai XW, Peleshanko S, Greco A, Shevchenko VV, Tsukruk VV (2001) Langmuir 17:5924
116. Bruening ML, Zhou Y, Aguilar G, Agee R, Bergbreiter DE, Crooks RM (1997) Langmuir 13:770
117. Franchina JG, Lackowski WM, Dermody DL, Crooks RM, Bergbreiter DE, Sirkar K, Russell RJ, Pishko MV (1999) Anal Chem 71:3133
118. Sheiko SS, Möller M (2001) Chem Rev (Washington DC) 101:4099
119. Schlüter AD (1998) Top Curr Chem 197:165
120. Schlüter AD, Rabe JP (2000) Angewandte Chemie Int Ed 39:864
121. Zhao B, Brittain WJ (2000) Prog Polym Sci 25:677
122. Pyun J, Matyjaszewski K (2001) Chem Mater 13:3436
123. Watson KJ, Zhu J, Nguyen ST, Mirkin CA (1999) J Am Chem Soc 121:462

124. Jordan R, West N, Ulman A, Chou Y-M, Nuyken O (2001) Macromolecules 34:1606
125. Bottcher H, Hallensleben ML, Nuss S, Wurm H (2000) Polym Bull (Berlin) 44:223
126. von Werne T, Patten TE (1999) J Am Chem Soc 121:7409
127. Perruchot C, Khan MA, Kamitsi A, Armes SP, von Werne T, Patten TE (2001) Langmuir 17:4479
128. von Werne T, Patten TE (2001) J Am Chem Soc 123:7497
129. Pyun J, Matyjaszewski K, Kowalewski T, Savin D, Patterson G, Kickelbick G, Huesing N (2001) J Am Chem Soc 123:9445
130. Beers KL, Gaynor SG, Matyjaszewski K, Sheiko SS, Möller M (1998) Macromolecules 31:9413
131. Börner HG, Beers K, Matyjaszewski K, Sheiko SS, Möller M (2001) Macromolecules 34:4375
132. Cheng G, Böker A, Zhang M, Krausch G, Müller AHE (2001) Macromolecules 34:6883
133. Zhang M, Breiner T, Mori H, Müller AHE (2003) Polymer 44:1449
134. Mori H, Chan Seng D, Zhang M, Müller AHE (2002) Langmuir 18:3682
135. Bharathi P, Moore JS (1997) J Am Chem Soc 119:3391
136. Bharathi P, Moore JS (2000) Macromolecules 33:3212
137. Reith D, Müller B, Müller-Plathe F, Wiegand S (2002) J Chem Phys 116:9100

Top Curr Chem (2003) 228: 39–63
DOI 10.1007/b11005

Metallodendrimers Composed of Organometallic Building Blocks

Kiyotaka Onitsuka · Shigetoshi Takahashi

The Institute of Scientific and Industrial Research, Osaka University, 8-1 Mihogaoka, Ibaraki, Osaka 567-0047, Japan
E-mail: onitsuka@sanken.osaka-u.ac.jp, takahashi@sanken.osaka-u.ac.jp

The incorporation of metallic species into dendritic molecules has been attracting much attention, because the addition of properties characteristic of metallic complexes, such as magnetic, electronic, and photo-optical properties, as well as reactivity, may lead to the realization of new functionalized dendrimers. Organometallic dendrimers offer several advantages in the design of a dendritic molecule with the desired functions due to the diversity of the structure and properties of the organometallic complexes. In the past, the focus was on organometallic dendrimers with metallic species only at specific positions of the molecules, such as the core and the periphery. In this chapter, we summarize recent developments in the synthesis of organometallic dendrimers, in which organometallic species act as building block in every generation. Such dendrimers are generated by successive reactions characteristic of organometallic complexes, and some of them show interesting properties.

Keywords. Metallodendrimer, Organometallic complex, Metal acetylide

1
Introduction

Recently, increasing attention has been focused on the functionalization of dendrimers, which are regularly hyper-branched macromolecules with well-defined nanostructures [1]. A number of interesting dendrimers with functional organic groups have been reported [2]. On the other hand, incorporation of metallic species into dendritic molecules has enabled access to highly ordered materials with attractive magnetic, electronic, and photo-optical properties as well as characteristic reactivity [3]. Balzani's group [4] and Newkome et al. [5] independently initiated such studies by using polypyridine Ru complexes in the early 1990s. Organometallic dendrimers offer several advantages for tailoring dendritic molecules with desirable functionalities due to not only the availability of a wide variety of organic compounds that coordinate to many kinds of metal atoms but also the flexibility of the coordination modes between metal atoms and organic ligands. The appropriate choice of organometallic complexes for the core and the building block allows us to design a novel dendritic molecule in which the number and position of the metallic species are precisely controlled, and specific applications such as catalysts, multi-redox system, and molecular sensors are realized [6].

From the view point of structural feature, organometallic dendrimers can be classified into two categories. One has metallic species at specific positions of the dendritic molecules, and the other has metallic species in every generation. Each of them is further classified into two types, as shown in Fig. 1. The former consists of (**A**) core type (Fig. 1A) and (**B**) surface type (Fig. 1B) dendrimers, both of which have been often applied to dendritic catalysts. The latter consists of (**C**) the side chain type (Fig. 1C), in which the skeleton of dendritic molecules is composed of organic units and metallic species are incorporated as structural auxiliaries, and (**D**) the main chain type (Fig. 1D), in which metallic species are used as a member of repeating units. Most of the organometallic dendrimers that have been reported so far belong to types **A**, **B**, or **C** because those dendrimers can be prepared by multi-step successive organic reactions, which are similar to those used for organic dendrimers, and a single organometallic reaction. In contrast, type **D** organometallic dendrimers have to be built up by multi-step successive organometallic reactions and therefore the reactions characteristic of organometallic complexes can be utilized for the preparation of such dendrimers. However, organometallic complexes suitable for use as repeating units of dendrimers are rare due to their low stability relative to organic and inorganic molecules. Type **D** dendrimer is a fascinating molecule not only as a synthetic target but also as a novel nano-sized material.

Some reviews on organometallic dendrimers have already appeared [6]. However, those review articles mainly focus on types **A**, **B**, and **C** organometallic dendrimers. In this review we summarize recent developments in the synthesis of type **D** organometallic dendrimers, including our contributions to this young and rapidly growing field. Although illustration of the primary structure is attempted for as many organometallic dendrimers as possible, for some molecules only one of the dendritic wedges around the core is shown and the other

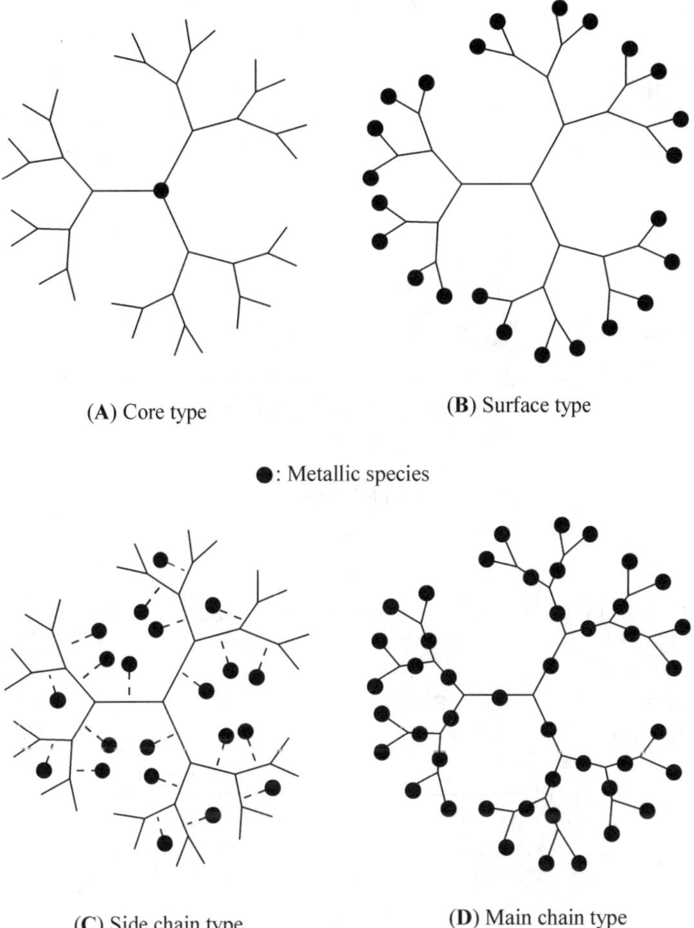

Fig. 1A–D. Classification of organometallic dendrimers by position of metal atoms

dendritic wedges are drawn using a sector with DW, because the full two-dimensional representation is fairly difficult for the higher generation dendrimers due to page space.

2
Dendrimers Based on Alkyl-Metal Complexes

In 1994, Achar and Puddephatt reported the first example of an organometallic dendrimer that involves metallic species in every generation [7]. Their synthetic strategy of alkylplatinum dendrimers was based on reactions characteristic of organometallic complexes, and is shown as follows: (1) oxidative addition of benzyl bromide to PtMe$_2$(bpy) (bpy=2,2′-bipyridyl) generating benzylplatinum(IV) species; and (2) ligand exchange reaction of [PtMe$_2$(μ-SMe$_2$)]$_2$ with bpy to give

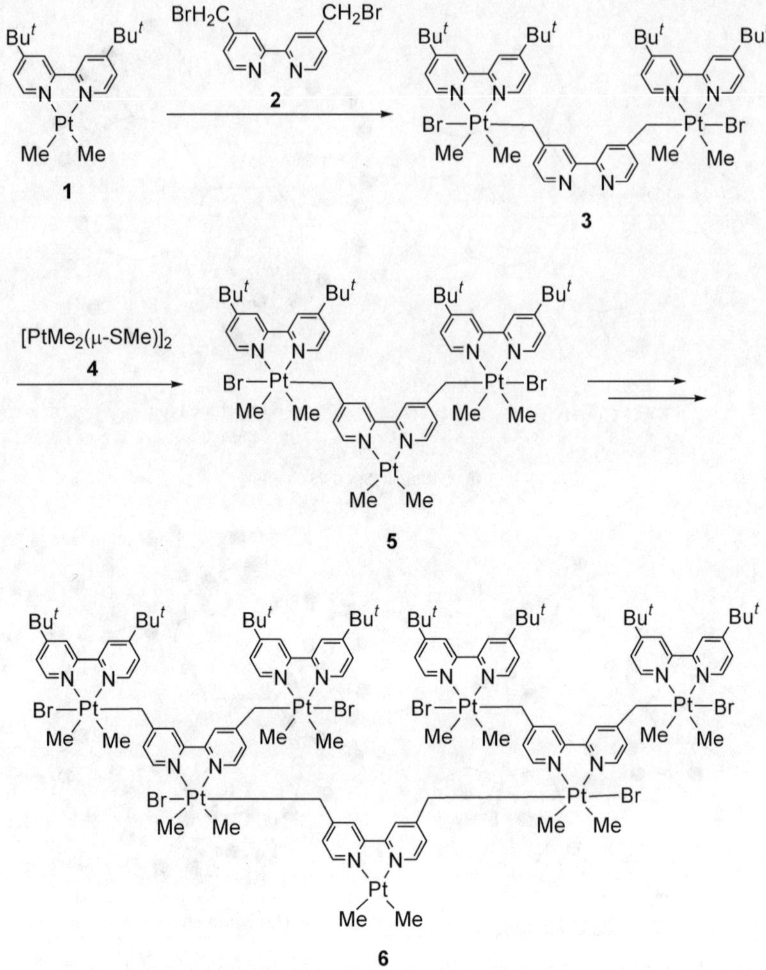

Scheme 1. Synthesis of alkylplatinum dendrons **5** and **6**

PtMe$_2$(bpy) complexes. Thus, they prepared organoplatinum dendrons from PtMe$_2$(4,4'-Bu$^t_2$bpy) (**1**) by repeating the reaction with 4,4'-di(bromomethyl)-2,2'-bipyridine (**2**) followed by the reaction with [PtMe$_2$(μ-SMe$_2$)]$_2$ (**4**) to give the first-generation dendron (**5**) (Scheme 1). Although the second-generation dendron (**6**) containing seven atoms of Pt in the molecule could be prepared, further growth of alkylplatinum dendrons was unsuccessful due to a decrease in the reactivity at the focal point. Treatment of the first- and second-generation dendrons with 1,2,4,5-tetra(bromomethyl)benzene resulted in the formation of the first- and second-generation dendrimers [8]. The second-generation dendrimer **7**, shown in Fig. 2, contains 28 Pt atoms in the molecule. A similar organoplatinum dendrimer modified by ferrocenyl groups at its periphery was also prepared by using PtMe$_2$[4,4'-(FcCH=CH)$_2$bpy] as the starting material [9].

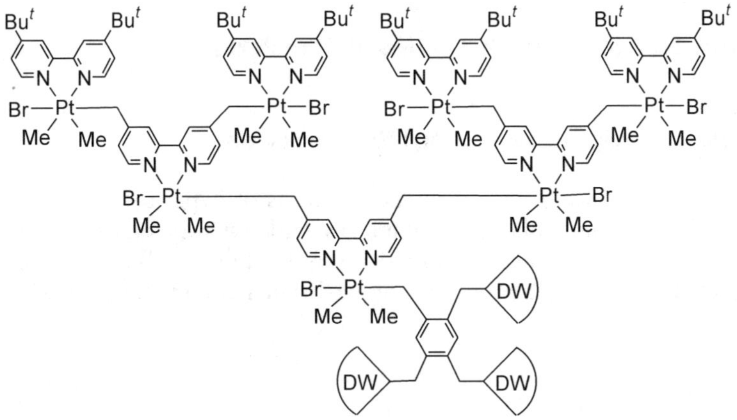

Fig. 2. Second generation alkylplatinum dendrimer 7

A divergent route for the synthesis of organoplatinum dendrimers has been developed by using a bis(bpy) compound as the building block [10]. Although a first-generation dendrimer was prepared by the reaction of a trinuclear platinum complex with the bis(bpy) compound followed by treatment with $[PtMe_2(\mu\text{-}SMe_2)]_2$, further growth was prevented due to low solubility of the resulting dendrimer. However, mixed-metal organometallic dendrimer (8) was prepared by treatment with $[PdMe_2(\mu\text{-}SMe_2)]_2$ instead of $[PtMe_2(\mu\text{-}SMe_2)]_2$ (Fig. 3).

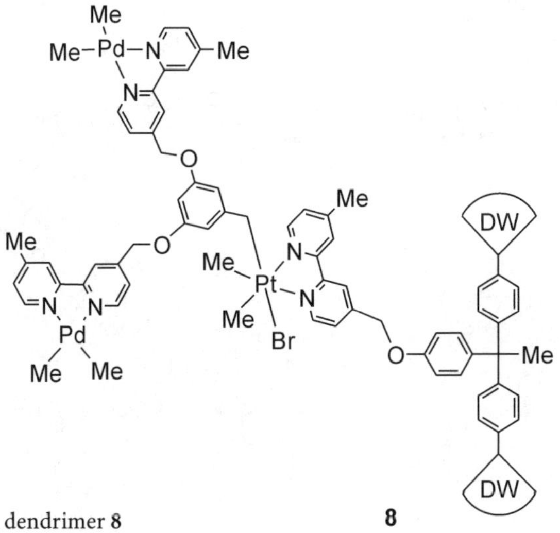

Fig. 3. Mix-metal dendrimer 8

3
Dendritic Sphere Based on Aryl-Metal Complexes

3.1
Hyper-Branched Polymers Based on Aryl-Metal Complexes

Reinhoudt and coworkers studied the synthesis of hyper-branched polymers composed of organopalladium complexes with an SCS pincer ligand [11]. Removal of acetonitrile ligands on palladium led to the self-assembly of dinuclear palladium complex (9) to give hyper-branched polymer (10), which was

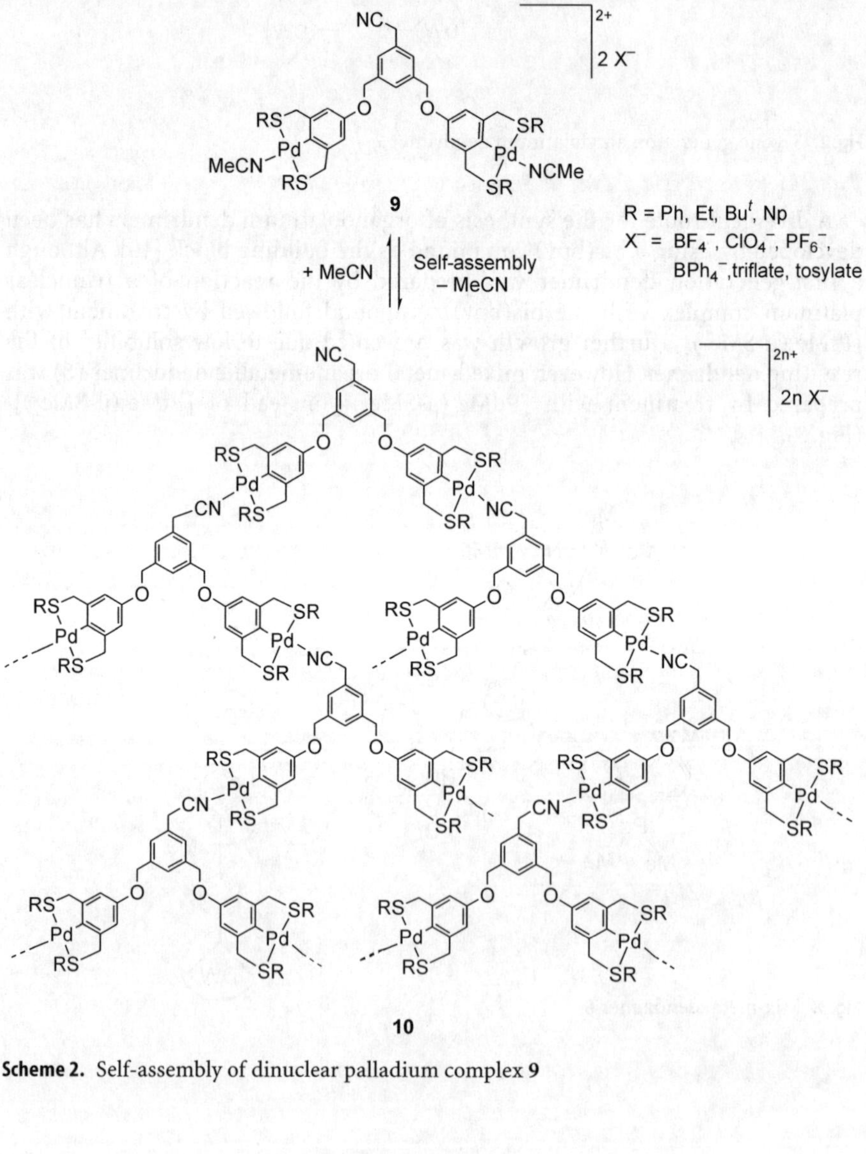

Scheme 2. Self-assembly of dinuclear palladium complex **9**

easily converted into the starting material by the addition of small amounts of acetonitrile (Scheme 2). Observation of the self-assembled sample by means of AFM and TEM revealed that the hyper-branched sphere is about 200 nm in diameter on average. The size of the hyper-branched polymer could be controlled by the substituents on the pincer ligand and the counter anion [12]. Larger substituents and/or counter anion tended to produce smaller assemblies.

3.2
Dendrimers Based on Aryl-Metal Complexes

Reinhoudt and coworkers reported the divergent synthesis of arylpalladium dendrimers, in which the assembly of the organopalladium building block was controlled by chloride ligands that prevent the coordination of cyano groups (Scheme 3) [13]. Treatment of trinuclear palladium complex (**12**), which is regarded as a zero generation dendrimer, with AgBF$_4$, followed by the addition of 3 equiv. of building block (**11**) produced the first-generation dendrimer (**13**) in a good yield. Repetition of the above procedure gave dendrimers up to the third generation (**14**). An organopalladium building block possessing a 4-pyridyl group at the focal point was also prepared and was suc-

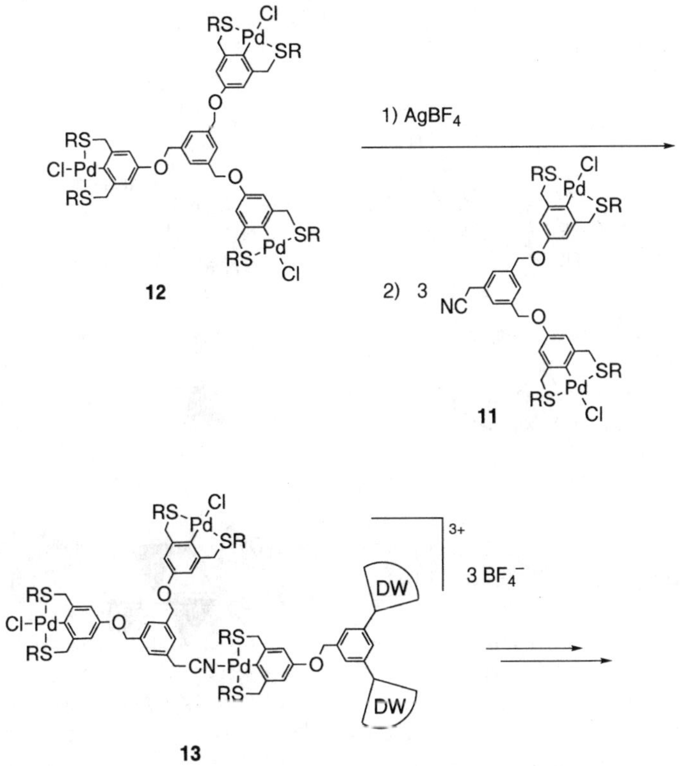

Scheme 3. Divergent synthesis of arylpalladium dendrimers

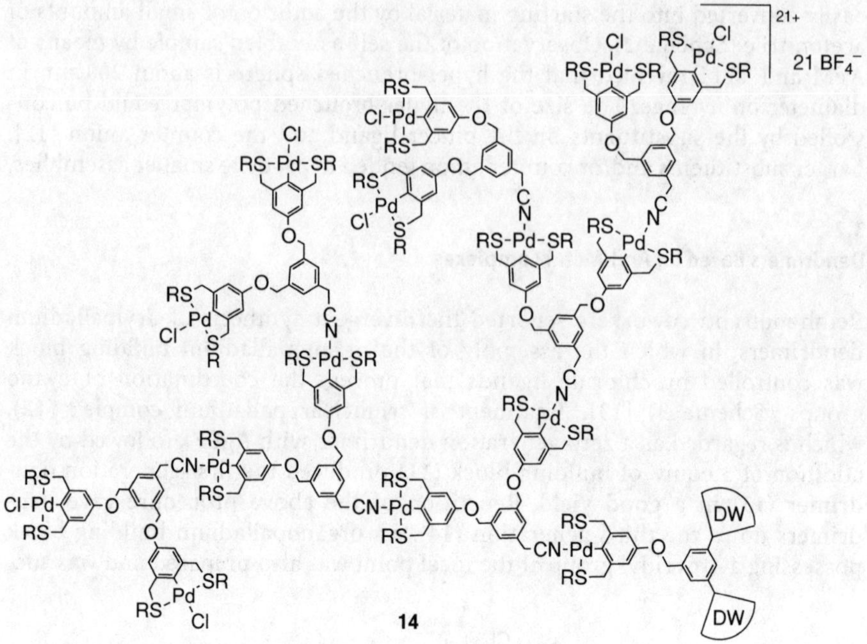

Scheme 3 (continued)

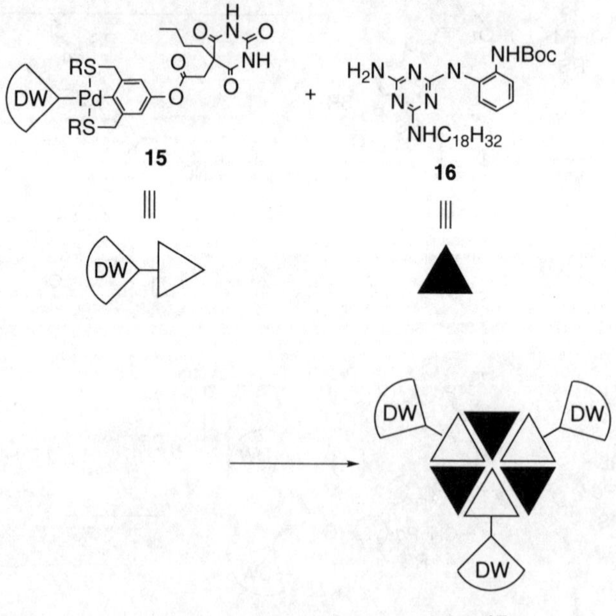

Scheme 4. Arylpalladium dendrimers 17 assembled by hydrogen bonds

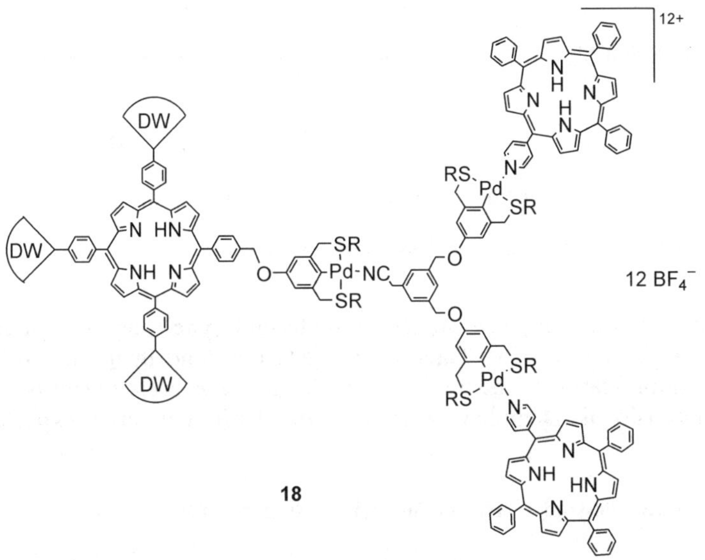

Fig. 4. Arylpalladium dendrimer **18** with porphyrin nuclei

cessfully applied to the synthesis of dendrimers by both divergent and convergent methods [14].

Organopalladium dendrons (**15**) equipped with barbituric acid residue were assembled around 2,4,6-triaminotriazine derivative (**16**) in a [3+3] fashion by hydrogen bonding to give dendrimers up to the third generation (**17**) (Scheme 4) [15]. Organopalladium dendrimer (**18**) having porphyrin nuclei at the center and/or periphery was also prepared by a convergent method (Fig. 4) [16]. However, no interaction between organopalladium moieties and porphyrin nuclei was observed. Organopalladium dendrimers containing hydrophobic dendrons at the periphery were also prepared [17].

4
Dendritic Sphere Based on Alkynyl-Metal Complexes

The alkynyl-metal (metal-acetylide) complex is one of the best building blocks for organometallic dendrimers, since it has some advantages compared to other organometallic complexes [18]. Most of the metal-acetylide complexes are thermally robust and stable, even when exposed to air and moisture. Metal-acetylide complexes are fairly accessible in high yields by well-established synthetic methodology [19]. These features are essential to the construction of dendrimers.

Metal-acetylide complexes have been used as a unit of organometallic polymers that have metallic species in the main chain [20]. Representative examples are metal-poly(yne) polymers (**19**) of group 10 metals depicted in Scheme 5. These polymers are easily prepared from $M(PR_3)_2Cl_2$ (M=Pt, Pd) and dialkynyl compounds catalyzed by Cu(I) salts in amine. Recently, this synthetic method was successfully applied to the construction of metal-acetylide dendrimers.

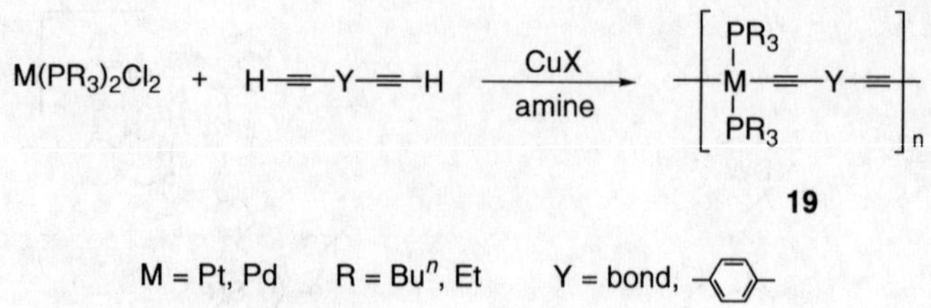

M = Pt, Pd   R = Bu^n, Et   Y = bond, —⟨benzene⟩—

Scheme 5. Synthesis of transition metal-poly(yne) polymers 19

Metal-acetylide complexes including metal-poly(yne) polymers often show unique properties [21–23]. Thus, metal-acetylide dendrimers are of interest because amplification of the functionality due to metal-acetylide units based on three-dimensional assembly with a regular dendritic structure is expected.

4.1
Hyper-Branched Polymers Based on Alkynyl-Metal Complexes

Lewis and coworkers were the first to report on the preparation of hyper-branched platinum-acetylide polymers [24]. The reaction of $Pt(PBu_3)_2Cl_2$ with triethynylbenzene in a 3:2 molar ratio gave polymer 20 that is insoluble in common organic solvents. Although the structure was not fully established, the polymer was proposed to have a highly cross-linked structure. When large amounts of p-diethynylbenzene were added to the reaction mixture (triethynylbenzene: p-diethynylbenzene=1:50), the resulting polymer 21 was fairly soluble in organic solvents due to the small number of branching units (Scheme 6).

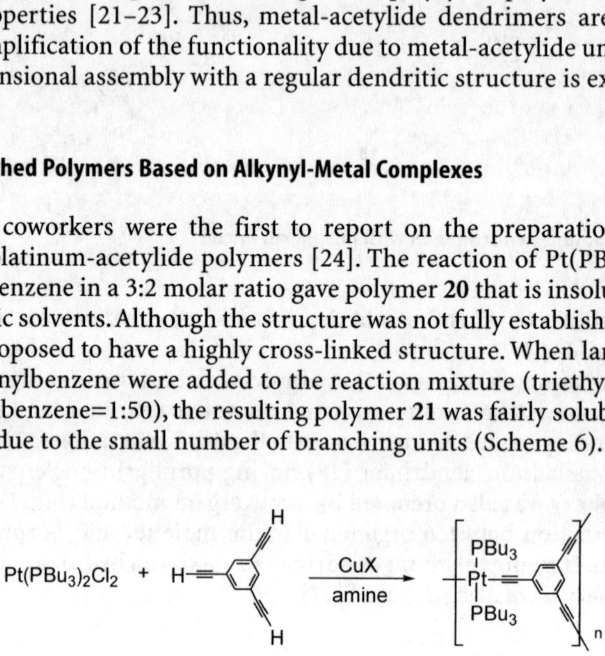

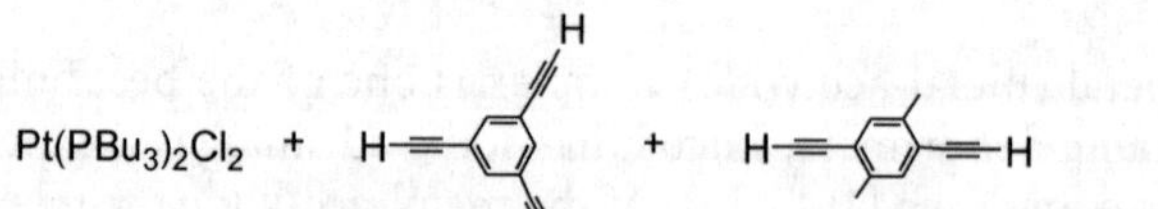

Scheme 6. Syntheses of hyper-branched platinum-acetylide polymers 20 and 21

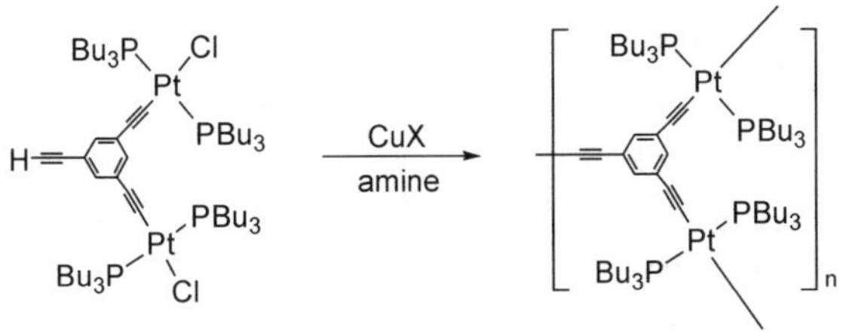

Scheme 7. Self-polycondensation of AB_2 type platinum-acetylide monomer

On the other hand, self-polycondensation of AB_2 type monomer **22** produced soluble polymer **23** (M_w=16,000, M_w/M_n=1.28) [25] (Scheme 7). Concentration of the monomer showed little influence on the molecular weight of the resulting polymers. ^{1}H and ^{31}P NMR spectra suggested that the polymer has a hyper-branched sphere partially involving a cyclic structure.

4.2
Approaches to the Synthesis of Platinum-Acetylide Dendrimers Without Protecting Groups

When metal-acetylide complexes of palladium and platinum are prepared from the reaction of $M(PR_3)_2Cl_2$ (M=Pt, Pd) with terminal acetylene catalyzed by CuX, the number of alkynyl groups on the metal can be controlled by the molar ratio of the starting materials [19a, 20]. The reaction using equimolar amounts of terminal acetylene relative to $M(PR_3)_2Cl_2$ produced a monoethynyl complex $M(PR_3)_2(C{\equiv}CR')Cl$ selectively, whereas the reaction with 2 equiv. of terminal acetylene gave a diethynyl derivative $M(PR_3)_2(C{\equiv}CR')_2$. In general, the M-C bonds of platinum-acetylide complexes are much more stable than those of palladium analogues. Cleavage of Pt-C bonds takes place in the presence of a CuX catalyst at approximately 100 °C, but not at room temperature. In contrast, Pd-C bonds are easily cleaved in the presence of the CuX catalyst even at room temperature. Thus, the first preparation of metal-acetylide dendrimers was realized by using platinum-acetylide complexes with non-protected terminal acetylene [26].

Trinuclear platinum-acetylide complex (**24**), which is a building block for the construction of dendrimers, was prepared from the reaction of triethynylmesitylene with 3.3 equiv. of $Pt(PBu_3)_2Cl_2$ in a good yield. Although the reaction of triethynylmesitylene with excess amounts of complex **24** produced the first generation dendrimer **25** having nine Pt atoms in the molecule (Scheme 8), separation of the resulting dendrimer from the unreacted trinuclear platinum-acetylide was very difficult. However, the first-generation dendrimer **26** was successfully isolated by alumina column chromatography after the introduction of phenylethynyl groups at the periphery by subsequent treatment of the reaction mixture with phenylacetylene.

When the preparation of the second-generation dendrimer was attempted by reacting trinuclear platinum acetylide complex (**27**) (Fig. 5), which is also a first

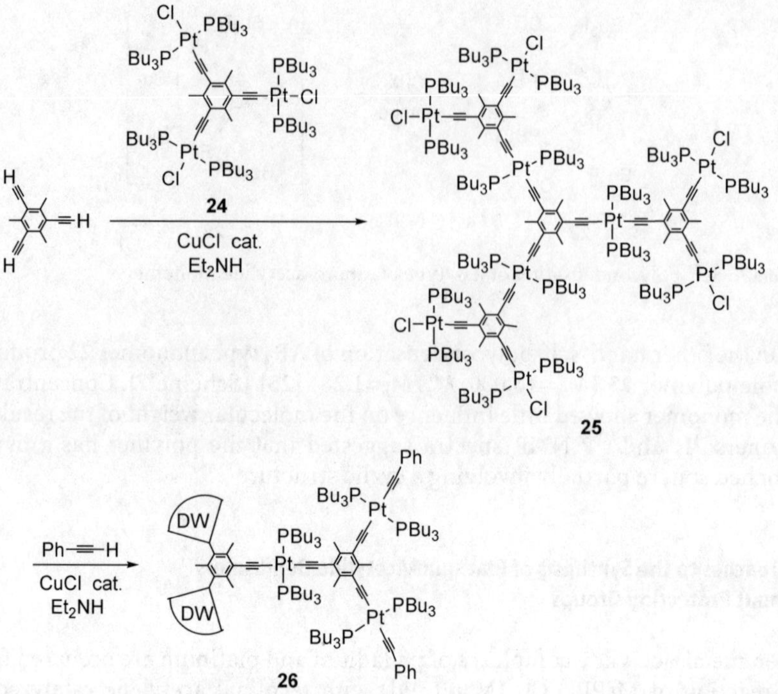

Scheme 8. Syntheses of first-generation platinum-acetylide dendrimer 26

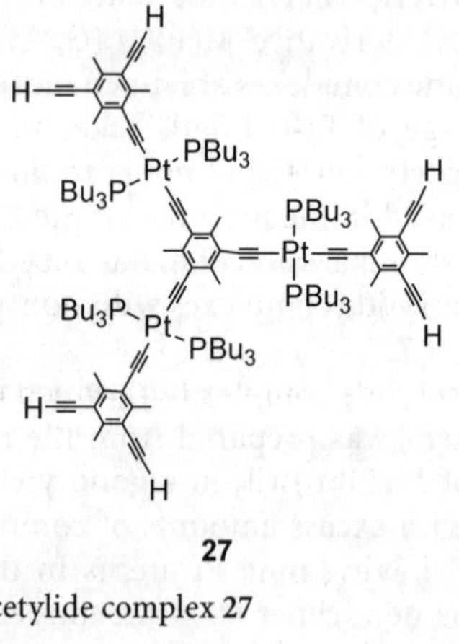

Fig. 5. Trinuclear platinum-acetylide complex 27

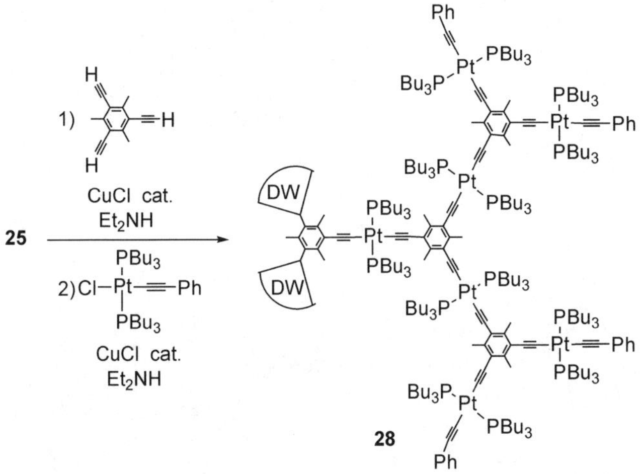

Scheme 9. Syntheses of second-generation platinum-acetylide dendrimer **28**

generation dendrimer having six ethynyl groups, with excess amounts of trinuclear platinum-acetylide complex **24**, a complex mixture of various molecular sizes was produced. However, modification of the synthetic procedure for the first-generation dendrimer resulted in the formation of the second generation dendrimer (Scheme 9). Thus, the first-generation dendrimer generated in situ was treated with excess amounts of triethynylmesitylene, followed by the reaction with excess amounts of Pt(PBu$_3$)$_2$(C≡CPh)Cl to give the second-generation dendrimer selectively. Isolation of the second-generation dendrimer was achieved by repetitive alumina column chromatography with much effort. Therefore, the preparation of higher generation dendrimers by this methodology is very difficult.

Stang and coworkers also reported the synthesis of platinum-acetylide dendrimers using 1,3,5-triethynylbenzene as the bridging ligand [27]. Their strategy was very similar to the above, and they prepared dendrimers up to the second generation.

4.3
Convergent Synthesis of Platinum-Acetylide Dendrimers

Recently, an efficient convergent synthesis of platinum-acetylide dendrimers has been reported [28]. The methodology involves the use of two kinds of trialkylsilyl protecting groups, trimethylsilyl and tri(isopropyl)silyl, of the terminal acetylene for the synthesis of platinum-acetylide dendrimers. In dendrimer synthesis, purification of the building blocks is very important for isolating pure samples of dendritic molecules up to high generations. To purify low molecular weight products by recrystallization, platinum complexes possessing PEt$_3$ ligands were used as starting materials for the synthesis of the desired dendrimers.

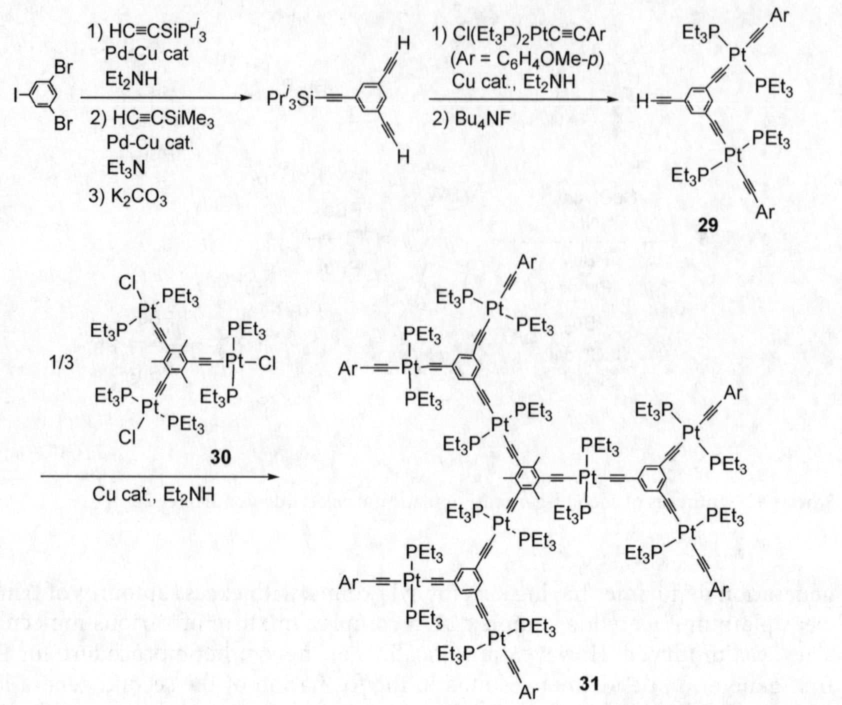

Scheme 10. Convergent synthesis of first-generation platinum-acetylide dendrimer **31**

The synthetic route for the first-generation dendrimer is illustrated in Scheme 10. Taking advantage of the difference in reactivity between aryl iodide and bromide, one triisopropylsilylethynyl group and two trimethylsilylethynyl groups were introduced by the successive reaction of 1-bromo-3,5-diiodobenzene with triisopropylsilylacetylene and trimethylsilylacetylene in the presence of Pd-Cu catalyst [29]. After selective removal of the trimethylsilyl groups by treatment with K_2CO_3, platinum-acetylide moieties were attached to the ethynyl groups. Finally, triisopropylsilyl group was removed by reacting with tetrabutylammonium fluoride (TBAF) to give the first-generation dendron (**29**) quantitatively.

The reaction of the first-generation dendron **29** with the core, which is a trinuclear platinum complex (**30**) bridged by 1,3,5-triethynylmesitylene, in a molar ratio of 3:1 resulted in the formation of the first-generation dendrimer (**31**). Since the reaction proceeded quantitatively and no side products were observed, the first-generation dendrimer was easily isolated by alumina column chromatography.

The first-generation dendron **29** was grown up to the second- and third-generation dendrons **33** and **34** by the successive reaction with dinuclear platinum complex **32** followed by treatment with TBAF for desilylation (Scheme 11). Reactions of the second- and third-generation dendrons **33** and **34** with the core **30** gave the second- and third-generation dendrimers, respectively. Although

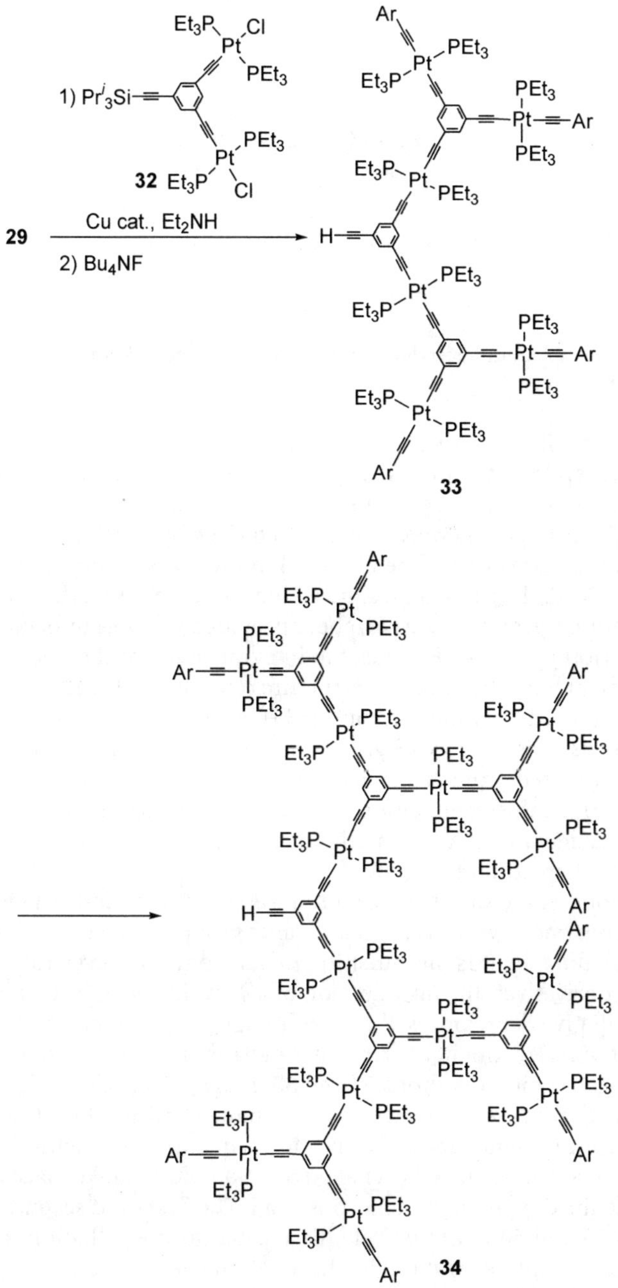

Scheme 11. Syntheses of second and first-generation platinum-acetylide dendrimers **33** and **34**

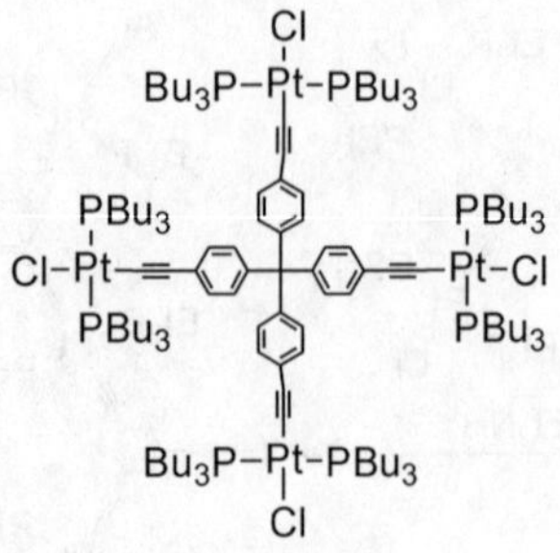

Fig. 6. Tetranuclear platinum-acetylide complex 35 used as the core for convergent synthesis

the fourth-generation dendron was successfully prepared, similar reaction of the fourth-generation dendron with the core led to the formation of not only the fourth-generation dendrimer but also defective dendrimers, which could not be separated. The dendrimers were fully characterized by NMR spectroscopy, while no molecular ion peaks could be observed in the mass spectrometry including MALDI-TOF MAS. In particular, signals due to the three methyl groups at the core and methoxy groups at the periphery provided valuable information for the characterization by ^{1}H NMR. Thus, the integral ratios of these signals were in good agreement with the expected structure of each dendrimer.

When tetranuclear platinum complex (35) bridged by tetra(4-ethynylphenyl)-methane was used as the core (Fig. 6), dendrimers having more platinum atoms than those prepared from trinuclear platinum core 30 in the molecule were obtained up to the third generation. GPC analysis of these dendrimers revealed that the dendrimers with a tetraplatinum core have a similar molecular size to those with the triplatinum core 30.

The development of the convergent route enabled us to design new platinum-acetylide dendrimers with functional organic groups at the core. Since it is well known that halide ligands on transition metal atoms are reversibly substituted by pyridine derivatives, the introduction of a 4-pyridyl group at a focal point of dendrons will give dendrimers, the morphology of which may be controlled by chemical stimuli. Treatment of trinuclear palladium complex 36 with the first- and second-generation dendrons 37 and 38 in the presence of NaBAr$_4$ (Ar=3,5-$(CF_3)_2C_6H_3$) resulted in complete ligand exchange and produced the first- and second-generation dendrimers 39 and 40 in quantitative yields (Scheme 12) [30]. However, a similar ligand exchange reaction did not take place in the reaction with the third-generation dendrons at all. The first- and second-generation dendrimers 39 and 40 quantitatively dissociated to the palladium core 36 and the first- and second-generation dendrons 37 and 38, respectively. Quantitative formation of dendrimers and dissociation into the core and dendrons were repeated up to three times by successive treatment of NaBAr$_4$ and Bu$_4$NCl, respectively. Although there are a few examples on the control of dendrimer morphology based on their photochemistry in the literature [31], this is the first example of control by using chemical stimuli.

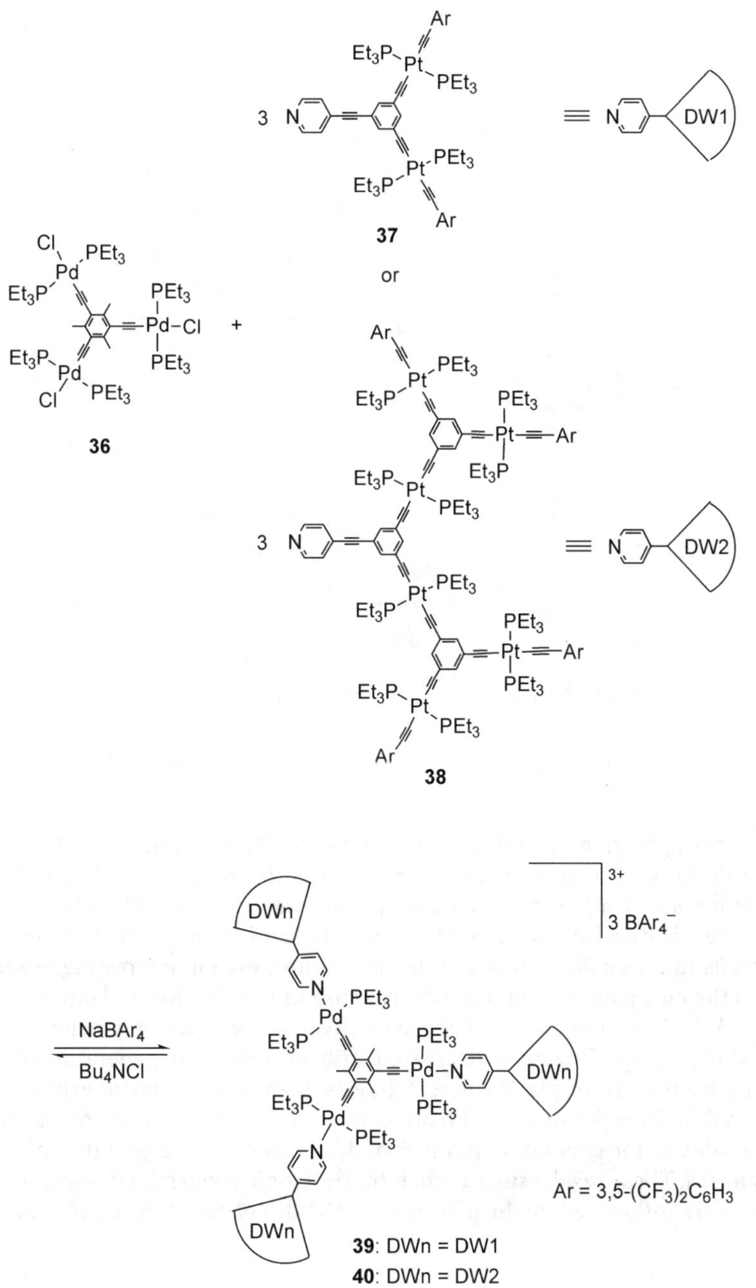

Scheme 12. Reversible formation of cationic metal-acetylide dendrimers **39** and **40**

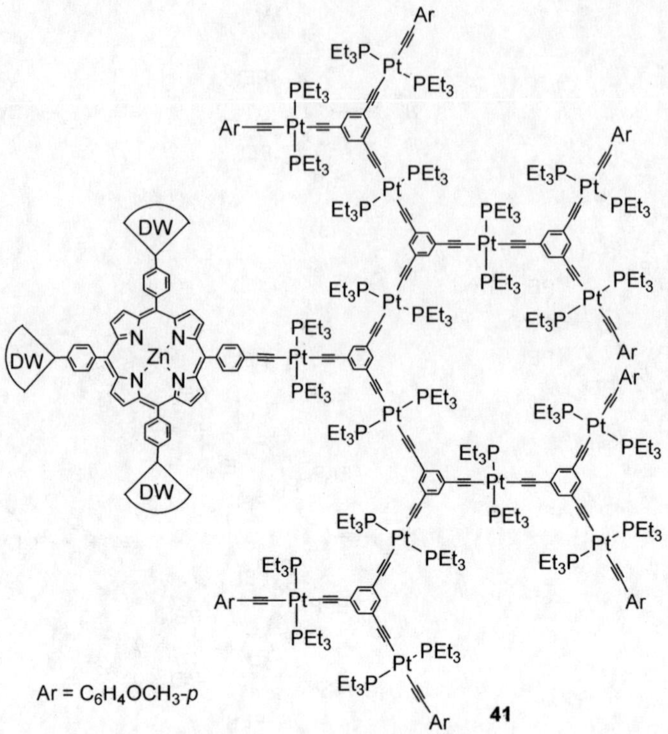

Fig. 7. Third-generation platinum-acetylide dendrimer **41** with a porphyrin core

Recently, platinum-acetylide dendrimers having a porphyrin core were also prepared by a convergent method up to the third generation **41** (Fig. 7) [32]. Dendrimers with a porphyrin core are promising molecules, because porphyrin is a representative functional organic molecule and the local environment often affects its functionalities. Some of the dendrimers exhibit interesting properties due to the encapsulation of a porphyrin core in the interior of dendritic structures [33]. Dendrimer **41** and its derivatives showed a Soret band at about λ_{max}=435 nm, and the ε value decreased with an increase in generation number. Energy transfer from platinum-acetylide dendrons to the porphyrin core was observed in these dendrimers. Fluorescence from the porphyrin core decreased appreciably as the generation number of dendrimers increased with a change of the spectra. These results suggest that the electronic properties of the porphyrin nucleus are influenced by the platinum-acetylide sphere of the dendrons.

4.4
Divergent Synthesis of Platinum-Acetylide Dendrimers

A divergent approach to the synthesis of platinum-acetylide dendrimers was also developed [34]. The building block for divergent synthesis is the mononuclear platinum-acetylide complex (**42**), which has been prepared from 1-bromo-3,5-diiodobenzene by a manner similar to that used for convergent synthesis, and the trimethylsilyl and triisopropylsilyl groups have been effectively employed as protecting groups of the terminal acetylene. Triethynylmesitylene, of which the methyl groups are very important for the characterization of the resulting dendrimers, was chosen as the core.

The reaction of triethynylmesitylene with 3 equiv. of the building block in the presence of CuCl catalyst in diethylamine resulted in the quantitative formation of the first-generation dendrimer (**43**). After the removal of the triisopropylsilyl groups by treatment with Bu$_4$NF, the dendrimer with terminal acetylenes was grown up to the second generation (**44**) by reacting with 6 equiv. of the building block, and further to the third-generation dendrimer (**45**) in a similar manner (Scheme 13). However, in the preparation of the fourth generation dendrimer, a

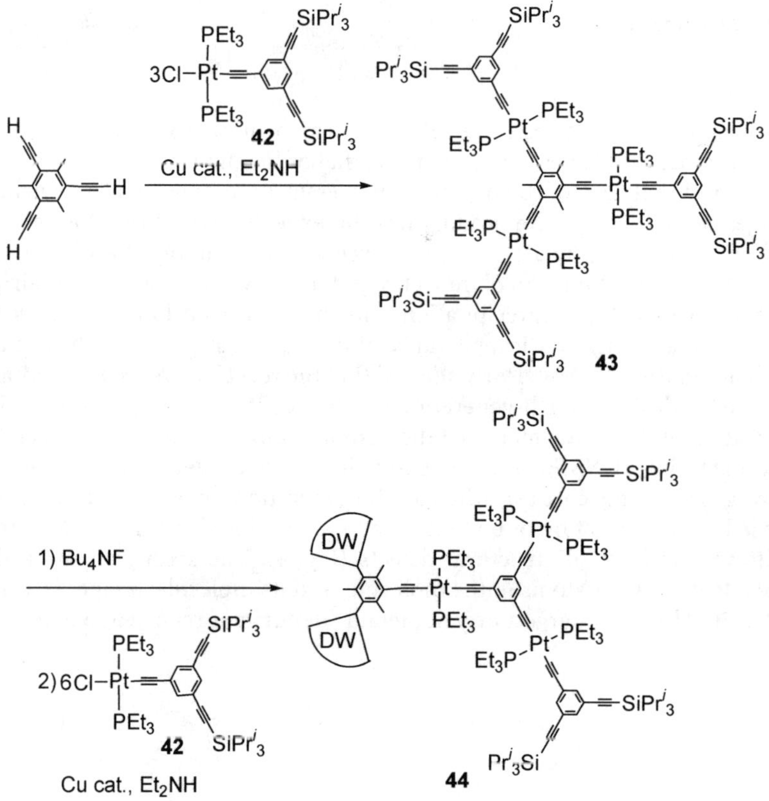

Scheme 13. Divergent synthesis of platinum-acetylide dendrimers

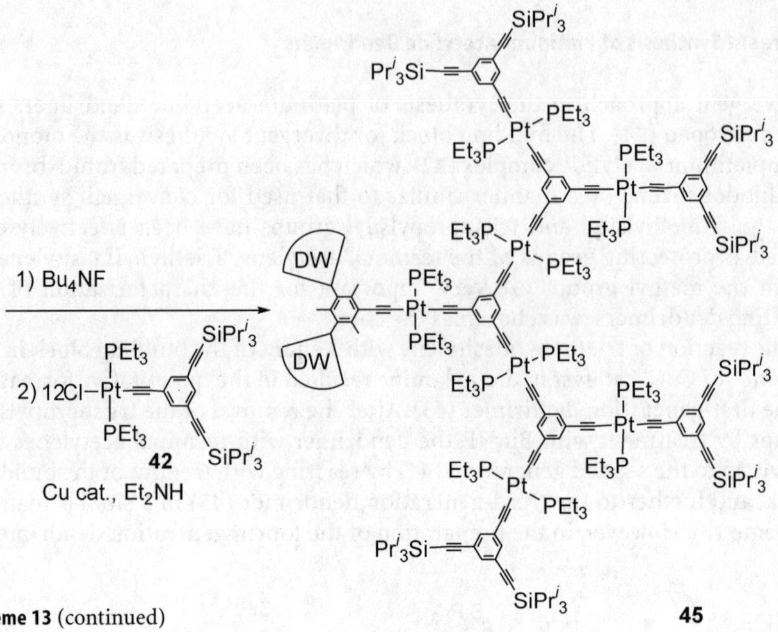

Scheme 13 (continued)

structural defect was detected by ^{1}H NMR, which showed signals due to unreacted terminal acetylenes. Based on the signal of the methyl groups at the core, it was found that 25% of the product has one unreacted terminal acetylene on average, though 75% of the product has the expected structure. However, this problem was resolved by using 25 mol% excess of the building block in the reaction. After purification by column chromatography, the resulting dendrimer was easily isolated by reprecipitation and the unreacted building block was recovered quantitatively. Disappearance of the signal assignable to the terminal acetylenes in the ^{1}H NMR clearly showed that the reaction proceeded completely to give the desired fourth-generation dendrimer. It is very important for characterization of the dendrimers that the number of unreacted terminal acetylene is estimated by ^{1}H NMR in every generation since all dendrimers have three methyl groups at the core regardless of the generation number. Successive reactions led to the formation of up to the sixth-generation dendrimer (**46**) with no significant amounts of structural defects (Fig. 8). The sixth-generation dendrimer **46** has 189 Pt atoms in the molecule, and its molecular weight is as high as 139,750. This is the largest organometallic dendrimer reported to date.

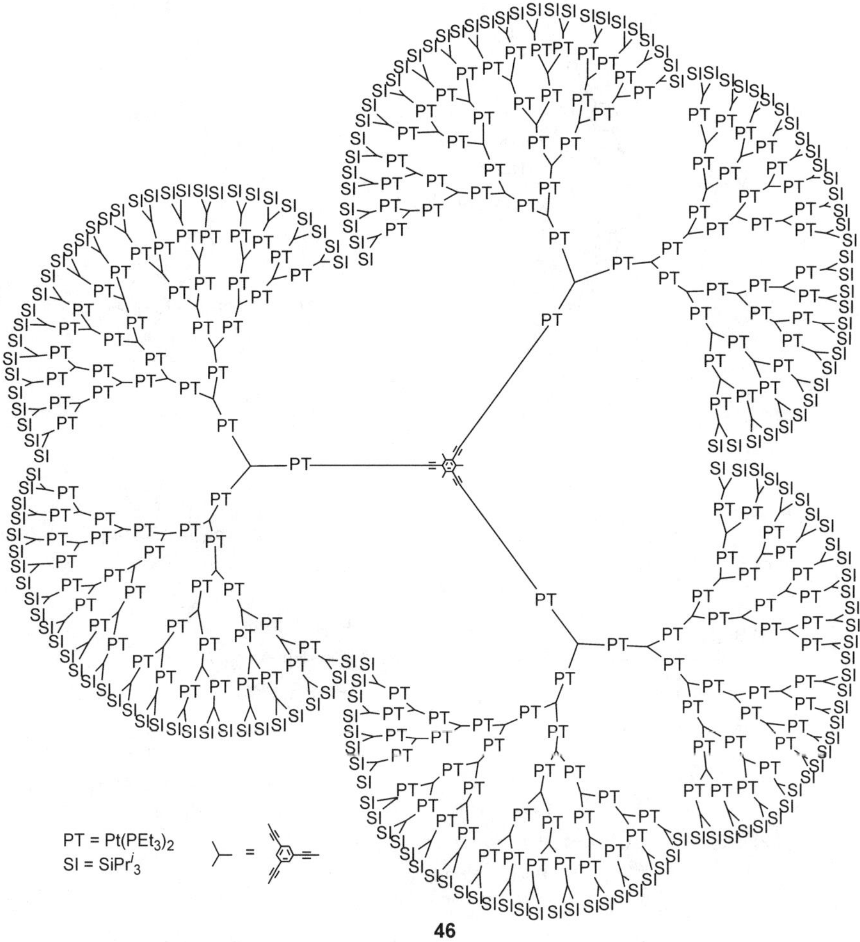

Fig. 8. Sixth generation platinum-acetylide dendrimer **46**

4.5
Synthesis of Ruthenium-Acetylide Dendrimers

Humphrey and coworkers reported the synthesis of ruthenium-acetylide dendrimers by a convergent method (Scheme 14). In their initial study, the first-generation dendrimer (**49**) was prepared from triruthenium complex (**48**) and the first-generation dendron (**47**) by using trimethylsilyl groups for protection of the terminal acetylene [35]. Recently, they developed a conventional route to the synthesis of ruthenium-acetylide dendrimers without having to use protecting groups [36]. Even if 1,3,5-triethynylbenzene was treated with excess amounts of *cis*-Ru(dppm)₂Cl₂, dinuclear ruthenium-acetylide complex (**50**) was selectively obtained due to the steric repulsion between the ruthenium moieties [37]. Complex **50** is an archetypal building block for ruthenium-acetylide dendrimers,

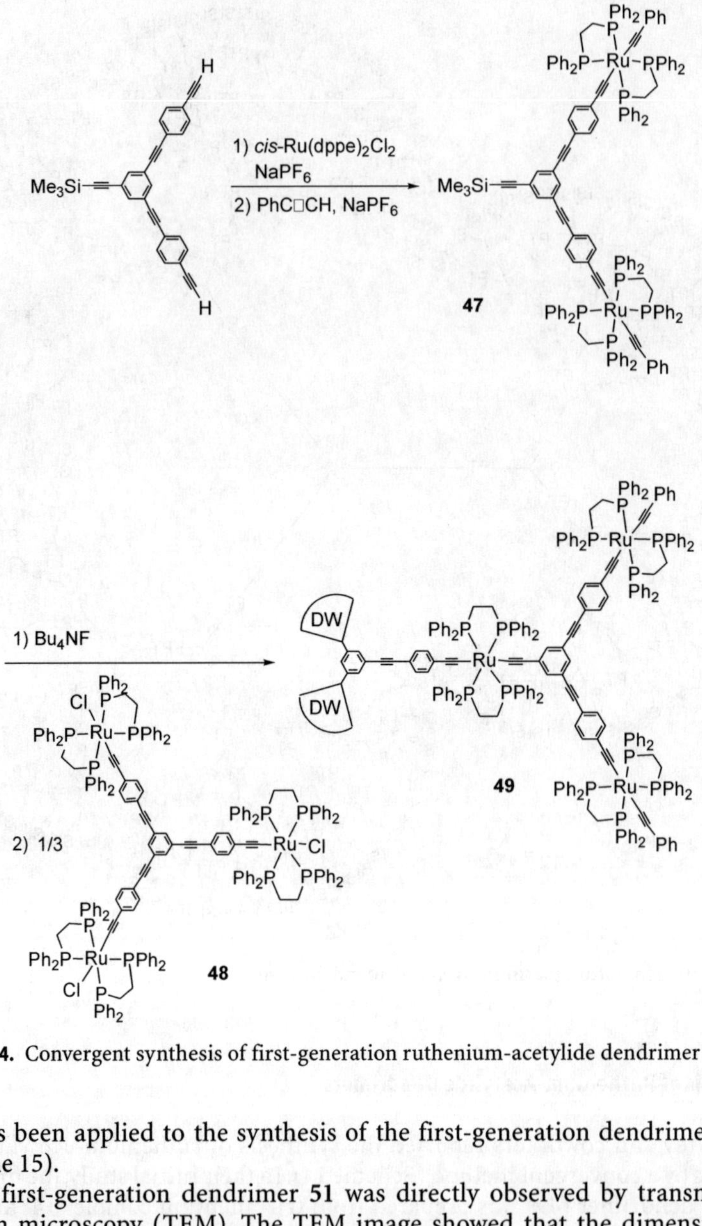

Scheme 14. Convergent synthesis of first-generation ruthenium-acetylide dendrimer **49**

and has been applied to the synthesis of the first-generation dendrimers (**51**) (Scheme 15).

The first-generation dendrimer **51** was directly observed by transmission electron microscopy (TEM). The TEM image showed that the dimensions of individual molecules are about 50 Å, which is consistent with the calculated one [36]. Third-order NLO measurements showed a significant enhancement of two-photon absorption upon proceeding from the constituent molecules to the dendritic complex [35].

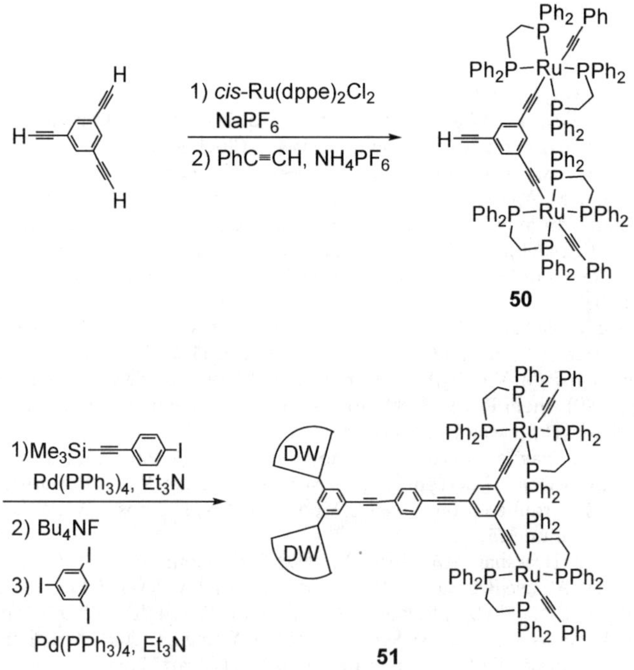

Scheme 15. Rapid convergent approach of the synthesis of first-generation ruthenium-acetylide dendrimer **51**

5
Concluding Remarks

The unique features of dendrimer architecture and the rich chemistry of organo-transition metal complexes have been combined in organometallic dendrimers that have potential for a wide range of applications. As evident from this review, dendrimers composed of organometallic building blocks have evolved recently following the development of organometallic dendrimers with metallic species at the core or periphery. A new family of organometallic dendrimers is expected to be developed through the design of the building blocks, considering the structural features of the organometallic complexes and the topological features of the dendrimers. The intramolecular interaction among metallic species in such dendritic molecules may provide a good opportunity for the synthesis of novel functional materials. However, the instability of organometallic complexes remains to be a barrier for the purification and characterization of the resulting dendrimer. Thus, the development and technical improvement of purification and characterization methods are crucial for the future chemistry of organometallic dendrimers that are expected to have more complicated structures. Since organometallic complexes have opened many new ways in supramolecular chemistry, organometallic dendrimers will play a very important role in not only organometallic chemistry and polymer science but also material science.

6
References

1. (a) Newkome GR, Moorefield CN, Vögtle F (2001) Dendrimers and dendrons – concepts, syntheses, applications. Wiley-VCH, Weinheim; (b) Fréchet JMJ, Tomalia DA (2001) (eds) Dendrimers and other dendritic polymers. Wiley, Chichester
2. For recent reviews: (a) Fischer M, Vögtle F (1999) Angew Chem, Int Ed 38:884; (b) Bosman AW, Janssen HM, Meijer EW (1999) Chem Rev 99:1665; (c) Adronov A, Fréchet JMJ (2000) Chem Commun: 1701; (d) Grayson SM, Fréchet JMJ (2001) Chem Rev 101:3819; (e) Hedrick JL, Magbitang T, Connor EF, Glauser T, Volksen W, Hawker CJ, Lee VY, Miller RD (2002) Chem Eur J 8:3308; (f) Tomalia DA, Fréchet JMJ (2002) J Polym Sci Part A Polym Chem 40:2719
3. For reviews: (a) Balzani V, Campagna S, Denti G, Juris A, Serroni S, Venturi M (1998) Acc Chem Res 31:26; (b) Gorman C (1998) Adv Mater 10:295; (c) Venturi M, Serroni S, Juris A, Campagna S, Balzani V (1998) Top Curr Chem 197:193; (d) Newkome GR, He E, Moorefield CN (1999) Chem Rev 99:1689; (e) Serroni S, Campagna S, Puntoriero F, Pietro CD, McClenaghan ND, Loiseau F (2001) Chem Soc Rev 30:367
4. (a) Denti G, Campagna S, Serroni S, Ciano M, Balzani V (1992) J Am Chem Soc 114:2944; (b) Campagna S, Denti G, Serroni S, Ciano M, Juris A, Balzani V (1992) Inorg Chem 31:2982
5. Newkome GR, Cardullo F, Constable EC, Moorefield CN, Thompson AMWC (1993) J Chem Soc Chem Commun 925
6. For reviews: (a) Hearshaw MA, Moss JR (1999) Chem Commun 1; (b) Cuadrado I, Moran M, Casado CM, Alonso B, Losada J (1999) Coord Chem Rev 193–195:395; (c) Astruc D, Blais J-C, Cloutet E, Djakovitch L, Rigaut S, Ruiz J, Sartor V, Valerio C (2000) Top Curr Chem 210:229; (d) Crooks RM, Zhao M, Sun L, Chechik V, Yeung LK (2001) Acc Chem Res 34:181; (e) Astruc D, Chardac F (2001) Chem Rev 101:2991; (f) van Manen H-J, van Veggel FCJM, Reinhoudt DN (2001) Top Curr Chem 217:121; (g) Kreiter R, Kleij AW, Gebbink RJMK, van Koten G (2001) Top Curr Chem 217:163
7. Achar S, Puddephatt RJ (1994) Angew Chem Int Ed Engl 33:847
8. (a) Achar S, Puddephatt RJ (1994) J Chem Soc Chem Commun 1895; (b) Achar S, Vittal JJ, Puddephatt RJ (1996) Organometallics 15:43
9. Achar S, Immoos CE, Hill MG, Catalano VJ (1997) Inorg Chem 36:2314
10. Liu G-X, Puddephatt RJ (1996) Organometallics 15:5257
11. Huck WTS, van Veggel FCJM, Kropman BL, Blank DHA, Keim EG, Smithers MMA, Reinhoudt DN (1995) J Am Chem Soc 117:8293
12. (a) Huck WTS, Snellink-Ruel BHM, Lichtenbelt JWT, van Veggel FCJM, Reinhoudt DN (1997) Chem Commun 9; (b) Huck WTS, van Veggel FCJM, Reinhoudt DN (1997) J Mater Chem 7:1213
13. Huck WTS, van Veggel FCJM, Reinhoudt DN (1996) Angew Chem Int Ed Engl 35:1213
14. Huck WTS, Prins LJ, Fokkens RH, Nibbering NMM, van Veggel FCJM, Reinhoudt DN (1998) J Am Chem Soc 120:6240
15. Huck WTS, Hulst R, Timmerman P, van Veggel FCJM, Reinhoudt DN (1997) Angew Chem Int Ed Engl 36:1006
16. Huck WT, Rohrer A, Anikumar AT, Fokkens RH, Nibbering NMM, van Veggel FCJM, Reinhoudt DN (1998) New J Chem 22:165
17. van Manen H-J, Fokkens RH, Nibbering NMM, van Veggel FCJM, Reinhoudt DN (2001) J Org Chem 66:4643
18. (a) Nast R (1982) Coord Chem Rev 47:89; (b) Lang H (1994) Angew Chem Int Ed Engl 33:547; (c) Manna J, John KD, Hopkins MD (1995) Adv Organomet Chem 38:79
19. (a) Sonogashira K, Yatake T, Tohda Y, Takahashi S, Hagihara N (1977) J Chem Soc Chem Commun 291; (b) Bruce MI, Humphrey MG, Matisons JG, Roy SK, Swincer AG (1984) Aust J Chem 37:1955; (c) Bruce MI, Horn E, Matisons JG, Snow MR (1984) Aust J Chem 37:1163; (d) Touchard D, Haquette P, Pirio N, Toupet L, Dixneuf PH (1993) Organometallics 12:3132
20. Hagihara N, Sonogashira K, Takahashi S (1981) Adv Polym Sci 41:149

21. (a) Kaharu T, Matsubara H, Takahashi S (1992) J Mater Chem 2:43; (b) Kaharu T, Ishii R, Adachi T, Yoshida T, Takahashi S (1995) J Mater Chem 5:687; (c) Whittall IR, Humphrey MG, Persoons A, Houbrechts S (1996) Organometallics 15:1935; (d) Whittall IR, Humphrey MG, Houbrechts S, Persoons A, Hockless DCR (1996) Organometallics 15:5738; (e) Whittall IR, Cifuentes MP, Humphrey MG, Luther-Davies B, Samoc M, Houbrechts S, Persoons A, Heath GA, Bogsányi D (1997) Organometallics 16:2631; (f) Hurst SK, Cifuentes MP, Morrall JPL, Lucas NT, Whittall IR, Humphrey MG, Asselberghs I, Persoons A, Samoc M, Luther-Davies B, Willis AC (2001) Organometallics 20:4664; (g) Yam VW-W (2001) Chem Commun 789; (h) Yam VW-W, Tang RP-L, Wong KM-C, Cheung K-K (2001) Organometallics 20:4476; (i) Garcia MH, Robalo MP, Dias AR, Duarte MT, Wenseleers W, Aerts G, Goovaerts E, Cifuentes MP, Hurst S, Humphrey MG, Samoc M, Luther-Davies B (2002) Organometallics 21:2107

22. (a) McDonagh AM, Humphrey MG, Samoc M, Luther-Davies B, Houbrechts S, Wada T, Sasabe H, Persoons A (1999) J Am Chem Soc 121:1405; (b) Cifuentes MP, Powell CE, Humphrey MG, Heath GA, Samo M, Luther-Davies B (2001) J Phy Chem A 105:9625

23. (a) Takahashi S, Takai Y, Morimoto H, Sonogashira K (1984) J Chem Soc Chem Commun 3; (b) Wilson JS, Dhoot AS, Seeley AJAB, Khan MS, Kohler A (2001) Nature 413:828; (c) Wong W-Y, Lu G-L, Choi K-H, Shi J-X (2002) Macromolecules 35:3506

24. Khan MS, Schwartz DJ, Pasha NA, Kakkar AK, Lin B, Raithby PR, Lewis J (1992) Z Anorg Allg Chem 616:121

25. Onitsuka K, Ohshiro N, Fujimoto M, Takei F, Takahashi S (2000) Mol Cryst Liq Cryst 342:159

26. (a) Ohshiro N, Takei F, Onitsuka K, Takahashi S (1996) Chem Lett 871; (b) Ohshiro N, Takei F, Onitsuka K, Takahashi S (1998) J Organomet Chem 569:195

27. Leininger S, Stang PJ, Huang S (1998) Organometallics 17:3981

28. Onitsuka K, Fujimoto M, Ohshiro N, Takahashi S (1999) Angew Chem, Int Ed 38:689

29. Takahashi S, Kuroyama Y, Sonogashira K, Hagihara (1980) Synthesis 627

30. Onitsuka K, Iuchi A, Fujimoto M, Takahashi S (2001) Chem Commun 741

31. (a) Smet M, Liao L-X, Dehaen W, McGrath DV (2000) Org Lett 2:511; (b) Takaguchi Y, Suzuki S, Mori T, Motoyoshiya J, Aoyama H (2000) Bull Chem Soc Jpn 73:1857; (c) Cao D, Meier H (2001) Angew Chem Int Ed 40:186

32. Onitsuka K, Kitajima H, Fujimoto M, Iuchi A, Takei F, Takahashi S (2002) Chem Commun 2576

33. (a) Bhyrappa P, Young JK, Moore JS, Suslick KS (1996) J Am Chem Soc 118:5708; (b) Sadamoto R, Tomioka N, Aida T (1996) J Am Chem Soc 118:3978; (c) Pollak KW, Leon JW, Fréchet JMJ, Maskus M, Abruna HD (1998) Chem Mater 10:30; (d) Jiang D-L, Aida T (1998) J Am Chem Soc 120:10,895; (e) Bhyrappa P, Vaijayanthimala G, Suslick KS (1999) J Am Chem Soc 121:262; (f) Weyermann P, Gisselbrecht J-P, Boudon C, Diederich F, Gross M (1999) Angew Chem Int Ed 38:3215; (g) Matos MS, Hofkens J, Verheijen W, De Schryver FC, Hecht S, Pollak KW, Fréchet JMJ, Forier B, Dehaen W (2000) Macromolecules 33:2967; (h) Kimura M, Saito Y, Ohta K, Hanabusa K, Shirai H, Kobayashi N (2002) J Am Chem Soc 124:5274

34. Onitsuka K, Shimizu A, Takahashi S (2003) Chem Commun 280

35. McDonagh AM, Humphrey MG, Samoc M, Luther-Davies B (1999) Organometallics 18:5195

36. Hurst SK, Cifuentes MP, Humphrey MG (2002) Organometallics 21:2353

37. (a) Whittall IR, Humphrey MG, Houbrechts S, Maes J, Persoons A, Schmid S, Hockless DCR (1997) J Organomet Chem 577:277; (b) Long NJ, Martin AJ, de Biani FF, Zanello P (1998) J Chem Soc Dalton Trans 2017

Top Curr Chem (2003) 228: 65–85
DOI 10.1007/b11006

An Adventure in Macromolecular Chemistry Based on the Achievements of Dendrimer Science: Molecular Design, Synthesis, and Some Basic Properties of Cyclic Porphyrin Oligomers to Create a Functional Nano-Sized Space

Ken-ichi Sugiura

Department of Chemistry, Graduate School of Science, Tokyo Metropolitan University, 1–1 Minami-Ohsawa, Hachi-Oji 192–0397 Tokyo Japan. *E-mail: sugiura@porphyrin.jp*

This review is dedicated to Emeritus Professor Soichi Misumi on his 77th birthday (NB the 77th birthday in Japan has special significance since the Chinese character for "77" resemble those for "happy").

Dendrimer chemistry has taught us that these molecules create a nano-sized closed space that, presumably, is the origin of the specific physical properties of this class of materials. As the next stage of dendrimer chemistry, a macromolecule capable of creating such a space inside its molecule is proposed. To create the nano-sized space, porphyrin is considered to be the best candidate for the component molecules, because it has versatile properties associated with its expanded π-electron system and the incorporated metal. The resultant multi-detectable properties of porphyrin, that is, a number of its properties are detectable by many physical methods, may reveal the function of the nanometer-sized space.

Keywords. Nano-sized space, Porphyrin, Cyclic oligomer, Multi-detectable molecule, Inverse-template effect

Abbreviations and Symbols

2HPor	porphyrin free base
CPO(s)	cyclic porphyrin oligomer(s)
GPC	gel permeation chromatography
HSAB	hard and soft acid and base theory
MALDI-TOF MASS	matrix-assisted laser desorption/ionization time-of-flight mass spectroscopy
MgPor	magnesium porphyrin
NMR	nuclear magnetic resonance
$Pd_2(dba)_2$	tris(dibenzylideneacetone)dipalladium(0)
QE	quantum efficiency
R_f value	valuerate of flow value for chromatography
$Ru^{II}(CO)Por$	ruthenium porphyrins
STM	scanning tunneling microscopy
VPO	vapor pressure osmometry
ZnPor	zinc porphyrin

1
What Dendrimer Chemistry has Taught Us

1.1
Achievements of Dendrimer Science and Another Novel Science Based on the Dendrimer Concept

The chemistry and physics of dendritic compounds started a decade ago [1–5]. Today, this science of uniquely shaped molecules, namely, dendrite-shaped molecules, is one of the most exciting topics of contemporary interdisciplinary research. The dendrimers and their related molecules have been investigated widely not only from the viewpoints of synthetic, physical, and material chemistries but also from that of mathematics. Accompanying the development of the science in this decade, research interest has shifted from the mere challenge of preparing molecules with unique shapes, via their excited state chemistries involving inter- and/or intramolecular photo-induced electron and/or energy transfer, to the nanoscience.

In this decade, all chemistry research fields have adopted and/or applied the dendrimers and/or dendrimer methodologies. The table of contents of this series, *Topics in Current Chemistry: Dendrimers Volumes I–IV*, clearly indicates this situation [1–4], that is, the concept of dendritic compounds has already been introduced in host–guest and/or supramolecular chemistry (Vol. I/Chap. 2, Vol. II/Chaps. 3,4, Vol. IV/Chap. 3), chiral chemistry (Vol. I/Chap. 4), electrochemistry (Vol. I/Chap. 6, Vol. III/Chap. 3), heteroatom and/or organometallic chemistries (Vol. I/Chap. 3, Vol. II/Chaps. 2,5, and Vol. IV/Chap. 4), and carbohydrate chemistry (Vol. IV/Chap. 6), as well as applied in the field of medicine (Vol. II/Chap. 6) and nanoscience (Vol. III/Chap. 4). The dendrimer methodology is expected to be used in future novel science as a conventional chemistry concept.

What research objectives should dendrimer chemistry therefore target as the second stage of this science? The answer is the progress of macromolecular chemistry while maintaining the dendrimer concept, even if we do not use molecules with the dendritic shape.

1.1.1
Wouldn't it be Possible for Another Novel Science Based on Dendrimer Science to Evolve?

During this decade, what has dendrimer chemistry taught us? Dendrimers exhibit various functions. Generally, most of the properties of materials are governed by their electrons and resultant molecular orbital interactions; however, several functions of dendrimers, such as their energy-harvesting property [6], are difficult to understand from simple molecular orbital interactions. These functions are, presumably, attributable to the three-dimensional spherical shape of the molecule, usually constructed by the covalent bonds of such typical elements as carbon, nitrogen, and oxygen. The molecules create a nano-sized closed space that, presumably, is the origin of the specific properties of the dendrimers. This hypothesis leads to another molecular design that the nano-sized closed space created inside the molecule may exhibit specific physical properties.

Based on the working hypothesis described above, the author focuses on the chemistry of cyclic porphyrin oligomers (CPOs). As described in the next section, porphyrin is a multi-functional compound and the nano-sized closed space created by these molecules should therefore be of interest to chemists.

1.2
Basic Properties of Metalloporphyrins

Porphyrin **1** is the parent system of the naturally occurring chlorophyll **2** and/or the industrially useful phthalocyanine **3** (Fig. 1) [7, 8]. Porphyrins play important roles in the field of biology, for example, in the light-harvesting process of green plants, in the photo-induced charge separation at the photosynthesis reaction center, as oxygen carriers, and as redox mediators. Reflecting these versatile functions, these molecules have been called the "pigments of life". In the field of material science, this expanded π-electron system is frequently used as a component molecule for molecule-based conductors [9], magnets [10], and light-emitting devices [11].

Porphyrin consists four pyrroles and each pyrrole is connected by a methyne carbon; therefore, the π-electrons of porphyrin are delocalized over the molecule. The eighteen π-electrons of the molecule obey the Hückel rule and therefore contribute to the aromaticity. This delocalization changes the nitrogen atoms of pyrrole into amine-type nitrogens with a hydrogen atom and imine-type nitrogen atoms, similar to pyridine's nitrogen. The two hydrogen atoms connected to the former amine-type nitrogens are weak acids; therefore, chelate complexes are formed in the presence of metal salts and/or complexes, $M^{n+}X_n$ (M=metal ion, X=counter anions). Porphyrin is formally a doubly charged ligand. Hence, the porphyrin ligand produces a neutral molecule with a divalent metal cation, such as Co^{2+}, Ni^{2+}, Cu^{2+}, Zn^{2+}, etc. These metalloporphyrins behave as pseudo-covalently bonded molecules that can be subjected to chromatogra-

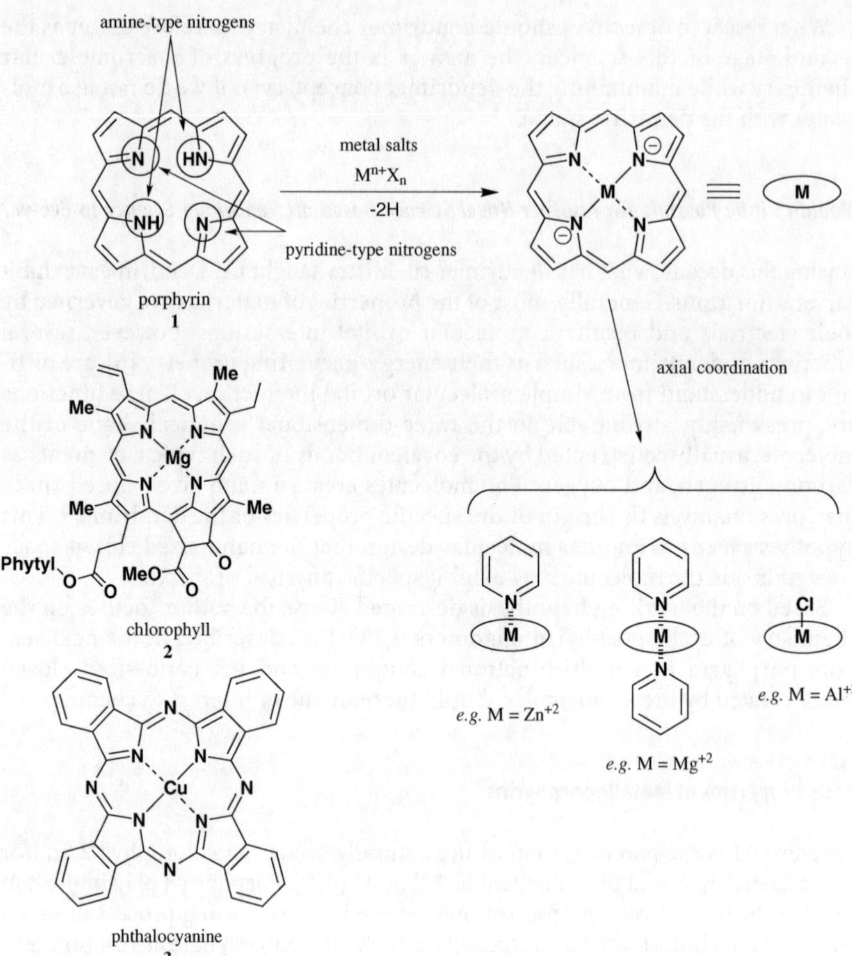

Fig. 1. Structures of porphyrin **1**, chlorophyll **2**, and phthalocyanine **3**. In the presence of metal salts $M^{n+}X_n$ (M=metal, X=counter anion, n=oxidation state or number of counter anions), porphyrins produce chelate complexes. Some metal chelates of the porphyrins, such as ZnPor, form further coordination bonds with other ligands such as pyridines

phy or vacuum sublimation purifications. Today, most elements form the corresponding metallocomplexes with porphyrin ligands and therefore, porphyrin acts as a "perfect ligand". Acid treatment of the complexes leads to demetalation reactions. The ability of the acid to remove the incorporated metal varies depending on the metal, for example, nickel porphyrin requires concentrated sulfuric acid and the pH of water is sufficiently strong to remove calcium from the corresponding metal complexes (vide infra).

The distance between the diagonal nitrogen atoms is approximately 4 Å [12]. The distortion induced by the incorporated metal is negligible and planar four-coordinated metallocomplexes are produced. Therefore, the metal–nitrogen distances of the metalloporphyrins are approximately 2 Å, and this ligand provides a similar coordination environment to the elements incorporated. Several incorporated metals require further axial ligation to produce pyramidal five-coordinated complexes, octahedral six-coordinated complexes, etc. (Fig. 1). For example, copper porphyrins ($Cu^{II}Pors$) require no axial ligand; however, zinc porphyrins ($Zn^{II}Pors$) produce pyramidal five-coordinate complexes with one pyridine-like ligand, and furthermore, magnesium porphyrin requires two axial ligands. Trivalent metals, such as Al^{3+}, Ga^{3+}, and In^{3+}, form pseudo-covalent bonds with anions such as halogens and five-coordinated complexes. The nature of the axial coordination phenomena is dependent on the metal and is difficult to discuss comprehensively.

As described before, the π-electrons of porphyrin are delocalized over the molecule and the energy levels of the HOMO and the LUMO are high and low, respectively. The resultant narrow intramolecular HOMO–LUMO gap causes absorption of the entire region of visible light. Usually, porphyrins are red to purple and phthalocyanines are blue to green. Furthermore, the long lifetime of their excited states is applicable to the construction of photo-induced electron and/or energy transfer systems.

1.3
Research Objectives of Cyclic Porphyrin Oligomers in the Context of Dendrimer Chemistry

Thousands of cyclic-shaped molecules are known. Why then must we focus on the cyclic porphyrin oligomers (CPOs) in the context of dendrimer chemistry? Are there any specific properties and functions associated with CPOs?

The most popular and important motivation for investigating CPOs is the mechanistic studies on the light-harvesting process in photosynthesis and the creation of novel photon–electron conversion devices aiming at artificial photosynthesis [13, 14]. The first stage of the photosynthesis is the light-harvesting process, and the combined photo-excited energy is used for charge separation and the subsequent chemical reaction that produces oxygen. The three-dimensional structures of the protein participating in the light-harvesting process are revealed by single-crystal diffraction studies and the resolution of the structure is increasing yearly [14]. Surprisingly, 27 porphyrins make up the ring in the protein. However, the reason why nature adopted the ring shape for the light-harvesting system is still unclear. Many methodologies are considered to reveal this perfect biological machine.

Many methodologies exist to elucidate the functions of complicated biological systems including photosynthesis. The direct observation of the phenomena using high-performance research facilities is one of the strategies. However, synthetic chemists adopt the method of design, synthesis, and evaluation of model compounds that serve as the essence of biological systems. They systematically design, prepare, and evaluate a series of model compounds by changing the structure little by little.

To reveal the light-harvesting process, synthetic studies on cyclic porphyrin oligomers, designed based on the information of the protein structure, are considered. Nature adopted the cyclic alignment of chlorophyll. Is the closed cyclic architecture needed for the efficient light harvest? For this query, we design a cyclic oligomer and a linear chain oligomer, both containing an identical number of porphyrins. Does the number of chlorophylls affect the efficiency of the light-harvesting process? For this query, we design a series of cyclic porphyrin oligomers constructed from two, three, four, or more porphyrins. A variety of CPOs are expected to be designed, synthesized, and evaluated to clarify the factors governing the light-harvesting process.

The second motivation of this research field is molecular recognition and enzyme reaction using the nano-sized space produced by CPOs [15]. The chemistry of crown ethers is the most representative example; however, the shape of the nano-sized space used in this chemistry is a "line" of ethylene glycol units. Calix-aromatic compounds form the space created by the planes of the substituted aromatic compounds. Porphyrin, a rigid and large square-shaped π-system of 10-Å length per side, is larger than benzene and acts as a panel [12]. Furthermore, porphyrin has important specific properties, namely, chelation of metals, further ligation with the incorporated metal, light absorption, long excited state lifetime, and porphyrin-centered and metal-centered redox reactions, and these properties are difficult to obtain from other π-electron systems. By using the nano-sized space constructed by this multi-functional molecular panel, novel chemistry is expected, which exceeds the previous host–guest chemistry.

Porphyrin is a multi-detectable molecule, that is, a number of its properties are detectable by many physical methods. Not only the most popular nuclear magnetic resonance and light absorption and emission spectroscopic methods, but also the electron spin resonance method for paramagnetic metalloporphyrins and Mössbauer spectroscopy for iron and tin porphyrins are frequently used to estimate the electronic structure of porphyrins. By using these multi-detectable properties of the porphyrins of CPOs, a novel physical phenomenon is expected to be found. In particular, the topology of the cyclic shape is an ideal one-dimensional state of the materials used in quantum physics [16]. The concept of aromaticity found in fullerenes, spherical aromaticity, will be revised using π-conjugated CPOs [17].

2
Synthetic Methodologies of CPOs

In this chapter, some experimental tips on the synthetic studies of CPOs are reviewed. Because this class of materials has specific properties associated with the large size of molecules, several special methodologies should be dealt with, which are applicable to the dendrimers.

2.1
High-Dilution and Template Methods

The synthesis of cyclic compounds is a "battle" for the yield. The general synthetic strategies of CPOs are illustrated in Fig. 2. The first one involves one-pot synthesis using a porphyrin unit (method A), and the second one is the ring-closing reaction of the chain-shaped PO (method B) [11].

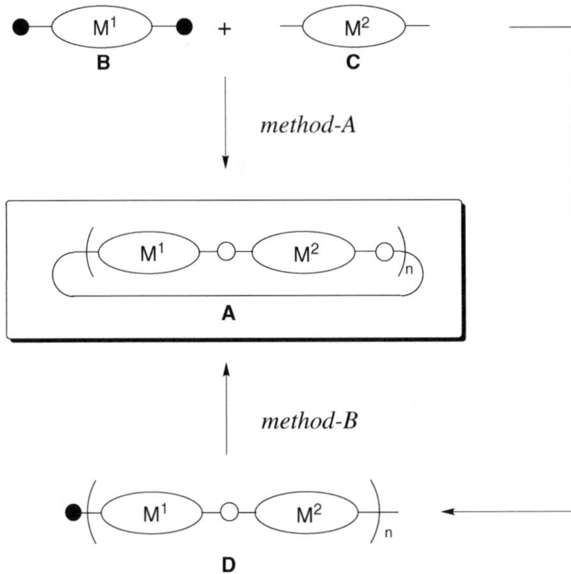

Fig. 2. Synthetic strategies, *method A* and *method B*, of cyclic porphyrin oligomers. *Black and white circles* indicate the reaction points and the resultant linkage moieties, respectively

The advantage of method A is that starting materials, such as **B** and **C**, are easily available. The disadvantage includes the difficulty of controlling the composition (n) of the products. Furthermore, the isolation of the desired CPOs from by-products of similar structure is difficult to carry out. The merit of method B is that the desired CPO is obtained selectively under the high-dilution condition. The disadvantage is that the preparation of **D** requires many synthetic steps. Although both methods are employed, the desired CPO is difficult to isolate as a major product with a high yield. In most cases, structurally unidentified polymeric products contaminate the reaction.

The high-dilution and template methods are frequently used in the synthesis of cyclic compounds with the aim of increasing the yield. The former method is carried out at substrate concentrations lower than 1 mM [18–20]. This reaction condition decreases the contact of the substrate molecules in the solution. The linear intermediate produced prefers the intramolecular cyclization reaction rather than the intermolecular reaction. Therefore, this reaction condition is useful for the intramolecular reaction, method B (Fig. 2).

The latter method, the template method, involves a reaction to produce a transition state similar to the desired product using a template. The template should have a shape similar to the space of the product. The template interacts with the substrate by forming noncovalent bonds such as coordination bonds (Fig. 3). The representative and most successful examples are found in crown ether chemistry. In the chemistry, alkali metals act as templates to create a crown-ether-like transition state with an ethylene glycol substrate by using metal–oxygen coordination bonds.

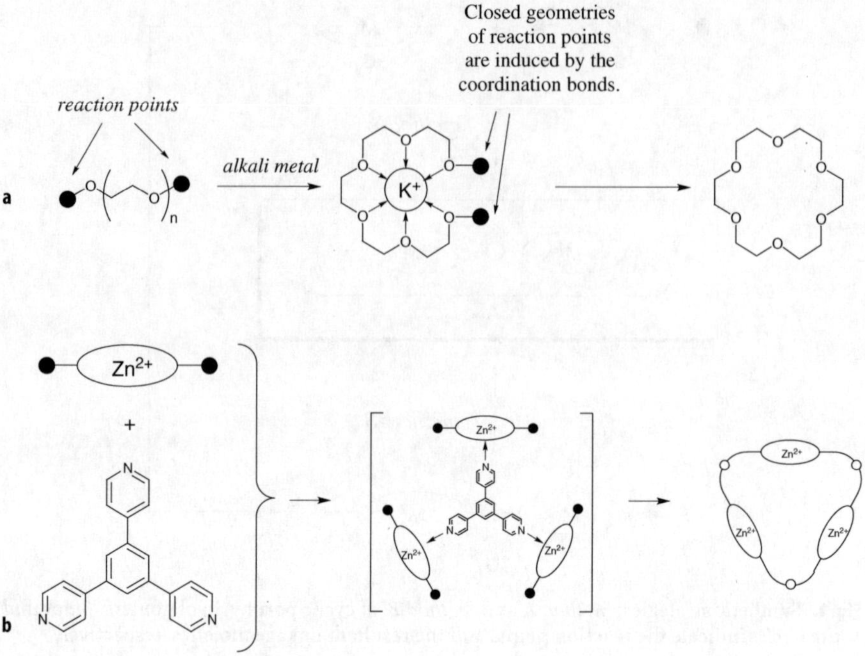

Fig. 3a,b. Template cyclization reactions of **a** crown ethers and **b** CPOs. The coordination bonds are illustrated by *black arrows*. In the crown ether synthesis, ethylene glycols coordinate toward the metal acting as the template (normal template reaction); however, the template coordinates to the incorporated metals of porphyrin in CPO synthesis (inverse-template reaction)

Is a similar method applicable to CPO synthesis? As described above (Fig. 1), some metalloporphyrins require axial ligand(s). For example, zinc porphyrin (ZnPor), magnesium porphyrin (MgPor), and ruthenium porphyrin ($Ru^{II}(CO)$ Por) require one (for ZnPor and $Ru^{II}(CO)$Por) or two (for MgPor) ligands. By using these properties, the existence of a template matching the shape of the space may produce the designed CPO selectively [21]. This method has the inverse-template effect, that is, the ligand acts as a template and the template coordinates to the substrate (i.e., metalloporphyrins): in contrast, the alkali metal acts as a template and the substrates (i.e., ethylene glycols) coordinate to the template in crown ether chemistry.

To select the metal to be incorporated into the substrate porphyrin unit, the following basic properties of metalloporphyrins should be considered. The stability constant of MgPor is too small to achieve the usual oligomeric reactions and purification by silica gel chromatography. The starting material ($Ru_3(CO)_{12}$) for $Ru^{II}(CO)$Por is expensive and the yield of the corresponding metalation reaction is low. Furthermore, the removal of ruthenium is difficult, and it is likewise difficult to remove the template from the obtained ruthenium CPOs. Therefore, ZnPor is frequently used as a substrate in this template reaction, because of the low prices of zinc sources (zinc acetate and/or zinc chloride), the high yield in the metalation reaction, the sufficient chemical stability of the ZnPor under con-

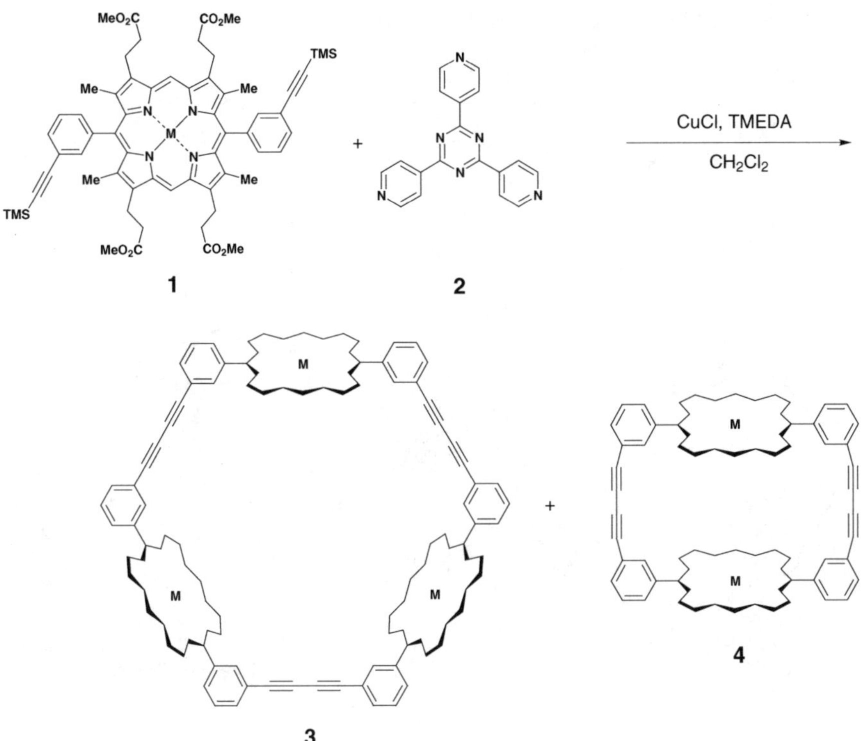

Fig. 4. Yields of CPOs 3 and 4 using inverse-template reactions (%).

Metal (M)	Template 2	3 (%)	4 (%)
Zn^{2+}	Present	55	6
Zn^{2+}	Absent	34	23
$Ru(CO)^{2+}$	Present	32	–

Synthesis of CPOs, 3 and 4, using inverse-template reaction. The substituents are omitted for clarity

densation and purification conditions, and the ease of the demetalation reaction for future transmetalations (Fig. 4).

The coupling reaction of 1 (M=Zn) affords CPO 3 (M=Zn) in 55% yield in the presence of template 2; however, the absence of 2 decreases the yield to 34% [22]. With the increase of yield of 3, template 2 induces the selectivity of the reaction: the yield of the by-product (cyclic dimer 4 (M=Zn)) was changed from 23% (with no template) to 6% (in the presence of template). A similar CPO formation reaction was reported for the corresponding ruthenium porphyrins (3, M=Ru(CO)), in which the stability constant of the Ru–N coordination bond is 10^2 larger than that of the Zn–N coordination bond [23]. Although the transition state of the CPO produced by the ruthenium-based substrate is expected to be more stable than that produced by ZnPor, the yield of 3 (M=Ru(CO)) is only

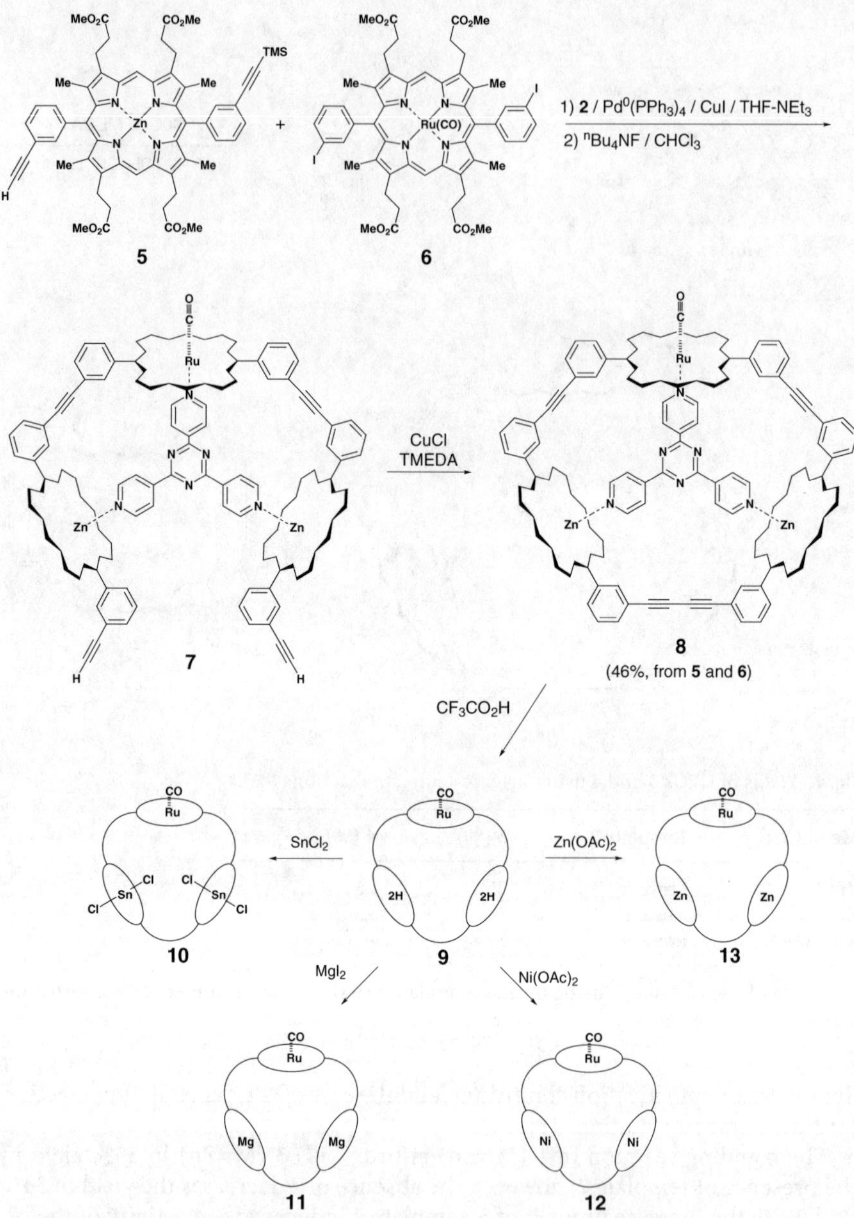

Fig. 5. Mixed cyclization reaction using two kinds of porphyrins. The introduced substituents of **7** and **8** are omitted for clarity and the structures of **9–13** are roughly drawn

32%. As described above, the removal of the template from the obtained ruthenium CPO **3** was unsuccessful. To obtain larger or smaller CPOs, another template design is required [24], the same as crown ether chemistry in which a series of alkali metals (Li$^+$, Na$^+$, K$^+$, etc.) produce crown ethers of different sizes.

These inverse-template reactions proceed using more than two kinds of metalloporphyrin units (Fig. 5). Webb and Sanders reported the stepwise cyclization reaction using a 2:1:1 mixture of ZnPor **5**, Ru^{II}(CO)Por **6**, and template **2**. Acyclic intermediate **7** gave CPO **8** containing two kinds of metals, zinc and ruthenium (46%, in two steps) [25]. The significance of this chemistry is as follows: first, the fine molecular design methodology is established, because each linkage moiety is selectable by stepwise synthesis. Compared to the number of acetylene moieties of CPO **3**, that of CPO **8** has two acetylenes less. This linkage manner produces a small nano-sized space that is perfect for enzymatic reaction (vide infra), compared to that produced by **3**; second, different kinds of metals can be incorporated into one CPO. Previously, porphyrin oligomers incorporating different metals were prepared under strictly limited reaction conditions and the yields were low [26]. As described in the following section, the creation of intramolecular energy transfer systems is one of the important research objectives of CPO chemistry. Such a phenomenon is usually observed in porphyrin oligomers containing porphyrin free base (2HPor) and ZnPor moiety, that is, the excited energy of ZnPor is transferred to the ground state of 2HPor, thereby intramolecularly producing the excited state of 2HPor. To achieve the synthesis of such heterometalated porphyrin oligomers, an understanding of the stability constant of metalloporphyrin is needed.

The qualitative stability constants of metalloporphyrins are summarized in Table 1. The metals classified in class I produce the most stable metalloporphyrins and the demetalation reaction does not proceed smoothly even under concentrated sulfuric acid condition. The incorporated metals classified in classes II and III are removed using mild acids such as hydrochloric acid. Calcium classified in class V is removed by (the pH of) water. The mixed cyclization reactions afford the heterometalated CPO, and the acid treatment of the CPO obtained produces the CPO containing 2HPor moiety. Further treatment of the metal salt classified in a class lower than that of the unremoved metal(s), which is classified in a class higher in Table 1, produces another heterometalated CPO. Representative examples are summarized in Fig. 5 [25]. The initial cyclization reaction is carried out by using Ru^{II}(CO)Por **6**, the metal of which is classified in class I, and ZnPor **5**, clas-

Table 1. Demetallation reaction conditions for metalloporphyrins based on the qualitative stability constant

	Condition	Example
Class I	Not completely demetallated by conc. H$_2$SO$_4$	Pd^{2+}, Pt^{2+}, Ru^{2+}, Sn^{4+}
Class II	conc. H$_2$SO$_4$	Ga^{3+}, Co^{2+}, Ni^{2+}, Cu^{2+}
Class III	conc. HCl	Zn^{2+}
Class IV	Glacial acetic acid	Mg^{2+}
Class V	Water	Ca^{2+}, Ba^{2+}

sified in class III. After the isolation of CPO **8** containing one ruthenium atom and two zinc atoms in one molecule, trifluoroacetic acid treatment removes only zinc metals and template **2** to produce another CPO **9**. Further treatment with metal salts of tin, magnesium, and nickel forms other heterometalated CPOs **10–12**, respectively. Treatment with zinc salt regenerates the ruthenium–zinc-containing CPO **13**; however, **13** does not contain the template molecule.

2.2
Chemical Reactions Used for Cyclization Reactions

The chemical reactions frequently used to form CPOs are summarized in this section (Fig. 6). Because of the low yields of CPOs, activated and selective reactions are preferred.

Because the Glaser coupling reaction, that is, the oxidative coupling reaction of hydrogen-terminated acetylene groups in the presence of copper(I) salt, proceeds smoothly under mild reaction conditions, it is frequently used in CPO chemistry as a key reaction (Fig. 6a). However, the most popular reaction conditions, such as pyridine/CuCl, are difficult to apply in the inverse-template reaction using a pyridine-based template such as **2**. The coordination of the large excess of pyridine used masks the template. Sanders and co-workers reported that the revised reaction conditions (CuCl/TMEDA/CH_2Cl_2) maximize the effect of the pyridine-based template [22].

Recently, great progress has been made in the aryl-coupling reactions using low-valence transition-metal catalysis. The revised novel conditions allow nearly quantitatively yields, and the reactions are widely applied in CPO chemistry (Fig. 6b). One of the linkage moieties formed, the diarylethene group (Ar–C≡C–Ar), is known to transfer photo-excited energy through the group by the through-bond-type electron-transfer process. Lindsey and co-workers used $Pd_2(dba)_2/AsPh_3$ conditions in the synthetic studies of intramolecular energy transfer systems mimicking the natural light-harvesting system. They designed the ZnPor–2HPor energy-transfer system, and for this purpose, any copper insertion reaction should be avoided at the 2HPor moiety by the copper catalyst usually used in this aryl coupling reaction to give the desired **14** [20]. As an extension of the reactions described above (Fig. 6a,b), the organometallic method was also reported, affording stable CPOs [24].

In this review, CPOs constructed by covalent bonds are mainly focused on; however, stable coordination bonds comparable to the stability of the covalent bonds have potential for future enhanced molecular design of novel CPOs. One representative is the bond between pyridine-type nitrogen and metal, which is widely used in supramolecular chemistry, that is, the cyclic supramolecular formation reaction between pyridine-substituted porphyrin and metal salts (Fig. 6d) [27, 28]. Palladium salts are frequently used as the metal salts. From the viewpoint of the hard and soft acid and base theory (HSAB), this N–Pd coordination bond is a well-balanced combination, because the bonds between nitrogen and other group X metals, N–Ni and Ni–Pt coordination bonds, are too weak and too strong to obtain the desired CPOs, respectively. For the former, the supramolecular architectures tend to dissociate into pieces in the solution state, and for the latter,

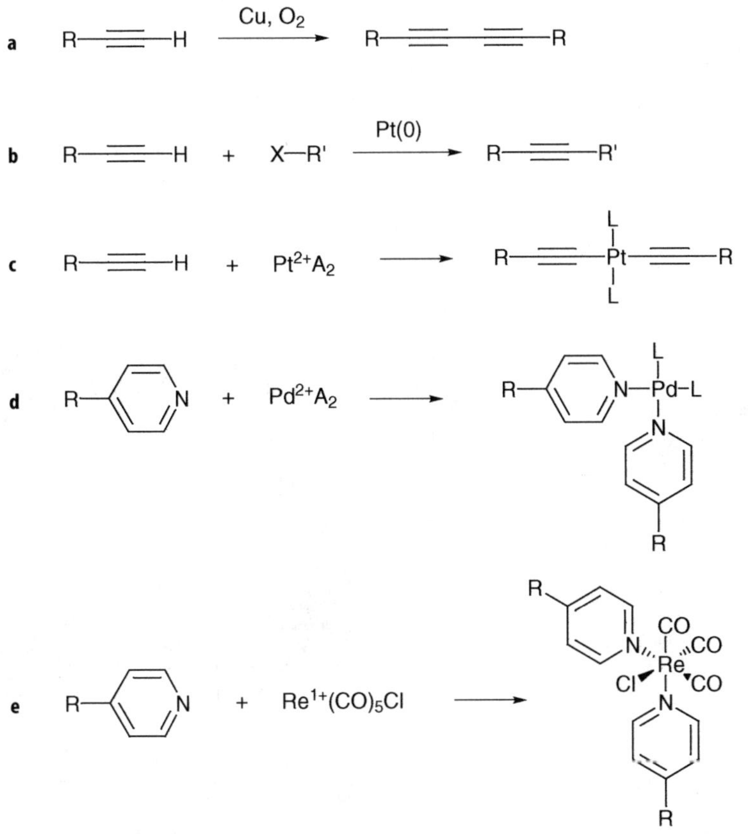

Fig. 6. Chemical reactions used for CPO formation reactions (R, R' porphyrin moieties, X halogen, A counter anion, and L ligand)

structurally unidentified polymeric products are formed and it is difficult to obtain the most thermodynamically stable single product. In contrast, the N–Pd bond has moderate stability and is believed to produce equilibrium between the coordinated species and the components. The repeated rearrangements and alternations of the bonds gradually produce the most thermodynamically stable supramolecular architecture stabilized by the cooperative multibonds. Therefore, the CPO constructed by twelve Pd–N bonds was isolated in a surprisingly high yield, >95%. Based on these strategies, N–Re coordination bonds are also used. A square-shaped cyclic tetramer was isolated in 86% yield from the pyridine-substituted porphyrin and $Re^+(CO)_3Cl$ (Fig. 6e) [29, 30]. The most interesting property of these coordination bonds is that they introduce a right angle into the architecture. By using these bonds, novel molecular designs are expected.

The yields of the CPOs tend to be inversely proportional to the size of the ring. Although the connection manners are different, square-shaped cyclic tetramer **14** was isolated in 7% yield [20], whereas the smaller square **15** was obtained in 22% yield [19] (Fig. 7).

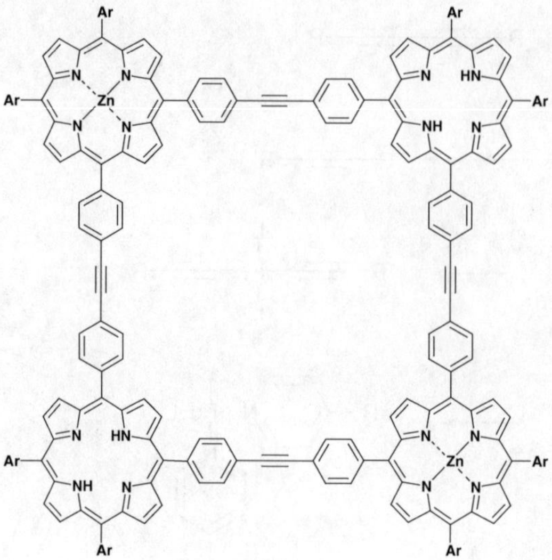

Fig. 7. Molecular structures of **14–17**

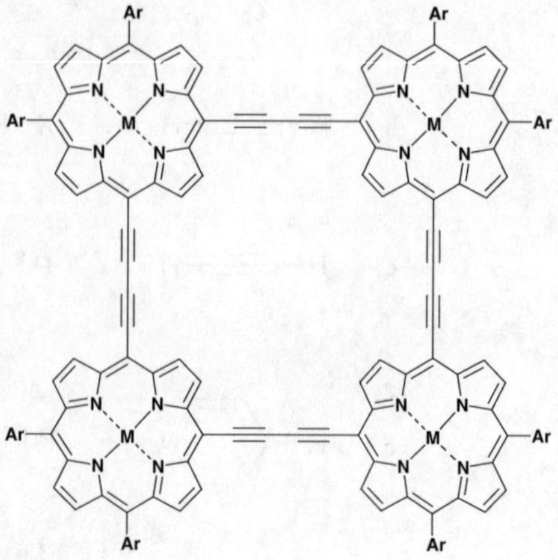

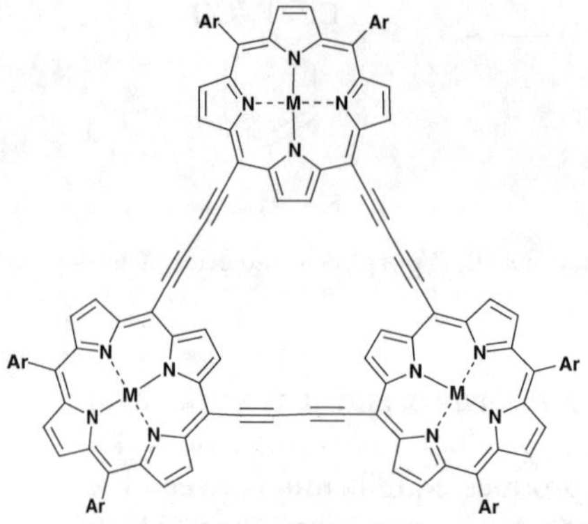

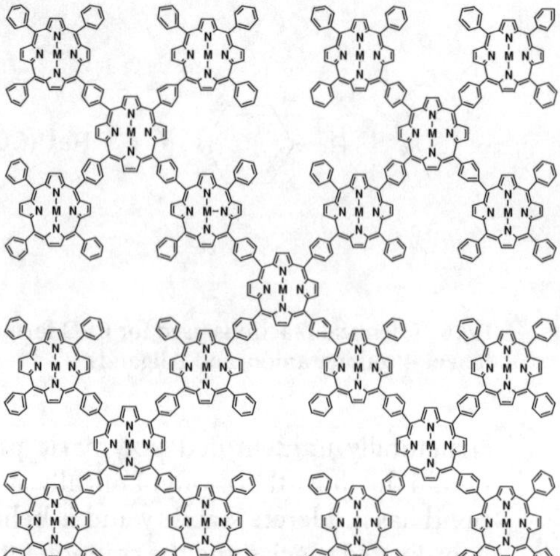

2.3
Purification

Although we choose neither method A nor B, difficulties in the purification
and/or isolation of CPOs are unavoidable. The component porphyrin unit with
polar substituents such as an ester group produces further polar products in
oligomerization reactions. The polarities of the products are proportional to
the degrees of oligomerization caused by the increase in the number of func-

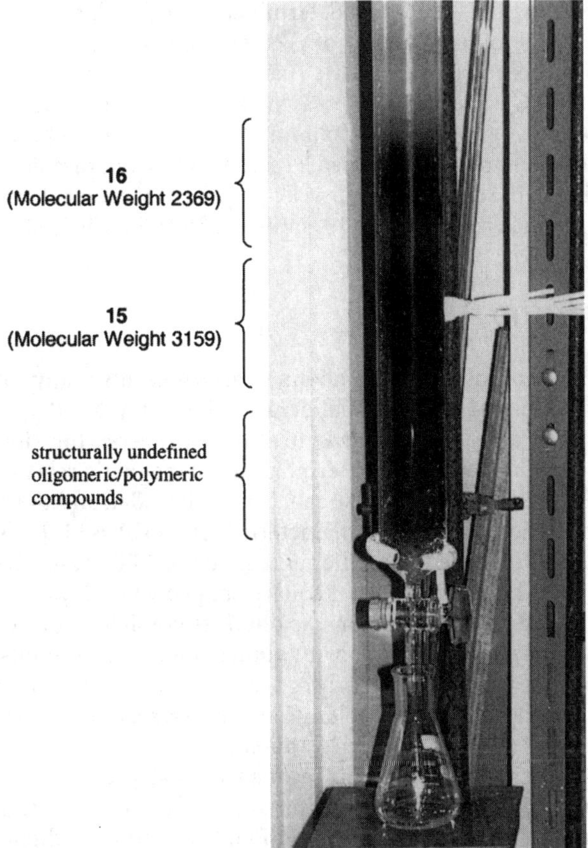

Fig. 8. GPC separation of square-shaped cyclic oligomer **15** from trimer **16** and structurally unidentified polymeric compounds

tional groups. In such cases, adsorption chromatography, such as silica gel chromatography, is useful. For example, cyclic trimer **3** has twelve ester groups and dimer **4** holds only eight ester groups, both of which are produced from the ZnPor unit having four ester groups by trimerization and dimerization reactions, respectively (Fig. 4). Therefore, the separation of **3** from **4** is achieved by silica gel chromatography using chloroform as the elute (R_f values of **4** and **3** are 0.11 and 0.13, respectively) [22, 31]. Because the purification and/or separation are easy, this reaction is convenient and affords a series of CPOs by one-pot synthesis.

However, because most of the CPOs contain nonpolar substituents, isolation of the desired CPO is difficult using adsorption chromatography techniques. In this regard, gel permeation chromatography (GPC) is useful. The gel contains small holes that can retain organic molecules. Molecules larger than the size of the hole pass through the gel. The retention time and/or volume of small molecules are longer/larger than those of large molecules. The maximum size of molecules separable by the GPC gel is determined as the exclusion size associated with the size of the hole. A variety of GPC gels are commercially available for open column chromatography as well as for high-performance liquid chromatography (HPLC). The author's research group usually uses BIO-BEANS S-X1 (Bio-Rad Co., Ltd.) gel for open column chromatography, because relatively large amounts of samples are purified. Figure 8 shows an example of our purification experiment of square-shaped cyclic oligomer **15** (molecular weight 3,159) from the by-product, cyclic trimer **16** (molecular weight 2,369) [19].

2.4
Characterization of CPOs

The characterization of the CPOs obtained also presents many difficulties. In the synthetic studies of conventional organic chemistry, NMR spectroscopy is the most popular and powerful analytical tool. However, the slow motion of oligomeric porphyrins in solution terminates the magnetic relaxation process. Therefore, the line widths of the resonance peaks broaden and the resultant few pieces of information prevent the characterization of CPOs [32]. A similar peak broadening is also observed upon the aggregation of CPOs associated with $\pi-\pi$ interactions. The characterization of square-shaped CPO **15** was performed by NMR under high dilution conditions, as dilute as possible to detect the signals [19]. Some oligomeric porphyrins with similar molecular weights show well-resolved resonance peaks, particularly rod- and/or wire-shaped oligomers, indicating that these molecules have relaxation processes associated with the rotation along the axis of the rod or wire. In marked contrast, sheet-shaped molecules, such as **17**, tend to broaden the resonance peaks [32].

Considering these situations, the observation of molecular weights, particularly by matrix-assisted laser desorption/ionization time-of-flight mass spectroscopy (MALDI-TOF MASS), is essential [33]. The operation is simple and enables us to observe the molecular ion peaks of CPOs with molecular weights exceeding 10,000. The quality of the measurement is strongly dependent on the choice of the matrix. Therefore, the search for the best matrix for each CPO should be pursued.

However, because of the high price, MALDI-TOF mass spectrometers have not come into wide use. Vapor pressure osmometry (VPO), an old and traditional method for estimating molecular weight, is useful in the field of CPO chemistry. The experimental error of this measurement is approximately 10%; however, the obtained data are sufficiently useful to estimate the number of porphyrins in a molecule.

The large size of CPOs allows their direct observation. For this purpose, scanning tunneling microscopy (STM) is the best method [32, 34]. Electron microscopic analysis is used for phthalocyanine **3** and its derivatives; however, most of the porphyrin derivatives are decomposed by electron beam irradiation. Presently, although only a limited number of researchers are able to perform atomic-scale resolution measurement, this powerful analytical method is expected to be used widely in the future. The author reported a summary of STM studies on porphyrins elsewhere [34].

3
Function of Cyclic Porphyrin Oligomers

CPOs are best characterized by the following three features: 1) axial coordination to the incorporated metals, 2) specific nano-sized space created by rigid porphyrin panels, and 3) specific (photo-induced) redox reactions associated with the porphyrin's π-electron system. In this chapter, some examples are reviewed based on these properties.

3.1
Reactions

3.1.1
Acceleration of Chemical Reactions

Attempts to realize enzymatic reactions have been reported over the past four decades in the context of host–guest chemistry, presently a well-established research field. In the field of CPOs, much attention has been paid to identical research objectives. The host–guest chemistry based on CPOs holds a special position, because specific selectivity and reactivity will be achieved using the coordination-bond-forming reactions between the substrate and the incorporated metals in the porphyrins, as well as the redox reaction associated with the porphyrin's π-electron system.

Sanders and his co-workers reported two enzymatic reactions using CPO **3** described above (Fig. 9). In the reactions, three zinc metals are incorporated into each porphyrin. The first reaction is the acylation reaction of pyridyl alcohol **18** using *N*-acetylimidazole (**19**) [35] to afford **20**. In the presence of **3**, the reaction is accelerated 16 times compared to the reaction without **3**. The replacement of **3** with **4** (M=Zn) gave no significant acceleration of the reaction; therefore, the enzymatic reaction proceeds through a supramolecular

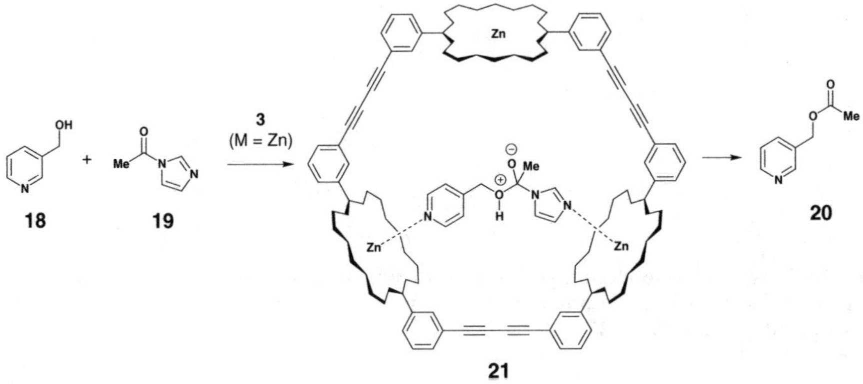

Fig. 9. Enzymatic acylation reaction of **18** and **19**. Structure **21** is proposed as the reaction intermediate

intermediate, such as **21**. The rigidity of catalyst **3**, that is, the stability of the nano-sized space produced by the three porphyrins, affects the enzymatic reaction. The reductive product of **3**, that is, the porphyrin trimer connected by $-CH_2CH_2CH_2CH_2-$ groups, accelerates the reaction by only five times.

3.1.2
Regioselectivity

The second example is the Diels–Alder reaction of **22** and **23** (Fig. 10) [36]. Usually, this reaction affords the *endo*-adduct **24**, the kinetically controlled product; however, the reversibility of the reaction produces the by-product, *exo*-adduct **25**, the thermodynamically controlled product. In the presence of CPOs **3** or **8** (M=Zn), the ratio of *exo*- to *endo*-adducts changes. The small nano-sized space produced by **8** affords *endo*-adduct **24** in 100% yield; in contrast, the large nano-sized space produced by **3** gives *exo*-adduct **25** in 100% yield. Catalysts **3** and **8** supply the perfect nano-sized spaces that best match the transition states for the *exo*- and *endo*-products. In other words, the formation of the coordination bonds of **22** and **23** with the zinc atoms of the catalyst produces the transition states to give the *exo*- and *endo*-products. The acyclic catalyst with a structure similar to that of **3** or **8** gives a mixture of **24** and **25**, indicating that the rigid nano-sized space is needed for the regioselective reaction.

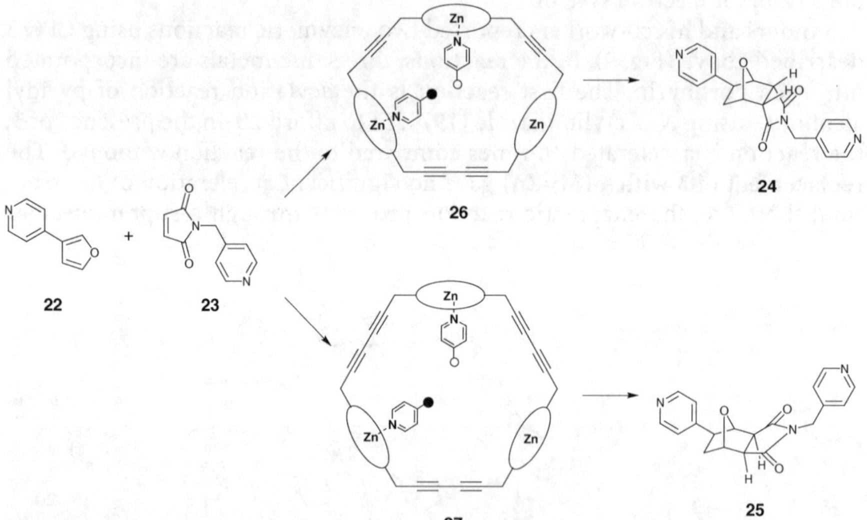

Fig. 10. Enzymatic regioselective Diels–Alder reaction of **22** and **23**. The catalyst with a small nano-sized space selectively produces *endo*-adduct, **24**, and the other catalyst with a large space produces *exo*-adduct **25**

3.2
Energy Transfer Systems

As described previously, the first stage of photosynthesis is the energy transfer. To elucidate the detailed mechanism of this complicated biological process, artificial porphyrin oligomers containing ZnPor and 2HPor moieties are frequently used as models, in which the energy is transferred from ZnPor to 2HPor. This energy transfer process is easily detectable from emission spectra as a single fluorescence from the 2HPor moiety. By using this ZnPor–2HPor system, several intramolecular energy transfer molecules are prepared. For example, Lindsey and co-workers designed, prepared, and evaluated cyclic tetramer **14** [20]. In the selective excitation of the ZnPor moieties of **14** by irradiation at 550 nm, which is the characteristic absorption of ZnPor, the excitation energy was transferred to the 2HPor moiety with 99.5% quantum efficiency (QE) by a through-bond mechanism. They discussed that this QE value is a significant increase from that of the 2HPor–ZnPor dimer having the identical linkage, 99.0%. One of the main reasons is that **14** has two energy transfer pathways, that is, the ZnPor moieties of **14** have two adjacent 2HPors. Presently, systematic synthetic studies that would enable discussions of the relationships between the QE and the cyclic molecular alignment of the chromophore are lacking.

Considering the situations described above, a theoretical study of the structure-dependent energy transfer efficiency was reported by Lindsey and co-workers [36]. They designed four kinds of porphyrin decamers using nine ZnPors and 2HPor (Fig. 11a–d). Type a mimics the alignment of the natural system that contains the cyclic molecular alignment. Type b is a structural analog of type a, and type d is designed based on dendrimer compounds. The energy transfer efficiency from ZnPors to 2HPor was theoretically estimated, and the QE of the natural type (a) was found to be 76%. This value is in marked contrast

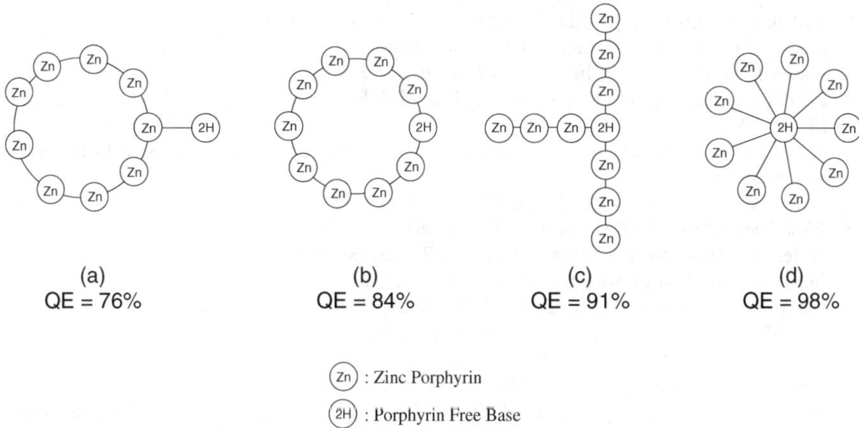

Fig. 11a–d. Designed porphyrin decamers. Theoretically obtained quantum efficiencies for the energy transfer (QE, %) are shown

to that of type c previously designed by the authors, 91%. The dendrite-type alignment, type d, showed the highest QE of 98%. Experiments on these are expected.

4
Conclusion

The maturity of dendrimer chemistry has led to confusion among researchers, because there are no specific research objectives in this field. It may be a good idea to design and synthesize a novel macromolecule, a non-dendrite-shaped molecule, keeping the concept of the dendrimer in mind. The nano-sized closed space inside the molecule is one of the key concepts for novel macromolecular chemistry.

Acknowledgements. The author appreciate the helpful discussions with Professor Masahiro Yamashita and Dr. Hitoshi Miyasaka, Tokyo Metropolitan University.

5
References

1. Vögtle F (1998) Dendrimers. In: Vögtle F (ed) Topics in Current Chemistry, Vol 187. Springer, Berlin Heidelberg New York
2. Vögtle F (2000) Dendrimers II. In: Vögtle F (ed) Topics in Current Chemistry, Vol 210. Springer, Berlin Heidelberg New York
3. Vögtle F (2001) Dendrimers III. In: Vögtle F (ed) Topics in Current Chemistry, Vol 212. Springer, Berlin Heidelberg New York
4. Vögtle F, Schalley CA (2001) Dendrimers IV; Topics in Current Chemistry Vol. 217; Vögtle F, Schalley CA, Eds.; Springer: Berlin
5. Newkome GR, Moorefield CN, Vögtle F (2001) Dendrimers and dendrons, concept, synthesis, applications. Wiley-VCH, Weinheim
6. Jiang D-L, Aida T (1997) Nature 388:454–456
7. Milgrom LR (1997) The colours of life. Oxford University Press, Oxford
8. Kadish KM, Smith KM, Guilard R (2000) In: Kadish KM, Smith KM, Guilard R (eds) The porphyrin handbook. Academic Press, San Diego
9. Marks TJ (1990) Angew Chem Int Ed Engl 29:857-879
10. Rittenberg DK, Sugiura K-i, Sakata Y, Mikami S, Epstein AJ, Miller JS (2000) Adv Mater 12:126–130
11. Baldo MA, O'Brien DF, You Y, Shoustikov A, Sibley S, Thompson ME, Forrest SR (1998) Nature 395:151–154
12. Chen BML, Tulinsky A (1972) J Am Chem Soc 94:4144–4151
13. Wasielewski MR (1992) Chem Rev 92:435–461
14. Pullerits T, Sundström V (1996) Acc Chem Res 29:381–389
15. Sanders JKM (2000) Pure App Chem 72:2265–2274
16. Susumu K, Maruyama H, Kobayashi H, Tanaka K (2001) J Mat Chem 11:2262–2270
17. Buhl M, Hirsch A (2001) Chem Rev 101:1153–1183
18. Anderson HL, Sanders JKM (1989) J Chem Soc Chem Commun:1714–1715
19. Sugiura K-I, Fujimoto Y, Sakata Y (2000) Chem Commun:1105–1106
20. Wagner RW, Seth J, Yang SI, Kim D, Bocian DF, Holten D, Lindsey JS (1998) J Org Chem 63:5042–5049
21. Anderson S, Anderson HL, Sanders JKM (1993) Acc Chem Res 26:469–475
22. Anderson S, Anderson HL, Sanders JKM (1995) J Chem Soc Perkin Trans 1:2247–2254

23. Marvaud V, Vidal-Ferran A, Webb SJ, Sanders JKM (1997) J Chem Soc Dalton Trans: 985–990
24. Mackay LG, Anderson HL, Sanders JKM (1995) J Chem Soc Perkin Trans 1:2269–2273
25. Webb SJ, Sanders JKM (2000) Inorg Chem 39:5912–5919
26. Sugiura K-I, Ponomarev G, Okubo S, Tajiri A, Sakata Y (1997) Bull Chem Soc Jpn 70:1115–1123
27. Fujita N, Biradha K, Fujita M, Sakamoto S, Yamaguchi K (2001) Angew Chem Int Ed Engl 40:1718–1721
28. Fujita M, Umemoto K, Yoshizawa M, Fujita N, Kusukawa T, Biradha K (2001) Chem Commun:509–518
29. Slone RV, Hupp JT (1997) Inorg Chem 36:5422–5423
30. Belanger S, Keefe MH, Welch JL, Hupp JT (1999) Coord Chem Rev 190–192:29–45
31. Anderson HL, Sanders JKM (1995) J Chem Soc Perkin Trans 1:2223–2229
32. Sugiura K-I, Tanaka H, Matsumoto T, Kawai T, Sakata Y (1999) Chem Lett:1193–1194
33. Stulz E, Mak CC, Sanders JKM (2001) J Chem Soc Dalton Trans:604–613
34. Sugiura K-I, Matsumoto T, Tada H, Nakamura T (2003) In: Sugiura K-I, Matsumoto T, Tada H, Nakamura T (eds) Chemistry of nanomolecular systems toward the realization of molecular devices. Springer, Berlin Heidelberg New York
35. Mackay LG, Wylie RS, Sanders JKM (1994) J Am Chem Soc 116:3141–3142
36. Van Patten PG, Shreve AP, Lindsey JS, Donohoe RJ (1998) J Phys Chem B 102:4209–4216

Top Curr Chem (2003) 228: 87–110
DOI 10.1007/b11007

Fullerodendrimers: Fullerene-Containing Macromolecules with Intriguing Properties

Jean-François Nierengarten

Institut de Physique et Chimie des Matériaux de Strasbourg, Groupe des Matériaux
Organiques, Université Louis Pasteur de Strasbourg et CNRS, 23 rue du Loess,
67034 Strasbourg Cedex, France
E-mail: niereng@ipcms.u-strasbg.fr

Fullerenes possess electronic and photophysical properties which make them natural candidates for the preparation of functional dendrimers. The attachment of a controlled number of dendrons on a C_{60} core provides a compact insulating layer around the carbon sphere, and the triplet lifetimes of the C_{60} chromophore can be used to evaluate its degree of isolation from external contacts. The fullerene core can also act as a terminal energy receptor in dendrimer-based light-harvesting systems. When a fullerodendrimer is further functionalized with a suitable electron donor, it may exhibit the essential features of a multicomponent artificial photosynthetic system in which photo-induced energy transfer from the antenna to the C_{60} core is followed by electron transfer. On the other hand, the preparation of dendrons with peripheral C_{60} subunits or containing a C_{60} sphere at each branching unit has been achieved. These fullerodendrons are not only interesting building blocks for the synthesis of monodisperse fullerene-rich macromolecules with intriguing properties, but they are also amphiphilic compounds capable of forming stable Langmuir films at the air–water interface.

Keywords. Dendrimers, Fullerenes, Optical limitation, Photophysical properties, Thin films

1
Introduction

In light of their unique molecular structure, dendrimers have attracted increasing attention in the past decade, and the design of functional dendrimers is an area with unlimited possibilities for fundamental new discoveries and practical applications [1]. Of the various electro- and photoactive chromophores utilized for dendrimer chemistry, C_{60} appears to be a versatile building block and at present a growing interest is developing in fullerene-functionalized dendrimers, that is, fullerodendrimers [2]. In particular, the unusual chemical and physical properties of fullerene derivatives [3] make fullerodendrimers attractive candidates for a variety of interesting features in supramolecular chemistry and materials science. In a recent review article, Hirsch and Vostrowski showed that C_{60} itself is a convenient core for dendrimer chemistry [4]. The functionalization of C_{60} with a controlled number of dendrons dramatically improves the solubility of the fullerenes [5–8]. Furthermore, variable degrees of addition within the fullerene core, especially from mono- up to hexaadducts, are possible and its almost spherical shape leads to globular systems even with low-generation dendrons [9–11]. The authors also highlighted specific advantages brought about the encapsulation of a fullerene moiety in the middle of a dendritic structure for the preparation of thin ordered films [12] or in the design of liquid crystalline derivatives [13–18]. In the present paper, the use of the fullerene sphere as a photoactive core unit will be emphasized. In particular, we will show that the special photophysical properties of C_{60} can be used to evidence dendritic shielding effects. On the other hand, a fullerene core can act as a terminal energy receptor in dendrimer-based light-harvesting systems. Following the description of dendrimers with a C_{60} core, the synthesis of dendrons with peripheral C_{60} subunits or containing a C_{60} sphere at each branching unit will be described. These fullerodendrons are not only versatile building blocks for the preparation of monodisperse fullerene-rich macromolecules with intriguing properties, but they are also amphiphilic compounds capable of forming stable Langmuir films at the air–water interface.

2
Dendrimers with a Photoactive Fullerene Core

2.1
A Fullerene Core to Probe Dendritic Shielding Effects

The ability of a dendritic shell to encapsulate a functional core moiety and to create a specific site-isolated microenvironment capable of affecting the molecular properties has been intensively explored in recent years [19]. A variety of experimental techniques have been employed to evidence the shielding of the core moiety and to ascertain the effect of the dendritic shell [19, 20]. Dendrimers with a fullerene core appear to be appealing candidates to evidence such effects resulting from the presence of the surrounding dendritic branches. Effectively, the lifetime of the first triplet excited state of fullerene derivatives

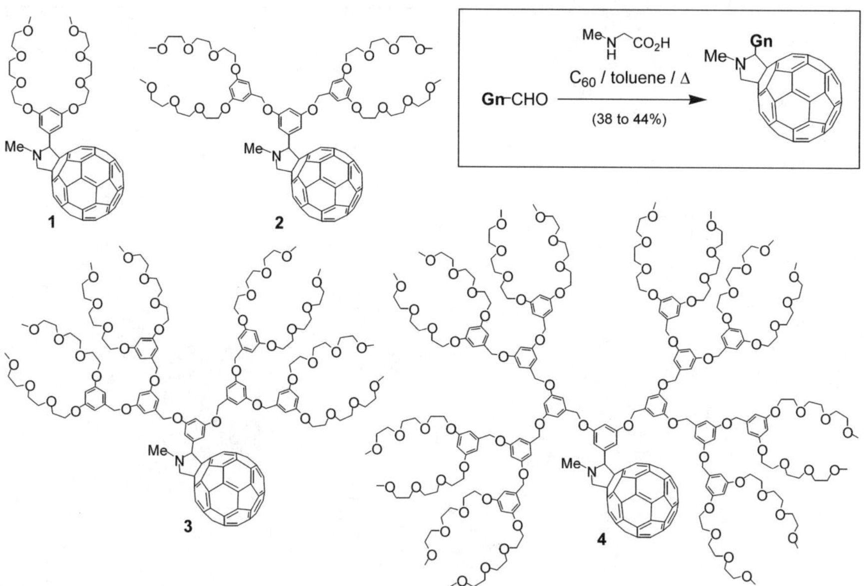

Fig. 1. Fullerodendrimers 1–4

is sensitive to the solvent [21]. Therefore, lifetime measurements in different solvents could be used to evaluate the degree of isolation of the central C_{60} moiety from external contacts. With this idea in mind, we have prepared fullerodendrimers 1–8 (Figs. 1 and 2). In the design of these compounds, it was decided to attach poly(aryl ether) dendritic branches terminated with peripheral triethyleneglycol chains to obtain derivatives soluble in a wide range of solvents [21, 22]. The synthetic approach to prepare compounds 1–4 relies upon the 1,3-dipolar cycloaddition of the dendritic azomethine ylides generated in situ from the corresponding aldehydes and *N*-methylglycine. This methodology has proven to be a powerful procedure for the functionalization of C_{60} due to its versatility and the ready availability of the starting materials [23]. Thus, reaction of the dendritic aldehydes with *N*-methylglycine and C_{60} in refluxing toluene gave the corresponding fulleropyrrolidines 1–4 in 38–44% isolated yield after column chromatography on silica gel followed by gel permeation chromatography.

Dendrimers 5–8 were obtained by taking advantage of the versatile regioselective reaction developed in the group of Diederich [24], which led to macrocyclic bis-adducts of C_{60} by a cyclization reaction at the C sphere with bis-malonate derivatives in a double Bingel cyclopropanation [25]. Reaction of the dendritic malonates with C_{60}, I_2, and 1,8-diazabicyclo[5.4.0]undec-7-ene (DBU) in toluene at room temperature afforded the corresponding cyclization products 5–8 (Fig. 2). The relative position of the two cyclopropane rings in 5–8 on the C_{60} core was determined based on the molecular symmetry deduced from the ^1H and ^{13}C NMR spectra (C_S) as well as on their UV/Vis spectra. It is well established that the absorption spectra of C_{60} bis-adducts are highly dependent on the addi-

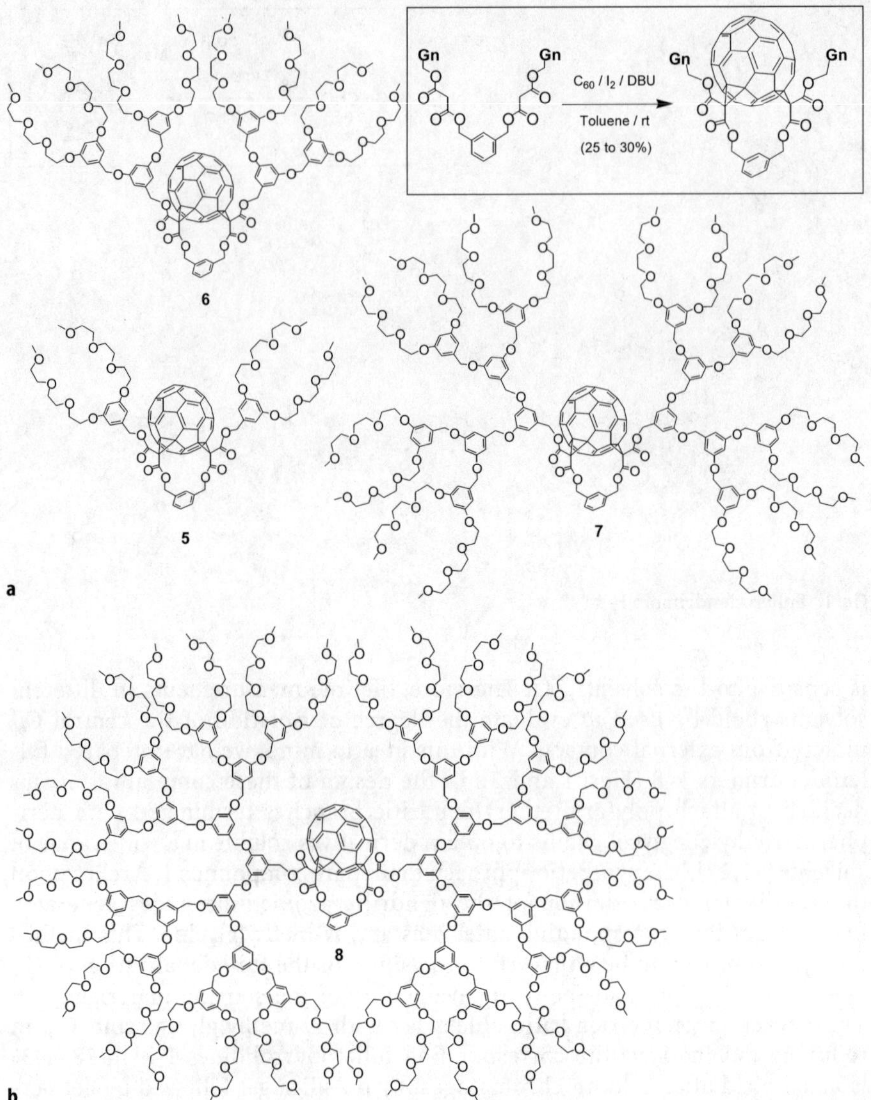

Fig. 2. Fullerodendrimers 5–8

tion pattern and characteristic for each regioisomers [24]. The UV/Vis spectra of **5–8** show all the characteristic features seen for previously reported analogous *cis*-2 bis-adducts [8]. In addition, it must also be added that 1,3-phenylene bis(methylene)-tethered bis-malonates produce regioselectively the *cis*-2 addition pattern at C_{60} [26–30].

The photophysical properties of **1–8** have been studied in different solvents (PhMe, CH_2Cl_2, and CH_3CN). The lifetimes of the lowest triplet excited states are summarized in Table 1.

Table 1. Lifetime of the first triplet excited state of **1–8** in air-equilibrated solutions determined by transient absorption at room temperature [31]

Compound	τ (ns) in toluene	τ (ns) in CH_2Cl_2	τ (ns) in CH_3CN
1	279	598	n/a[a]
2	304	643	330
3	318	732	412
4	374	827	605
5	288	611	314
6	317	742	380
7	448	873	581
8	877	1,103	1,068

[a] Not soluble in this solvent.

For both series of dendrimers interesting trends can be obtained from the analysis of triplet lifetimes in air-equilibrated solutions (Table 1) [21]. A steady increase of lifetimes is found by increasing the dendrimer size in all solvents, suggesting that the dendritic wedges are able to shield, at least partially, the fullerene core from external contacts with the solvent and from quenchers such as molecular oxygen. For compounds **1–4**, the increase is particularly marked in polar CH_3CN, where a better shielding of the fullerene chromophore is expected as a consequence of a tighter contact between the strongly nonpolar fullerene unit and the external dendritic wedges; in this case a 45% lifetime prolongation is found in passing from **2** to **4** (23% and 28% only for $PhCH_3$ and CH_2Cl_2, respectively). It must be emphasized that the triplet lifetimes of **4** in the three solvents are rather different from each other, probably reflecting specific solvent–fullerene interactions that affect excited state deactivation rates. This suggests that, albeit a dendritic effect is evidenced, even the largest wedge is not able to provide a complete shielding of the central fulleropyrrolidine core in **4** [21]. The latter hypothesis was confirmed by computational studies. As shown in Fig. 3, the calculated structure of **4** revealed that the dendritic shell is unable to completely cover the fullerene core (it must be noted that the calculations have been performed in the absence of solvent, our aim being only to estimate the possible degree of isolation). In contrast, the triplet lifetimes of **8** [31] in the three solvents tend to a similar value suggesting that the fullerene core is in a similar environment whatever the nature of the solvent is. In other words the C_{60} unit is, to a large extent, not surrounded by solvent molecules but substantially buried in the middle of the dendritic structure which is capable of creating a specific site-isolated microenvironment around the fullerene moiety. The latter hypothesis is quite reasonable based on the calculated structure of **8** (Fig. 3) showing that the dendritic branches are able to fully cover the central fullerene core.

The dendritic effect evidenced for **1–8** might be useful to optimize the optical limiting properties characteristic of fullerene derivatives. Effectively, the intensity dependant absorption of fullerenes originates from larger absorption cross sections of excited states compared to that of the ground state [32], therefore the

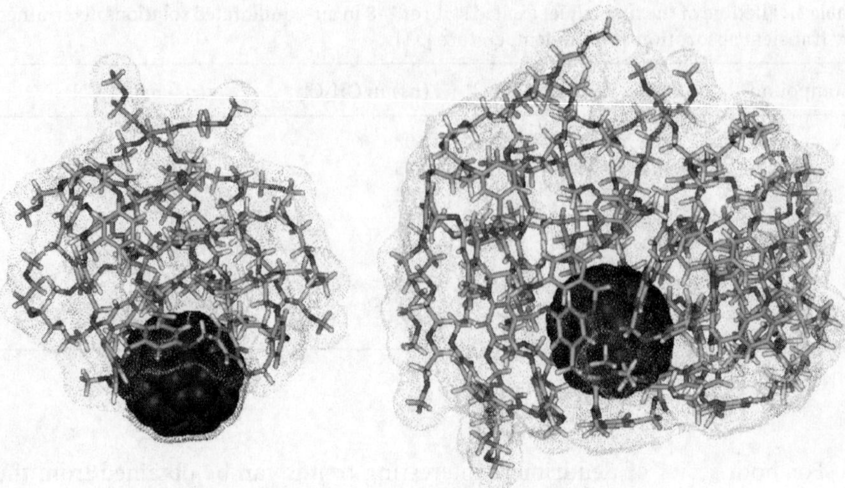

Fig. 3. Calculated structure of fullerodendrimers **4** (*left*) and **8** (*right*)

increased triplet lifetime observed for the largest fullerodendrimers may allow for an effective limitation on a longer time scale. For practical applications, the use of solid devices is largely preferred to solutions and inclusion of fullerene derivatives in sol–gel glasses has shown interesting perspectives [33]. However, faster de-excitation dynamics and reduced triplet yields are typically observed for fullerene-doped sol–gel glasses when compared to solutions [34]. These observations are mainly explained by two factors: (i) perturbation of the molecular energy levels due to the interactions with the sol–gel matrix and (ii) interactions between neighboring fullerene spheres due to aggregation [34]. Therefore, the encapsulation of the C_{60} core evidenced by the photophysical studies for both series of fullerodendrimers might also be useful to prevent such undesirable effects. The incorporation of fullerodendrimers **1–8** in sol–gel glasses has been easily achieved by soaking mesoporous silica glasses with a solution of **1–8** [21]. For the largest compounds, the resulting samples only contain well-dispersed fulleroden-drimer molecules. Preliminary measurements on the resulting doped samples have revealed efficient optical limiting properties [21]. For example, the optical transmission as a function of the fluence of the laser pulses is shown in Fig. 4 for a sol–gel sample containing compound **4**. The transmission remains nearly constant for fluences lower than 5 mJ cm^{-2}. When the intensity increases above this threshold, the effect of induced absorption appears, and the transmission diminishes rapidly, thus showing the potential of these materials for optical limiting applications. Further studies are underway to determine the influence of the dendritic branches on the optical limiting behavior of these composite materials.

Fullerodendrimers also allow an evaluation of the accessibility of the C_{60} core unit by studying bimolecular deactivation of its excited states by external quenchers. Recently Ito, Komatsu, and co-workers have used this approach to investigate a series of fullerodendrimers (**9–11**) in which Fréchet-type dendrons have been connected to a fullerene moiety via an acetylene linker (Fig. 5) [35].

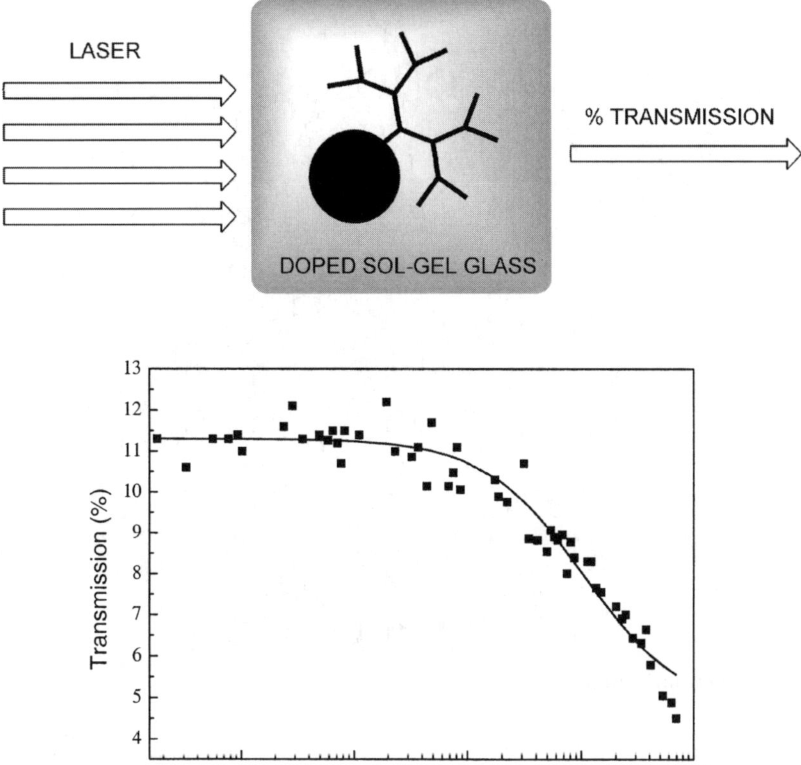

Fig. 4. Transmission versus incident fluence at 532 nm of a sol–gel sample containing dendrimer **4**

Both energy and electron transfer quenchers have been employed to show that the quenching rates of the fullerene triplet state are decreased as a function of the size of the dendrimer shell [36]. These results further demonstrate that fullerene is an excellent functional group to probe the accessibility of a dendrimer core by external molecules.

2.2
Light-Harvesting Dendrimers with a Fullerene Core

The synthesis and the study of dendrimer-based light-harvesting structures have attracted increased attention in recent years [37]. In such systems, an array of peripheral chromophores is able to transfer the collected energy to the central core of the dendrimer thus mimicking the natural light-harvesting complex in which antenna molecules collect sunlight and channel the absorbed energy to a single reaction center [37]. The C_{60} sphere is an attractive functional core for the preparation of light-harvesting dendrimers. Effectively, its first singlet and

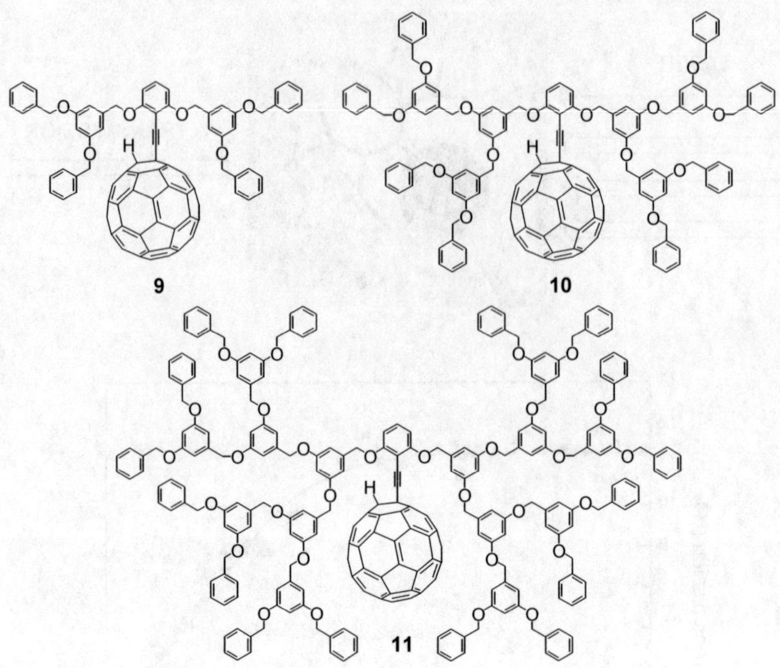

Fig. 5. Fullerodendrimers 9–11

Fig. 6. Fullerodendrimers 12–14

triplet excited states are relatively low in energy and photo-induced energy transfer events have been evidenced in fullerene-based dyads [38]. In particular, photophysical investigations of some fulleropyrrolidine derivatives substituted with oligophenylenevinylene (OPV) moieties revealed a very efficient singlet–singlet $OPV{\rightarrow}C_{60}$ photo-induced energy transfer [39, 40]. Based on this observation, dendrimers **12–14** comprising a fullerene core and peripheral OPV subunits (Fig. 6) have been prepared [41].

The photophysical properties of fullerodendrimers **12–14** have been investigated in CH_2Cl_2 solutions. Upon excitation at the OPV band maximum, dramatic quenching of OPV fluorescence is observed for all fullerodendrimers. This is attributed to an $OPV{\rightarrow}C_{60}$ singlet–singlet energy-transfer process [41]. At 394 nm (corresponding to OPV band maxima) the molar absorptivities (ε) of these fullerodendrimers are 95,800 for **12**, 134,800 for **13**, and 255,100 M^{-1} cm^{-1} for **14**. Since the ε of the ubiquitous N-methyl-fulleropyrrolidine at 394 nm is only 7,600 a remarkable light-harvesting capability of the peripheral units relative to the central core is evidenced along the series.

More recently, we have succeeded in the preparation of the next generation compound **15** (Fig. 7) for which the molar absorptivity is approximately 450,000 M^{-1} cm^{-1}. The photophysical properties of **15** have not been yet investi-

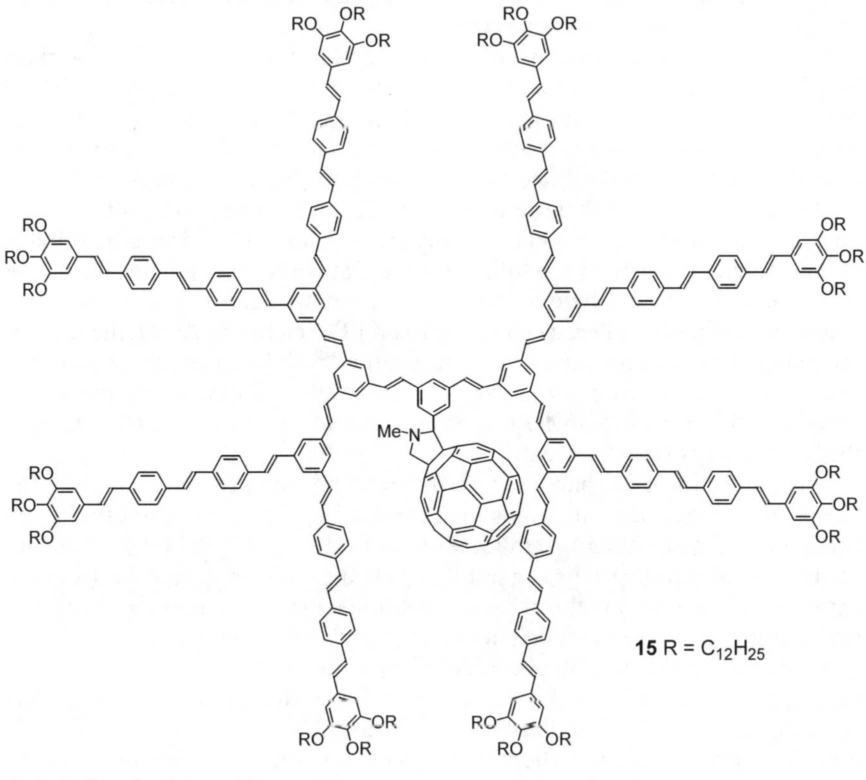

Fig. 7. Fullerodendrimer **15**

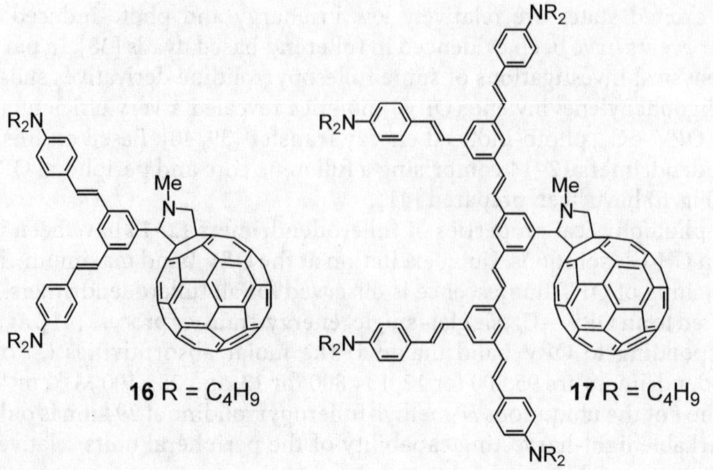

Fig. 8. Fullerodendrimers 16–17

gated in detail; however, preliminary measurements in CH_2Cl_2 have revealed a strong quenching of the OPV fluorescence upon excitation at the OPV band maximum suggesting that the dendritic wedge is still capable of channeling the absorbed energy to the fullerene core.

Related compounds have been reported by Martin, Guldi, and co-workers [42]. The end-capping of the dendritic moiety with dibutylaniline units yielded the multicomponent photoactive systems 16 and 17 in which the dendritic wedge plays the dual role of antenna, capable of channeling the absorbed energy to the fullerene core, and electron-donating unit (Fig. 8). Photophysical investigations in benzonitrile solutions have shown for both compounds that, upon photoexcitation, efficient and fast energy transfer takes place from the initially excited antenna moiety to the fullerene core. This process populates the lowest fullerene singlet excited state which is able to promote electron transfer from the dendritic unit to the fullerene core. For 16 and 17, relative to 12–14, the charge-separated state is significantly lower in energy than the fullerene singlet, as a result of the increased donating ability of the terminal dialkylaniline units [43]. Therefore, the electron transfer is thermodynamically possible in 16 or 17 after the initial energy transfer event.

Langa and co-workers have prepared fullerodendrimers 18 and 19 in which the phenylenevinylene dendritic wedge is connected to a pyrazolino [60] fullerene core rather than to a fulleropyrrolidine one as for 12–17 (Fig. 9) [44]. Preliminary photophysical investigations suggest that the efficient energy transfer from the excited antenna moiety to the pyrazolino[60]fullerene core is followed by an electron transfer involving the fullerene moiety and the pyrazoline N atom.

Methanofullerene 20 with phenylacetylene dendrimer addends has also been reported [45] (Fig. 10). The UV absorption of fullerodendrimer 20 is particularly strong and is mainly attributed to transitions located on the two dendritic branches of the molecule. The photophysical investigations revealed that the large poly(aryl)acetylene branches act as photon antennae [46].

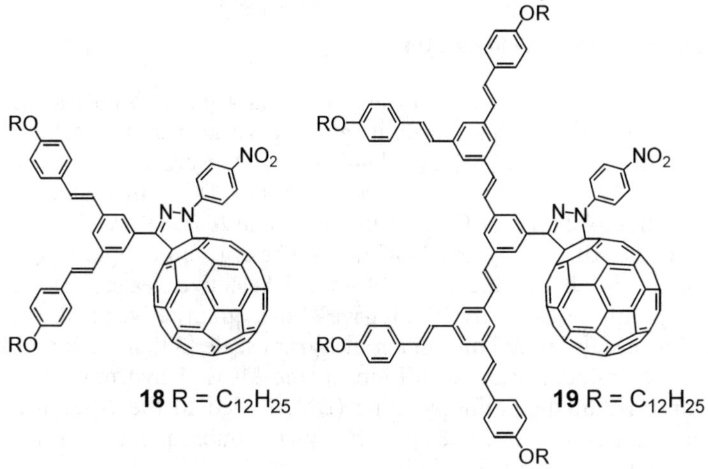

18 R = $C_{12}H_{25}$ **19** R = $C_{12}H_{25}$

Fig. 9. Fullerodendrimers **18–19**

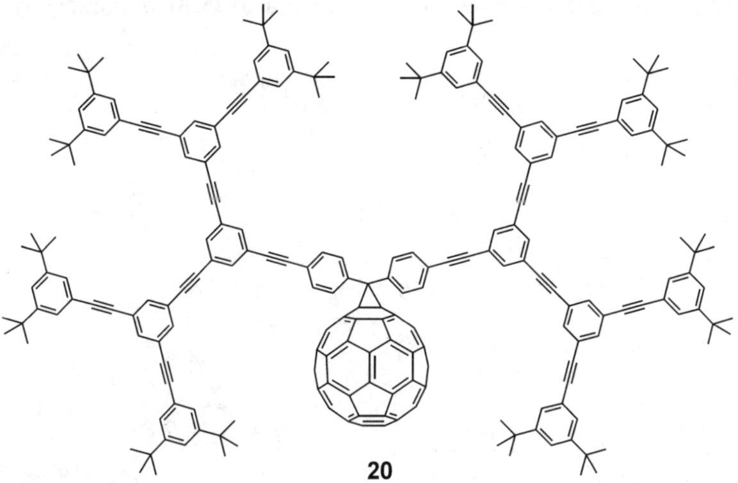

20

Fig. 10. Fullerodendrimer **20**

3
Fullerene-Functionalized Dendrons

All the fullerene-containing dendrimers reported to date have been prepared with a C_{60} core but never with fullerene units at their surface or with C_{60} spheres in the dendritic branches. We have recently started a research program on the synthesis of dendrons substituted with fullerene moieties. These fullerodendrons are interesting building blocks for the preparation of monodisperse fullerene-rich macromolecules. In addition, they are also amphiphilic compounds capable of forming stable Langmuir films at the air–water interface.

3.1
Dendrons with Peripheral Fullerene Units

We have developed an efficient convergent preparation of highly soluble dendritic branches with fullerene subunits at the periphery and a carboxylic acid function at the focal point [47]. The dendrons **21–23** were prepared as depicted in Figs. 11 and 12. *N,N'*-Dicyclohexylcarbodiimide (DCC)-mediated esterification of **25** with diol **24** in CH_2Cl_2 gave bis-malonate **26** in 83% yield. Reaction of bis-malonate **26** with C_{60}, I_2, and DBU in toluene at room temperature gave the C_s-symmetric *cis*-2 bis-adduct **27** in 59% yield. Selective cleavage of the *t*-butyl ester moiety with CF_3CO_2H in CH_2Cl_2 gave **21** in a quantitative yield. The preparation is thus easily carried out on a multi-gram scale [30]. Reaction of **21** with diol **24** under esterification conditions using DCC, 1-hydroxybenzotriazole (BtOH), and 4-dimethylaminopyridine (DMAP) led to the *t*-butyl-protected second-generation dendron **28** in 90% yield. Subsequent treatment with CF_3CO_2H in CH_2Cl_2 afforded **22** in high yield (99%).

Esterification of **22** with diol **24** (DCC/DMAP/BtOH) gave **29** in 84% yield (Fig. 12) and subsequent hydrolysis of the *t*-butyl ester group under acidic conditions afforded the third-generation carboxylic acid **23**. All of the spectroscopic

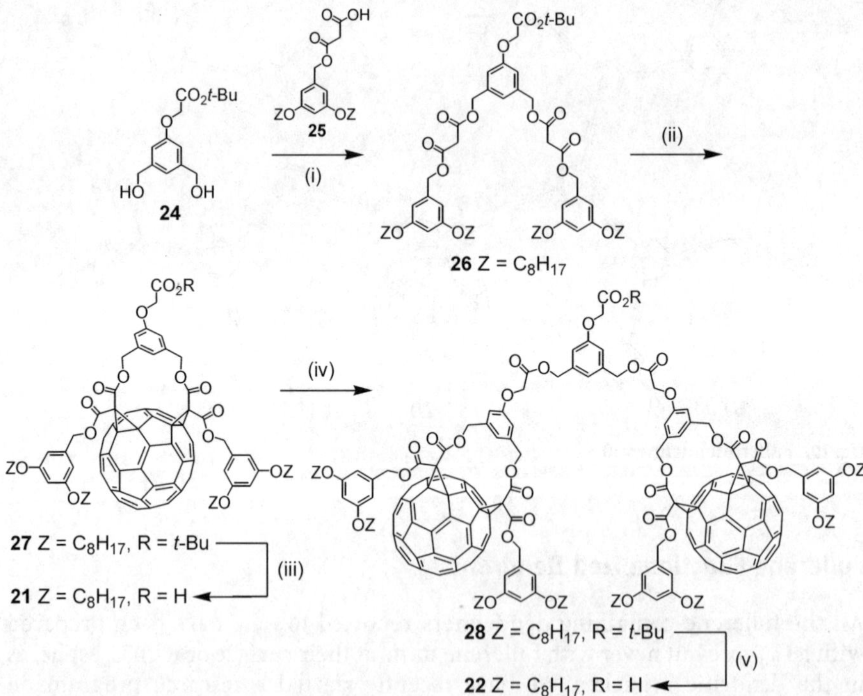

Fig. 11. Synthesis of fullerodendrons **21** and **22**. Reagents and conditions: (i) DCC, DMAP, CH_2Cl_2, 0 °C to room temperature (83%); (ii) C_{60}, DBU, I_2, toluene, room temperature (59%); (iii) CF_3CO_2H, CH_2Cl_2, room temperature (99%); (iv) **24**, DCC, DMAP, BtOH, CH_2Cl_2, 0 °C to room temperature (90%); (v) CF_3CO_2H, CH_2Cl_2, room temperature (99%)

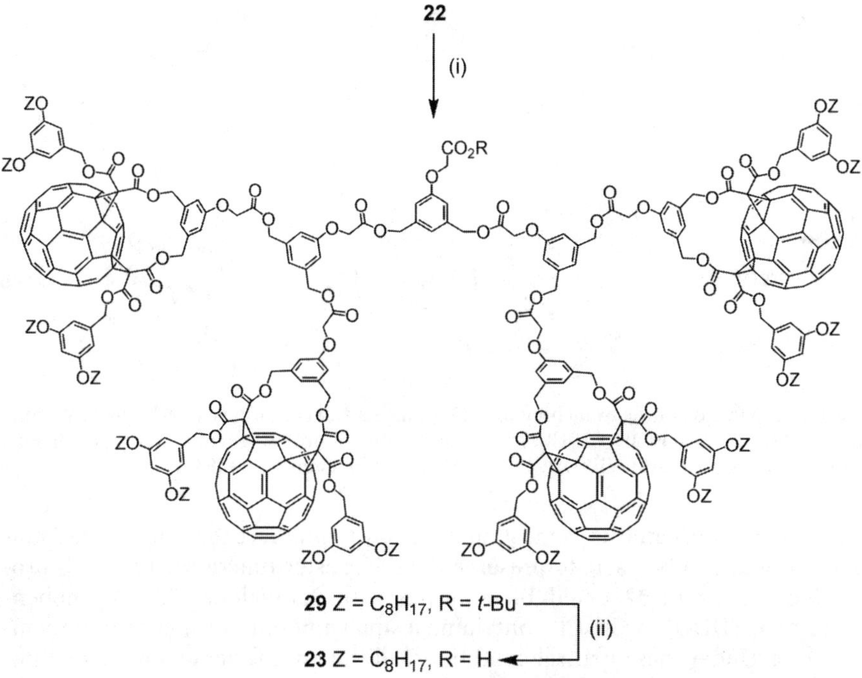

Fig. 12. Synthesis of fullerodendron **23**. Reagents and conditions: (i) **24**, DCC, DMAP, BtOH, CH_2Cl_2, 0 °C to room temperature (90%); (ii) CF_3CO_2H, CH_2Cl_2, room temperature (99%)

studies and elemental analysis results were consistent with the proposed molecular structures assigned to the dendrons of each generation.

Dendrons **21–23** are easily prepared on a multi-gram scale and are highly soluble in common organic solvents thanks to the presence of the four long alkyl chains per peripheral fullerene unit. Therefore, they appear to be good candidates for the preparation of fullerene-rich macromolecules, for example, as shown by their attachment to a phenanthroline diol derivative and the preparation of the corresponding copper(I) complexes (see below).

3.2
Dendrons Containing a C_{60} Sphere at Each Branching Unit

The preparation of fullerodendrons with a C_{60} group at each branching units was carried out by a convergent approach using successive esterification and deprotection steps [48]. The preparation of the key building block **33** is depicted in Fig. 13. This diol containing one *t*-butyl ester function appears to be the branching unit for the construction of the fullerodendrons. DCC-mediated esterification of **30** with *t*-butyl 2-hydroxyacetate in CH_2Cl_2 yielded malonate **31**, which after treatment with C_{60}, I_2, and DBU in toluene at room temperature gave methanofullerene **32** in 30% yield. The choice of the appropriate protecting group for the two alcohol groups in **32** was the key to this synthesis. Effectively,

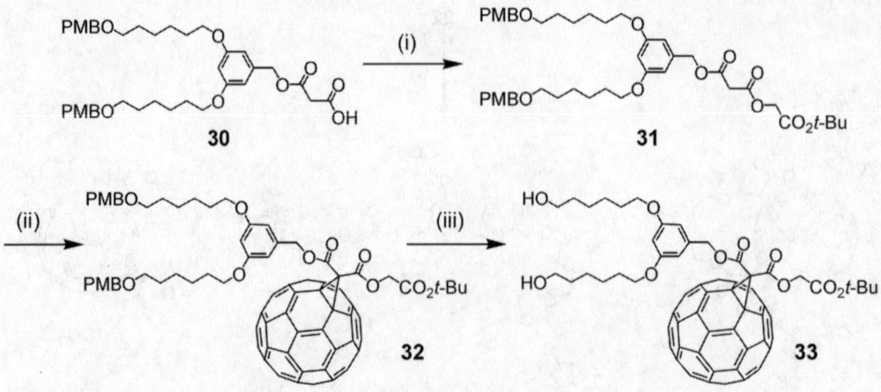

Fig. 13. Preparation of the branching unit 33. Reagents and conditions: (i) t-Butyl 2-hydroxy-acetate, DCC, DMAP, BtOH, CH_2Cl_2, 0 °C to room temperature (91%); (ii) C_{60}, DBU, I_2, toluene, room temperature (30%); (iii) DDQ, H_2O, CH_2Cl_2, room temperature (84%)

the deprotection conditions must not be acidic to preserve the t-butyl ester moiety and may not be basic to preserve the other ester functions. The PMB protecting groups in 32 could be removed with 2,3-dichloro-5,6-dicyanobenzoquinone (DDQ) in CH_2Cl_2 containing a small amount of water at room temperature. Under these neutral conditions, all the ester functions remained unchanged and compound 33 was thus obtained in a good yield (84%).

Fullerene derivative 34 substituted with two long alkyl chains (solubilizing groups) and a carboxylic function was used as peripheral subunit for the constructions of the dendrons (Fig. 14).

Esterification of acid 34 with diol 33 under conditions using DCC, BtOH, and DMAP afforded the t-butyl-protected fullerodendron 35 in 70% yield. Selective hydrolysis of the t-butyl ester under acidic conditions afforded acid 36 in 98% yield. Subsequent reaction of 36 with the branching unit 33 in the presence of DCC, BtOH, and DMAP afforded fullerodendron 37, which after treatment with CF_3CO_2H gave 38. The ^1H and ^{13}C NMR, IR, UV/Vis, and elemental analysis data were consistent with the proposed molecular structures assigned to the fullerodendrons of each generation. The electrochemical properties of fullerodendrons 35 and 37 have also been investigated [49]. Compound 35 shows the characteristic behavior previously reported for methanofullerenes [50]. Effectively, up to three one-electron reduction steps are seen and no oxidation could be observed at potentials below +1 V versus Fc/Fc^+. The reduction potentials (−1.05, −1.41, and −1.85 V versus Fc/Fc^+) are also quite similar. This seems to indicate that the three methanofullerene units in 35 behave as independent redox centers. By CV, fullerodendron 37 behaves like compound 35 and a simultaneous reduction of the seven methanofullerene independent redox centers has been observed. Whereas the FAB-MS spectra of methanofullerene 34 showed the expected molecular ion peak, no characteristic peaks could be observed for the fullerodendrons 35–38. In fact, due to condensation reactions resulting probably from fullerene–fullerene interactions under the high energy of the FAB gun or due to fullerene aggregation in the matrix, the FAB-MS analysis yields only frag-

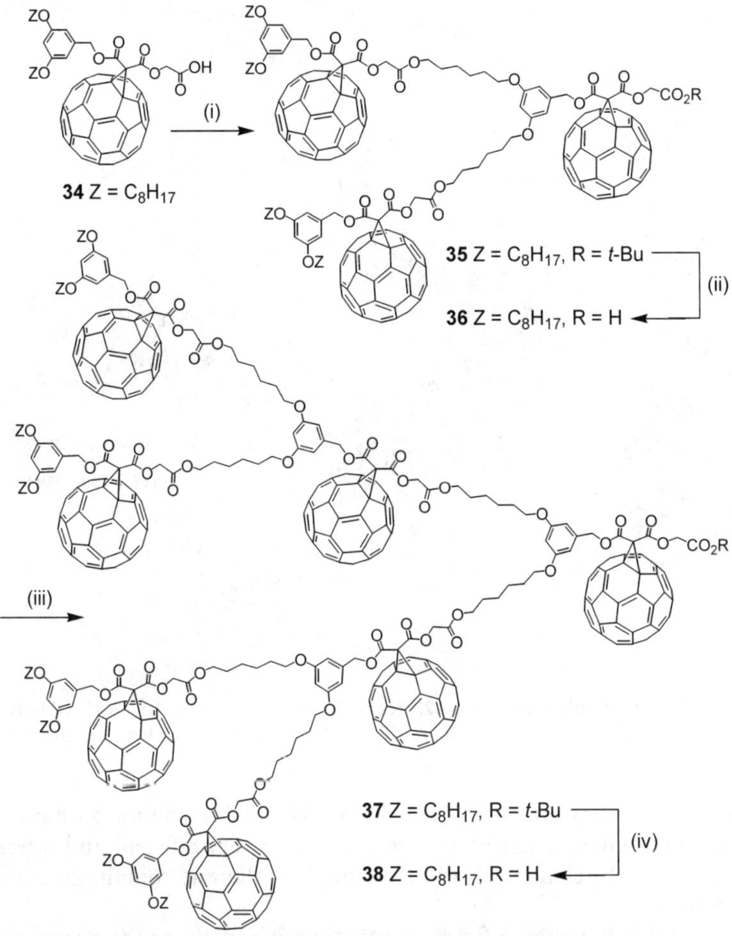

Fig. 14. Synthesis of fullerodendrons **36** and **38**. Reagents and conditions: (i) **33**, DCC, DMAP, BtOH, CH_2Cl_2, 0 °C to room temperature (70%); (ii) CF_3CO_2H, CH_2Cl_2, room temperature (98%); (iii) **33**, DCC, DMAP, BtOH, CH_2Cl_2, 0 °C to room temperature (37%); (iv) CF_3CO_2H, CH_2Cl_2, room temperature (80%)

mentations. The mass spectrometric characterization of **35–38** was, however, possible by using an approach based on their redox properties. Radical anions (negative mode) can be generated during the electrospray (ES) process owing to the ability of the ES source to behave like an electrolysis cell [49]. For example, the ES mass spectrum obtained in the negative mode with fullerodendron **35** shows the peak at m/z 1226.3 corresponding to the radical tri-anion (calculated m/z 1,225.95). The mass spectrum obtained with **37** under similar conditions displays the peak corresponding to the radical hepta-anion at m/z 1,209.6 (calculated m/z 1,209.55). The MS analysis of the reduced species have not only confirmed the structures of **35–38**, but also show that all the methanofullerene subunits in the fullerodendrons of highest generation behave as independent redox

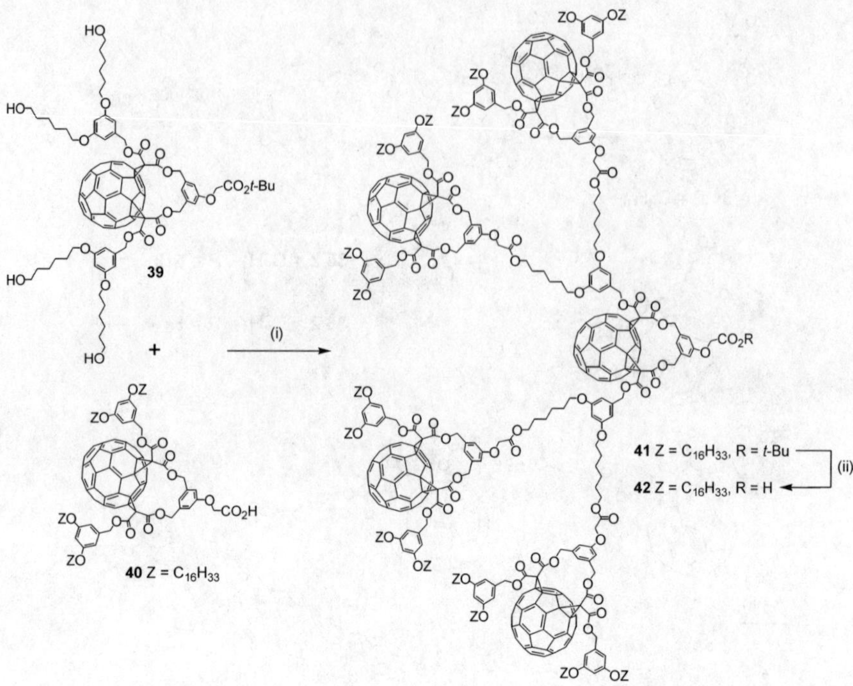

Fig. 15. Synthesis of fullerodendron **42**. Reagents and conditions: (i) DCC, DMAP, BtOH, CH$_2$Cl$_2$, 0 °C to room temperature (65%); (ii) CF$_3$CO$_2$H, CH$_2$Cl$_2$, room temperature (96%)

centers as deduced from their electrochemical behavior. This ionization strategy of uncharged fullerene derivatives appears to be very efficient and offers new perspectives in the characterization of complex fullerene-containing molecular architectures.

We have developed an efficient synthetic route allowing the preparation of fullerene-functionalized dendrons containing a C$_{60}$ sphere at each branching unit. However, the solubility of compounds **36** and **38** is low in common organic solvent. Therefore, the attachment of these dendrons on a core for the preparation of fullerene-rich macromolecules appears difficult. This prompted us to design new dendritic branches containing an increased number of solubilizing groups [51]. The synthesis of this fullerodendron is depicted in Fig. 15.

Reaction of acid **40** with **39** under esterification conditions using DCC, DMAP, and BtOH afforded **41** in 65% yield. Finally, selective cleavage of the *t*-butyl ester moiety with CF$_3$CO$_2$H in CH$_2$Cl$_2$ gave fullerodendron **42** in 96% yield. Thanks to the presence of the four hexadecyloxy substituents per peripheral fullerene subunits, compound **42** is highly soluble in common organic solvents such as CH$_2$Cl$_2$, CHCl$_3$, toluene, or THF. Therefore, this dendron appears to be an interesting building block for the functionalization of a core moiety.

3.3
Incorporation into Langmuir and Langmuir–Blodgett Films

A growing attention is currently devoted to large dendritic structures for applications in nanotechnology and materials science [52]. In this respect, the incorporation of such compounds into thin ordered films appears as an important issue and one of the most widely pursued approaches towards structurally ordered dendrimer assemblies has been the preparation of Langmuir films at the air–water interface [52]. As a part of this research, incorporation of the amphiphilic fullerodendrimers **21–23** into Langmuir films has been investigated [53]. Compounds **21–23** form good quality Langmuir films at the air–water interface and the isotherms taken at 20 °C are depicted in Fig. 16. Dendrons **21** and **23** can withstand pressures up to $\Pi \approx 20$ mN m^{-1}, the collapse of the films being indicated by the rounding of the curve, while the film prepared with **22** begins to collapse around 30 mN m^{-1}. The molecular areas for **21–23** extrapolated at zero pressure are 140±7 Å^2, 310±15 Å^2, and 560±30 Å^2, respectively. They are in the expected 1:2:4 proportion given the structure of the dendrimers and are in good agreement with the values estimated by molecular modeling. Compound **42** is also able to form Langmuir films at the air–water interface and the general shape of the isotherm is similar to the one obtained with **23**.

We also succeeded in forming Langmuir–Blodgett (LB) films by transferring monolayers of **21–23** and **42** onto solid substrates. However, due to the difference in size between the hydrophobic and hydrophilic groups, the Langmuir films of the largest derivatives are not sufficiently stable to withstand the pressure over long period of time. Therefore, the preparation of multilayered films was found to be difficult. In order to stabilize the films, it was decided to functionalize

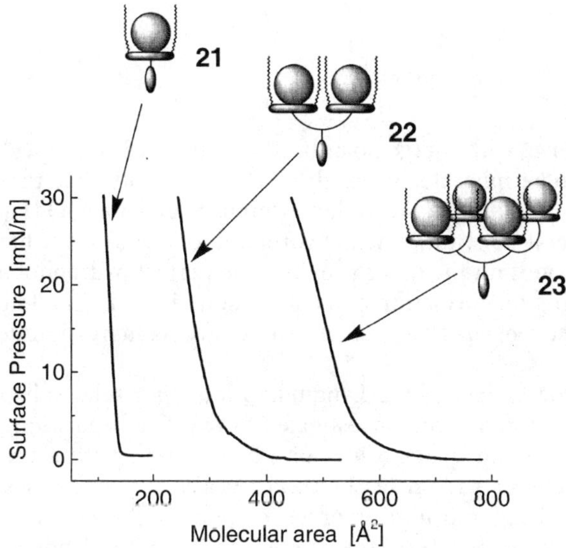

Fig. 16. Pressure-area isotherms for **21**, **22**, and **23** at 20°C

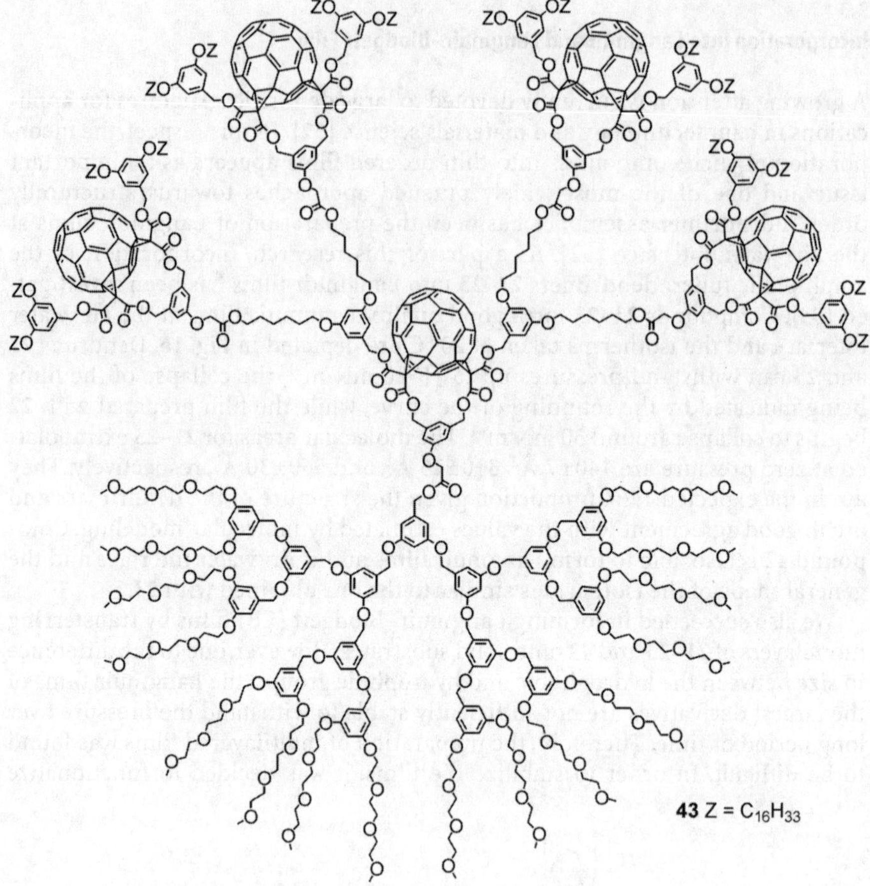

Fig. 17. Amphiphilic fullerodendrimer **43**

fullerodendron **42** with a large polar head group. A Fréchet-type dendron functionalized with peripheral ethyleneglycol chains was attached to the focal point of compound **42** leading to the diblock globular dendrimer **43** (Fig. 17) [51].

The peripheral substitution with hydrophobic chains on one hemisphere and hydrophilic groups on the other provides the perfect hydrophobic/hydrophilic balance allowing the formation of stable Langmuir films. In addition, a perfect reversibility has been observed in successive compression/decompression cycles (Fig. 18).

Transfer experiments of the Langmuir films onto solid substrates and the preparation of LB films were investigated for **43**. The deposition of films of **43** occurred regularly on quartz slides or silicon wafers with a transfer ratio of 1±0.05. The diblock structure of dendrimer **43** also appeared crucial for efficient transfers of the Langmuir films in order to obtain well-ordered multilayered LB films. Effectively, the transfer of the Langmuir films of the dendrimer **42** with the small polar head group was found to be difficult with a transfer ratio of about

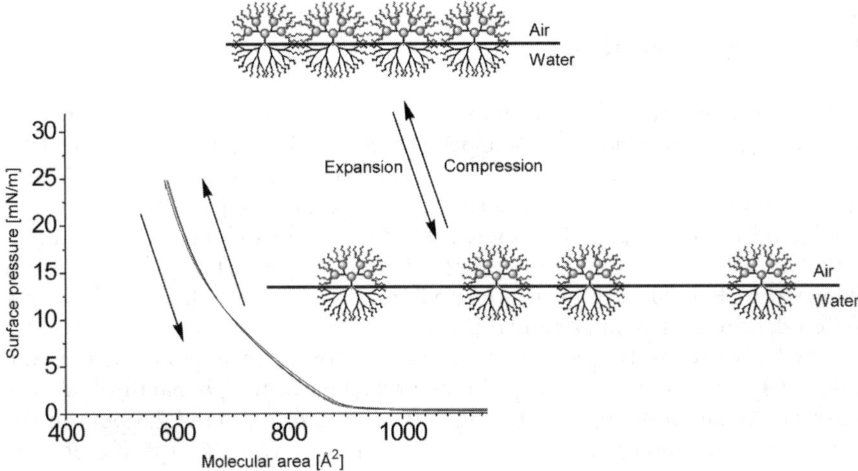

Fig. 18. Four successive compression/expansion cycles with a monolayer of **43** showing the perfect reversibility of the process

0.5–0.7. The structural quality of mono- and multi-layered films of **42** and **43** was investigated by grazing incidence X-ray diffraction. The quality of the LB films made with **42** was not too good and only allowed an estimation of their thickness, the roughness being always in the 3 Å range. Monolayers of **42** were about 20 Å thick and the average value of the layer thickness was found to be somewhat smaller than expected for the multilayer films (approximately 18 Å). This smaller value is probably the result of a partial interpenetrating of the successive layers within the film. For LB films of **43**, the presence of low-angle Kiessig fringes in the grazing X-ray patterns indicates that the overall quality of the films is good. The best fit of the grazing X-ray pattern obtained for a monomolecular film gives a thickness of 36±1 Å and a roughness of about 2 Å. For the multilayer films, the average layer thickness was found to be about 36 Å, indicating no or little interpenetration of successive layers. The excellent quality of the LB films prepared with **43** is also deduced from the plot of their UV/Vis absorbance as a function of the layer number: a straight line is obtained, indicating an efficient stacking of the layers. It is worth stressing the quality of the stacking and, as a consequence, the quality of the multilayered films obtained with such a megamolecule.

The peripheral substitution of a globular dendrimer with hydrophobic chains on one hemisphere and hydrophilic groups on the other provides the perfect hydrophobic/hydrophilic balance allowing the formation of stable Langmuir films. On the one hand, this approach shows some of the fundamental architectural requirements for obtaining stable films with amphiphilic dendrimers. On the other hand, functional groups not well adapted for the preparation of Langmuir and LB films such as fullerenes can be attached into the branching shell of the dendritic structure and, thus, efficiently incorporated in thin ordered films.

4
Fullerene-Rich Dendrimers

Fullerodendrons **21–23** have been used for the construction of new dendrimers with a bis(1,10-phenanthroline) copper(I) core [54]. The preparation of the first-generation complex is depicted in Fig. 19. Diol **44** was allowed to react with dendron **21** under DCC-mediated esterification conditions to give the corresponding ligand **L1**. The copper(I) complex **(L1)$_2$Cu** was then obtained by treatment with Cu(CH$_3$CN)$_4 \cdot$ BF$_4$. The highest generation complexes **(L2)$_2$Cu** and **(L3)$_2$Cu** depicted in Fig. 20 have been prepared from diol **44** and the corresponding fullerodendron following a similar route.

The ^{1}H NMR spectra provide good evidence for the formation of **(L1–3)$_2$Cu**. Effectively, the methylene group directly attached to the phenanthroline core observed at about 3.2 ppm in the ligands is shifted to around 2.6 ppm in the corresponding complexes. This particular behavior is highly specific of such copper(I) complexes and is the result of the ring-current effect of one phenanthroline subunit on the 2,9-substituents of the second one in the complex. Furthermore, the FAB-MS confirmed the structure of **(L1)$_2$Cu** with signals at m/z 7,872.7 and 3,967.6 corresponding to [M–BF$_4^-$]$^+$ (calculated m/z 7,872.6) and [M-L1-BF$_4^-$]$^+$ (calculated m/z 3,968.1), respectively. In the FAB-MS spectra of **(L2)$_2$Cu**, only the peak corresponding to [M-L2-BF$_4^-$]$^+$ could be observed at m/z 7,908.1 (calculated m/z 7,908.5). In the FAB-MS of **(L3)$_2$Cu**, no characteristic peak could be observed. Due to the presence of the 64 surrounding long alkyl chains, **(L3)$_2$Cu** aggregates strongly and high energy is required for dissociation

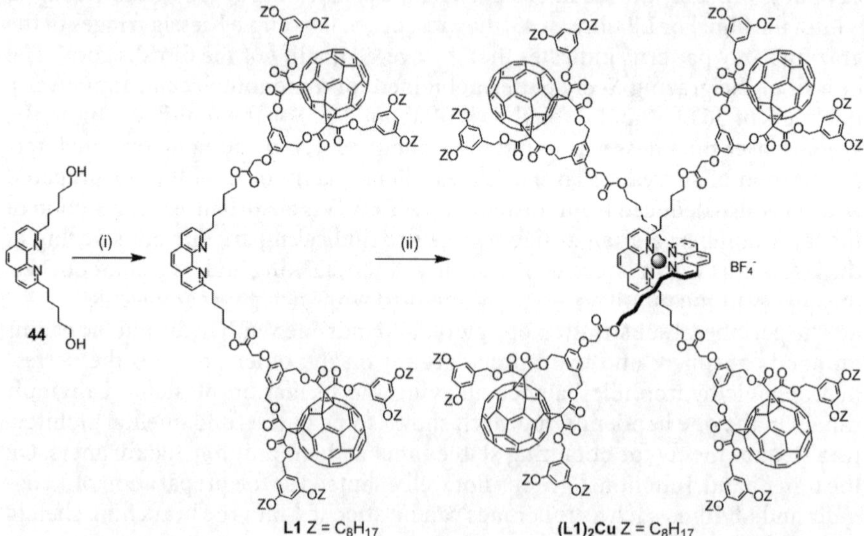

Fig. 19. Preparation of fullerodendrimer **(L1)$_2$Cu**. Reagents and conditions: (i) **21**, DCC, DMAP, BtOH, CH$_2$Cl$_2$, 0 °C to room temperature (75%); (ii) Cu(CH$_3$CN)$_4$BF$_4$, CH$_3$CN, CH$_2$Cl$_2$, room temperature (56%)

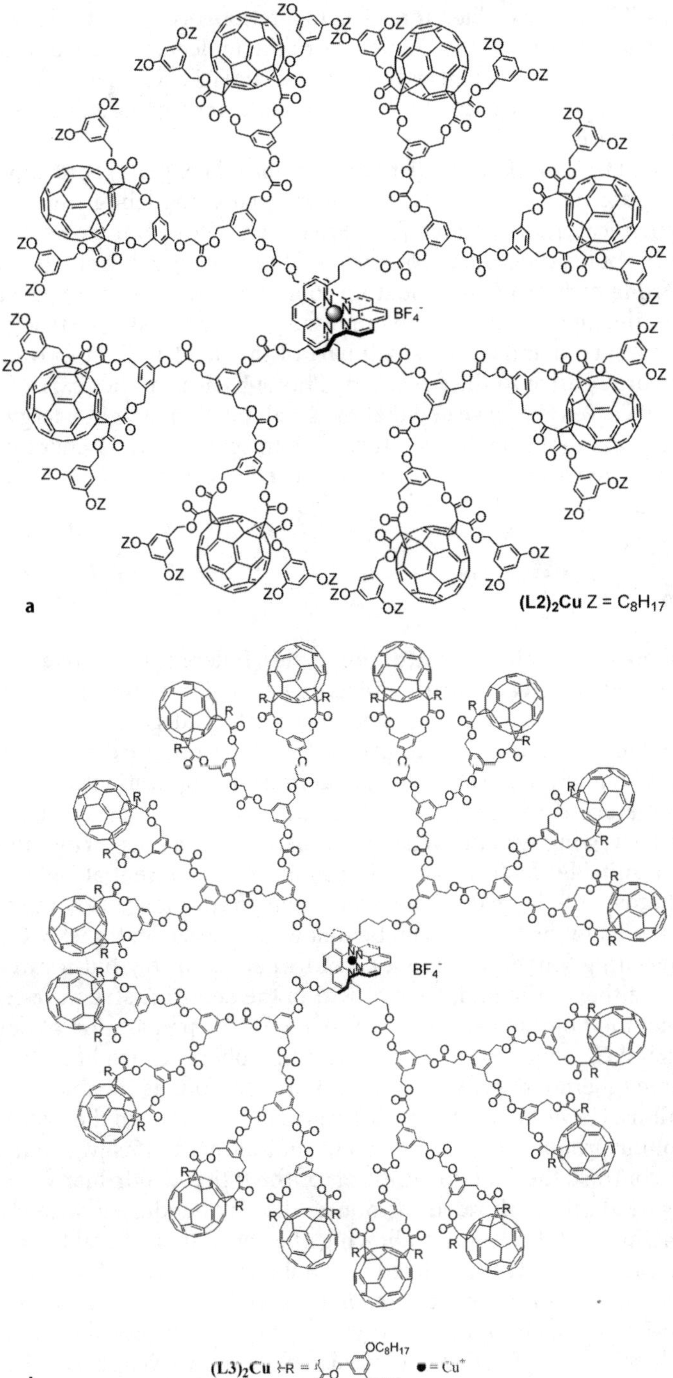

Fig. 20. Fullerodendrimers (L1)₂Cu and (L1)₂Cu

during FAB-MS analysis. Therefore, fragmentation occurs, especially on the fragile benzylic ester functions; furthermore the molecular mass of (L3)$_2$Cu is quite high (31,601.8). Nevertheless, the NMR and UV/Vis data obtained for (L3)$_2$Cu and comparison with (L1–2)$_2$Cu provide very good evidence for the proposed structure.

For the two largest fullerodendrimers, the central copper(I) core appears inaccessible to external contacts as shown by molecular modeling and confirmed by electrochemical investigations [55]. Effectively, Gross and co-workers have shown that the bulky fullerodendrons around the Cu center prevent its approach on the electrode surface and its oxidation could no longer be observed. Furthermore, due to the high number of fullerene subunits in (L3)$_2$Cu, a strong shielding effect is observed and only a small part of the incident light is available to the central core relative to the periphery. Photophysical studies carried out by Armaroli and co-workers revealed that the small portion of light energy able to reach the central Cu(I) complex is returned to the external fullerenes by energy transfer [55]. Therefore, one can conclude that the central core is buried in a dendritic black box.

5
Conclusions

Owing to their special photophysical properties, fullerene derivatives are good candidates for evidencing dendritic effects. In particular, we have shown that the triplet lifetimes of a C$_{60}$ core can be used to evaluate its degree of isolation from external contacts. In addition, the protective effect observed for fullerodendrimers 4 and 8 might be useful for optical limiting applications. On the other hand, the fullerene core can act as a terminal energy receptor in dendrimer-based light-harvesting systems. When the fullerodendrimer is further functionalized with a suitable electron donor, it may exhibit the essential features of an artificial photosynthetic system where an initial photo-induced energy transfer from the antenna to the C$_{60}$ core can be followed by electron transfer. C$_{60}$ is not only an interesting functional core for dendrimer chemistry, but it can also be incorporated either at the periphery or within the dendritic structure. Efficient synthetic methodologies have been developed for the preparation of dendrons with peripheral C$_{60}$ subunits or containing a C$_{60}$ sphere at each branching unit. Some of these fullerodendrons are amphiphilic compounds capable of forming Langmuir films. However, due to the difference in size between the hydrophobic and hydrophilic groups, the preparation of multilayered films was found to be difficult. In contrast, the functionalization of the fullerodendrimer with a dendritic polar head group afforded a globular diblock dendrimer with a perfect hydrophobic/hydrophilic balance allowing the formation of stable Langmuir films which can be readily transferred onto solid substrates, yielding high quality LB films. Incorporation of fullerenes into well-ordered structures can be easily achieved and such fullerodendrons appear to be promising compounds for materials science applications. It has also been shown that these fullerodendrons are versatile building blocks for the preparation of fullerene-rich macromolecules with intriguing properties. Particularly, a dendritic black box effect has been observed. The electrochemical and photophysical studies of the dendrimers with a bis(1,10-phenanthroline)Cu(I) core and fullerene π chromophores at the periphery have shown that the surrounding fullerene-functionalized dendritic branches are able to isolate the central Cu(I) complex. In the future, due to the high number of C_{60} subunits, such fullerodendrons could also be useful as antennas for light harvesting when attached to a suitable functional group able to act as a terminal receptor of the excitation energy. This could be achieved if the lowest excited state of the central core is lower in energy than that of the surrounding fullerene spheres.

Acknowledgements. This work was supported by the CNRS, the French Ministry of Research (ACI Jeunes Chercheurs) and ECODEV. I would like to warmly thank all my co-workers and collaborators for their outstanding contributions, their names are cited in the references.

6
References

1. Newkome GR, Moorefield CN, Vögtle F (2001) Dendrimers and dendrons: concept, syntheses, applications. VCH, Weinheim
2. Nierengarten J-F (2000) Chem Eur J 20:3667
3. Diederich F, Thilgen C (1996) Science 271:317
4. Hirsch A, Vostrowsky O (2001) Dendrimers with carbon rich-cores. In: Vögtle F (ed) Dendrimers. Top Curr Chem 217:51
5. Wooley KL, Hawker CJ, Fréchet JMJ, Wudl F, Srdanov G, Shi S, Li C, Kao M (1993) J Am Chem Soc 115:9836
6. Hawker CJ, Wooley KL, Fréchet JMJ (1994) Chem Commun 925
7. Brettreich M, Hirsch A (1998) Tetrahedron Lett 39:2731
8. Nierengarten J-F, Habicher T, Kessinger R, Cardullo F, Diederich F, Gramlich V, Gisselbrecht J-P, Boudon C, Gross M (1997) Helv Chim Acta 80:2238
9. Camps X, Hirsch A (1997) J Chem Soc Perkin Trans 1 1595
10. Camps X, Schönberger H, Hirsch A (1997) Chem Eur J 3:561
11. Herzog A, Hirsch A, Vostrowsky O (2000) Eur J Org Chem 171
12. Cardullo F, Diederich F, Echegoyen L, Habicher T, Jayaraman N, Leblanc RM, Stoddart JF, Wang S (1998) Langmuir 14:1955
13. Dardel B, Deschenaux R, Even M, Serrano E (1999) Macromolecules 32:5193
14. Dardel B, Guillon G, Heinrich B, Deschenaux R (2001) J Mater Chem 11:2814
15. Tirelli N, Cardullo F, Habicher T, Suter UW, Diederich F (2000) J Chem Soc Perkin Trans 2 193
16. Campidelli S, Deschenaux R (2001) Helv Chim Acta 84:589
17. Campidelli S, Deschenaux R, Eckert J-F, Guillon D, Nierengarten J-F (2002) Chem Commun 656
18. Chuard T, Deschenaux R (2002) J Mater Chem 12:1994
19. Hecht S, Frechet JMJ (2001), Angew Chem Int Ed 40:74
20. Gorman CB, Smith JC (2001) Acc Chem Res 34:60
21. Rio Y, Accorsi G, Nierengarten H, Rehspringer J-L, Hönerlage B, Kopitkovas G, Chugreev A, Van Dorsselaer A, Armaroli N, Nierengarten J-F (2002) New J Chem 26:1146
22. Rio Y, Nicoud J-F, Rehspringer J-L, Nierengarten J-F (2000) Tetrahedron Lett 41:10207
23. Prato M, Maggini M (1998) Acc Chem Res 31:519
24. Nierengarten J-F, Gramlich V, Cardullo F, Diederich F (1996) Angew Chem Int Ed Engl 35:2101
25. Bingel C (1993) Chem Ber 126:1957
26. Nierengarten J-F, Schall C, Nicoud J-F (1998) Angew Chem Int Ed Engl 37:1934
27. Nierengarten J-F, Oswald L, Nicoud J-F (1998) Chem Commun 1545

28. Felder D, Carreon MP, Gallani J-L, Guillon D, Nierengarten J-F, Chuard T, Deschenaux R (2001) Helv Chim Acta 84:1119
29. Bourgeois J-P, Woods CR, Cardullo F, Habicher T, Nierengarten J-F, Gehrig R, Diederich F (2001) Helv Chim Acta 84:1207
30. Felder D, Gutiérrez Nava M, del Pilar Carreon M, Eckert J-F, Luccisano M, Schall C, Masson P, Gallani J-L, Heinrich B, Guillon D, Nierengarten J-F (2002) Helv Chim Acta 85:288
31. Nierengarten J-F, Armaroli N, Accorsi G, Rio Y, Eckert JF (2003) Chem Eur J 9:37
32. Tutt LW, Krost A (1992) Nature 356:225
33. Felder D, Guillon D, Levy R, Mathis A, Nicoud J-F, Nierengarten J-F, Rehspringer J-L, Schell J (2000) J Mater Chem 10:887
34. Schell J, Felder D, Nierengarten J-F, Rehspringer J-L, Lévy R, Hönerlage B (2001) J Sol-Gel Sci Technol 22:225
35. Murata Y, Ito M, Komatsu K (2002) J Mater Chem 12:2009
36. Kunieda R, Fujitsuka M, Ito O, Ito M, Murata Y, Komatsu K (2002) J Phys Chem B 106:7193
37. Adronov A, Fréchet JMJ (2000) Chem Commun 1701
38. Armaroli N, Barigelletti F, Ceroni P, Eckert J-F, Nicoud J-F, Nierengarten J-F (2000) Chem Commun 599
39. Eckert J-F, Nicoud J-F, Nierengarten J-F, Liu S-G, Echegoyen L, Barigelletti F, Armaroli N, Ouali L, Krasnikov V, Hadziioannou G (2000) J Am Chem Soc 122:7467
40. Armaroli N, Accorsi G, Gisselbrecht J-P, Gross M, Krasnikov V, Tsamouras D, Hadziioannou G, Gomez-Escalonilla MJ, Langa F, Eckert J-F, Nierengarten J-F (2002) J Mater Chem 12:2077
41. Accorsi G, Armaroli N, Eckert J-F, Nierengarten J-F (2002) Tetrahedron Lett 43:65
42. Segura JL, Gomez R, Martin N, Luo CP, Swartz A, Guldi DM (2001) Chem Commun 707
43. Guldi DM, Swartz A, Luo C, Gomez R, Segura JL, Martin N (2002) J Am Chem Soc 124:10875
44. Langa F, Gomez-Escalonilla MJ, Diez-Barra E, Garcia-Martinez JC, de la Hoz A, Rodriguez-Lopez J, Gonzalez-Cortes A, Lopez-Arza V (2001) Tetrahedron Lett 42:3435
45. Avent AG, Birkett PR, Paolucci F, Roffia S, Taylor R, Wachter NK (2000) J Chem Soc Perkin Trans 2 1409
46. Schwell M, Wachter NK, Rice JH, Galaup J-P, Leach S, Taylor R, Bensasson RV (2001) Chem Phys Lett 339:29
47. Nierengarten J-F, Felder D, Nicoud J-F (1999) Tetrahedron Lett 40:269
48. Nierengarten J-F, Felder D, Nicoud J-F (2000) Tetrahedron Lett 41:41
49. Felder D, Nierengarten H, Gisselbrecht J-P, Boudon C, Leize E, Nicoud J-F, Gross M, Van Dorsselaer A, Nierengarten J-F (2000) New J Chem 24:687
50. Boudon C, Gisselbrecht J-P, Gross M, Isaacs L, Anderson HL, Faust R, Diederich F (1995) Helv Chim Acta 78:1334
51. Nierengarten J-F, Eckert J-F, Rio Y, Carreon MP, Gallani J-L, Guillon D (2001) J Am Chem Soc 123:9743
52. Tully DC, Fréchet JMJ (2001) Chem Commun 1229
53. Felder D, Gallani J-L, Guillon D, Heinrich B, Nicoud J-F, Nierengarten J-F (2000) Angew Chem Int Ed Engl 39:201
54. Nierengarten J-F, Felder D, Nicoud J-F (1999) Tetrahedron Lett 40:273
55. Armaroli N, Boudon C, Felder D, Gisselbrecht J-P, Gross M, Marconi G, Nicoud J-F, Nierengarten J-F, Vicinelli V (1999) Angew Chem Int Ed Engl 38:3730

Top Curr Chem (2003) 228: 111–140
DOI 10.1007/b11008

Rotaxane Dendrimers

Jae Wook Lee[1] · Kimoon Kim[2]

[1] Department of Chemsitry, Dong-A University, Busan 604-714, Republic of Korea
[2] National Creative Research Initiative Center for Smart Supramolecules and Department
of Chemistry, Division of Molecular and Life Sciences, Pohang University of Science
and Technology, San 31 Hyojadong, 790–784 Pohang, Republic of Korea
E-mail: kkim@postech.ac.kr

The synthesis, properties, and potential applications of rotaxane dendrimers, dendritic molecules containing rotaxane-like mechanical bonds to link their components are described.
Rotaxane dendrimers are classified into three types depending on where rotaxane-like features are introduced – Type I, II, and III rotaxane dendrimers which incorporate rotaxane-like
features at the core, termini, and branches, respectively. Several different types of macrocycles
are employed as the ring component in the templated synthesis of rotaxane dendrimers. In the
synthesis of rotaxane dendrimers, several aspects should be carefully considered, including
the binding affinity of the macrocycle (ring) and guest (rod). The properties of these rotaxane
dendrimers are quite different from those of the individual rotaxanes or dendrimers and often
a blend of both. Potential applications of rotaxane dendrimers include molecular nanoreactors, drug delivery, and gene delivery.

Keywords. Rotaxane dendrimers, Host-guest interaction, Recognition, Self-assembly, Supramolecular chemistry

1
Introduction

The chemistry of dendrimers, a class of highly branched molecules that combine the properties of polymers as well as small discrete molecules, is blossoming into an exciting field of research [1–6]. During the past two decades, a wide variety of dendrimers with different cores, branches, and end groups have been synthesized and applied in different areas including drug-delivery, catalysis, light energy harvesting, sensors, etc. Concurrently with this, another fascinating development in chemistry has been the efficient synthesis of mechanically interlocked molecules such as rotaxanes and catenanes, which have also attracted considerable attention not only due to their aesthetic appeal but also their potential applications such as molecular machines [7–9].

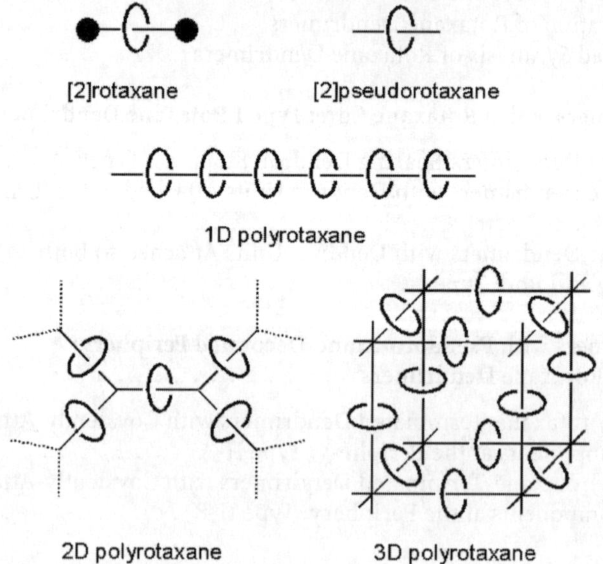

Scheme 1. Schematic representations of a [2]rotaxane, a [2]pseudorotaxane, and one-, two-, and three-dimensional polyrotaxanes

Rotaxanes are a class of compounds that consist of linear and cyclic species bound together in a threaded structure by mechanical bonds (Scheme 1). In the simplest rotaxanes, [2]rotaxanes, a ring (wheel) is threaded on a rod (axle) having two bulky stoppers at the ends, which prevent dethreading. Without stoppers, the ring and rod components can dissociate; such a supermolecule is called a [2]pseudorotaxane. One-dimensional (1D) polyrotaxanes contain many rings threaded on a long rod. This idea can be extended into two and three dimensions to define 2D and 3D polyrotaxanes [10, 11]. Rotaxane dendrimers can be broadly defined as the dendritic molecules containing rotaxane-like mechanical bonds to link their components. Combining many attractive features of conventional dendrimers with properties that result from the mechanical bonds, they are not only structurally appealing but also hold great promise for applications in many areas including materials science. In this review, we will describe the synthesis, properties, and potential applications of rotaxane dendrimers.

1.1
Classification of Rotaxane Dendrimers

Since dendrimers contain three distinct structural parts, the core, end-groups, and branching units connecting core and periphery, rotaxane dendrimers can be classified into three types depending on where rotaxane-like features are introduced – Type I, II and III rotaxane dendrimers which incorporate rotaxane like features at the core, termini and branches, respectively (Table 1 and Scheme 2). Type I rotaxane dendrimers may be called rotaxane-core dendrimers. Depending on where dendron units are attached, Type I rotaxane dendrimers can be further classified into I-A, I-B and I-C in which dendron units are attached to rod, ring, and both ring and rod components, respectively. Type II pseudorotaxane dendrimers which incorporate pseudorotaxane-like features at the *periphery* can be further classified as Type II-A and Type II-B depending on whether the terminal units of the dendrimers serve as rod components or ring

Table 1. Classification of rotaxane dendrimers

Type I	**Dendrimers with a rotaxane core**
Type I-A:	Rotaxane dendrimers with dendron units attached to the rod component
Type I-B:	Rotaxane dendrimers with dendron units attached to the ring component
Type I-C:	Rotaxane dendrimers with dendron units attached to both the ring and rod components
Type II	**Dendrimers with (pseudo)rotaxane-decorated periphery**
Type II-A:	(Pseudo)rotaxane-terminated dendrimers with covalently-attached rod components at the periphery
Type II-B:	(Pseudo)rotaxane-terminated dendrimers with covalently-attached ring components at the periphery
Type III	**Dendritic polyrotaxanes**
Type III-A:	Dendritic polyrotaxanes incorporating ring components *on* the branches
Type III-B:	Dendritic polyrotaxanes incorporating ring components *at* the branching points

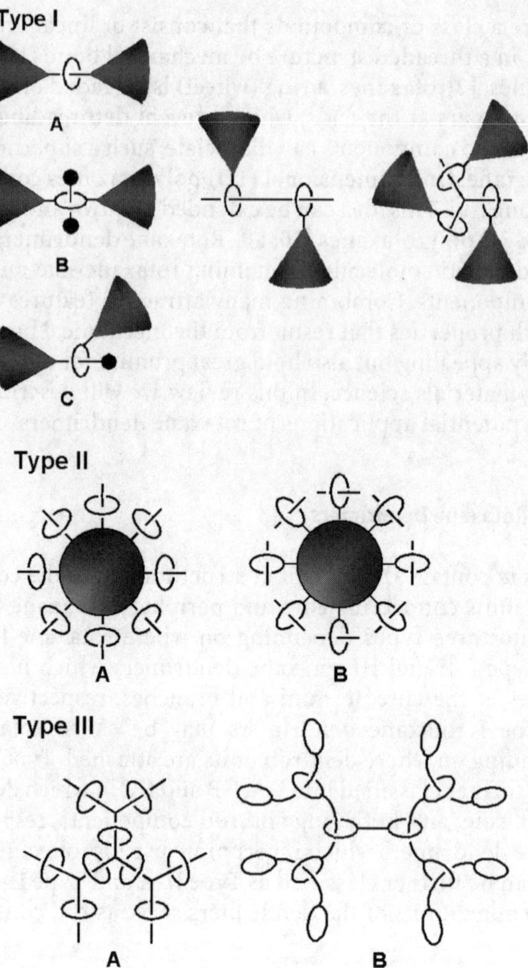

Scheme 2. Classification of rotaxane dendrimers: Type I, II and III rotaxane dendrimers incorporating rotaxane-like features at the core, termini, and branches, respectively

components, respectively. Type III rotaxane dendrimers may be viewed as dendritic polyrotaxanes in which rotaxane building units grow like a dendrimer. Depending on whether ring components are located *on the branches* or *at the branching points*, Type III rotaxane dendrimers are further classified as III-A and III-B, respectively (Scheme 2). This review will focus mainly on the synthesis of rotaxane dendrimers according to the above classification. Throughout the review, generations of dendrons or dendrimers will be designated by **G-*n*** (*n*=0, 1, 2, 3, 4, and 5).

1.2
Templated Synthesis of Rotaxane Dendrimers

Template effects have been used in rotaxane synthesis to direct threading of the axle through the wheel. Since macrocyclic compounds such as cyclodextrins, crown ethers, cyclophanes, and cucurbiturils form stable complexes with specific guest molecules, they have been widely used in the templated synthesis of rotaxanes as ring (wheel) components. Here, we briefly discuss macrocycles used in the synthesis of rotaxane dendrimers and their important features.

Crown ether type macrocyclic polyethers have been extensively used in the synthesis of interlocked molecules such as catenanes and rotaxanes by Stoddart and others. The characteristic chemistry of crown ethers involves complexation of the ether oxygens with various ionic species. Depending on which ion is targeted, selectivity for certain metal cations over others is achieved by manipulating the size of the cavity within the macrocyclic compound, the choices among the heteroatoms or groups and their relative sequence. These receptors form fairly stable inclusion complexes utilizing several noncovalent interactions. For example, in nonprotic organic media the crown ether bis-p-phenylene-34-crown-10 (BPP34C10) can form 1:1 complexes with bipyridinium ions ($K_a=$ $\sim 10^3\,M^{-1}$) mainly by the charge-transfer interaction between the π-electron-deficient bipyridinium unit and the π-electron-rich aromatic rings [12]. A deep red color builds up in the reaction mixture which signifies considerable charge-transfer interaction between the host and the guest. Bis-dibenzo[24]crown-8 (DB24C8) can form 1:1 complexes with the secondary ammonium ions by the hydrogen bonding ($K_a=10^3-10^4\,M^{-1}$) [13]. The crown ether oxygen atoms bind the guest using multiple hydrogen bonds with the amine as well as with one of the methylene hydrogen atoms. An additional π-π stacking interaction arises in case of the dibenzylammonium complex between a phenyl group in the cation and a catechol group of the crown.

Another class of host molecules extensively studied in host-guest chemistry is cyclophanes, a term originally applied to compounds having two p-phenylene groups held face to face by $- [CH_2]_n -$ bridges. It now designates compounds having cyclic systems consisting of ring(s) or ring system(s) having the maximum number of noncumulative double bonds connected by saturated and/or unsaturated chains. Most cyclophanes employ the recognition strategy of substrate-induced organization of the conformation [14]. Suitable hydrophilic functions appended to the ring can lead to water-soluble cyclophanes, which can form inclusion complexes stable in aqueous media. For example, water-soluble cyclophanes, in which ether groups and methylene units link benzene or naphthalene, can form inclusion complexes with naphthalene-2,7-diol, 6-(p-toluidino)naphthalene-2-sulfonate and steroids in aqueous media. The major driving force of the host-guest complexation is hydrophobic and π-π interactions. The association constants vary in the range $10^2-10^4\,M^{-1}$ depending on both the cavity size of the cyclophanes and the size of guests.

Macrocyclic isophthalamides (the tetralactam macrocycles) have been used as a host to build rotaxanes and catenanes by Vögtle and others [15]. Structurally analogous to cyclophanes, their initial usage as a synthetic receptor was for recognition of p-benzoquinone. Recognition sites include hydrogen bonds and a variety of π-π interactions. A key feature of the macrocyclic isophthalamides is the selective stabilization of the quinone radical anion over the parent neutral quinone. The macrocycles bind anions such as chloride, bromide, or phenolate in nonpolar media through hydrogen bonding [16]. The tetralactam-phenolate anion complex (K_a~3×10^2 M^{-1}) has been employed as a "wheeled" nucleophile. Thus the phenolate ion while still inside the macrocycle can react with bromide compounds. Such a reaction has been incorporated in a rotaxane synthesis, where once reacted, the complex becomes a stopper for the rotaxane.

One of the most widely used hosts in the study of host-guest chemistry as well as in the synthesis of rotaxanes is cyclodextrins (CDs) [17]. The most common cyclodextrins contain six, seven and eight glucopyranose rings (α-CD, β-CD, and γ-CD, respectively). Having a rigid, bucket-shaped structure with a hydrophobic cavity, they can bind a wide range of rod-like guest molecules in the cavity. Most of the host-guest complexes of cyclodextrins are studied in aqueous media. Frequently, this complexation is used as a means to solubilize the hydrophobic guests in water. Depending on the nature of the guest, the association constants vary over a wide range. For example, the association constants of β-CD with phenolphthalein, ferrocene, admantane, sodium cholate, and 2-naphthalenesulfonate vary in the range of 10^2–10^5 M^{-1}. For the inclusion complexes of low binding affinity to form in an appreciable concentration in solution often it is practice to use an excess amount of the host. The ability to form stable host-guest complexes with a very wide range of guest species in aqueous solutions makes these readily available molecules one of the most attractive building blocks in the synthesis of rotaxanes and polyrotaxanes [18].

Cucurbituril (CB[6]), a macrocyclic cavitand comprising six glycoluril units, has a cavity accessible through two identical carbonyl-fringed portals [19]. The hydrophobic cavity, whose size is almost comparable to that of α-CD, provides a potential site for inclusion of hydrocarbon molecules. Unlike CDs, however, the polar carbonyl groups at the portals allow CB[6] to bind cationic species through charge-dipole and hydrogen bonding interactions. For example, protonated aminoalkanes, particularly diaminoalkanes form exceptionally stable inclusion complexes with CB[6] ($K=10^5$–10^7) [20]. CB[6] also forms a stable inclusion complex with 1,6-di(pyridinium)hexane (log $K=4.40$: $K=2.5\times10^4$ at 25 °C [21]). This ability to form very stable inclusion complex makes CB[6] attractive as a building block for construction of interlocked molecules, in particular rotaxanes and polyrotaxanes [22]. Larger cucurbituril homologues, cucurbit[n]uril (CB[n], n=7 and 8) containing seven and eight glycoluril units, respectively, were synthesized recently [23] and their inclusion behavior was studied [24–30]. The driving force for inclusion is similar to that in CB[6], but the larger cavities of CB[7] and CB[8] allow them to bind larger guest molecules. For example, CB[7] forms a stable 1:1 inclusion complex ($K>10^5$) with methyl viologen (dimethylbipyridinium ions) [24, 27]. Moreover, CB[8] can accommodate two guest molecules to form ternary complexes. For example, CB[8] can form a stable 1:1:1 complex with methylviologen and 2,6-dihydroxynaphthalene, [24] which is driven by charge-transfer interaction between the electron-deficient and electron-rich guest molecules inside the hydrophobic cavity of CB [8].

As enumerated above, the macrocycle as a ring (wheel) component can form an inclusion complex with a suitable ligand as an axle utilizing mostly hydrophobic, hydrogen bonding, electrostatic, and charge-transfer interactions. Often, stable pseudorotaxanes do not result from these interactions, particularly when only one such interaction is available for complexation. Frequently, in such cases of low host-guest interaction strength, a large excess of the macrocycle is used to stabilize the complex in solution. Yet, low binding affinity between the macrocycle and ligand can be utilized in the template-directed synthesis of rotaxane dendrimers if multiple interactions are achievable. Rotaxane dendrimers stable even in aqueous medium can result from the synergistic effect of multiple interactions. Conversely, if the intrinsic binding affinity between the macrocycle and the guest is very high, such as that between cucurbituril and akyldiammonium ions, stable pseudorotaxanes can result from stoichiometric mixtures, and hence an easy access to the rotaxane dendrimers can be realized.

A stable pseudorotaxane resulting from the cooperative effect of weak multiple interactions or from a strong host-guest interaction does not necessarily always mean that the resulting rotaxane dendrimer can be established in solution. A great variety of other factors, such as the nature of binding among the pseudorotaxane units, the sizes of the host and the guest, steric congestion with successive dendrimer generations, solvent, pH of solution, temperature, etc., are as crucial. In the following section we shall illustrate the importance of these factors operating in tandem while discussing the synthesis of the rotaxane dendrimers of diverse kinds.

2
Dendrimers with a Rotaxane Core: Type I Rotaxane Dendrimers

2.1
Rotaxane Dendrimers Bearing Dendron Stoppers: Type I-A

Type I-A rotaxane dendrimers are the dendrimers having a core comprising a ring and a rod, to the end of which dendrons are attached. Here, the dendrons act as the stoppers to prevent dethreading of the ring. This type of rotaxane dendrimer was first synthesized by Stoddart and coworkers by threading followed by capping [31]. They incorporate a crown ether type macrocycle as the ring component(s), bipyridinium derivatives as the rod components and Fréchet-type dendrons as the stopper. Reaction of bipyridine with Fréchet-type dendron wedges [G-3]-**Br** in the presence of four equiv of **1** in DMF under high pressure conditions at 25 °C afforded the [2]rotaxane-core dendrimer **2** in 27% yield (Fig. 1). By increasing the number of bipyridinium recognition sites incorporated within the dumbbell-shaped components, higher order rotaxanes can be self-assembled using the same synthetic procedure. Reaction of bis(bipyridinium)-*p*-xylene with the stopper [G-3]-**Br** in the presence of eight equiv of **1** in DMF under similar conditions affords the [2]rotaxane-core dendrimer **3** and [3]rotaxane-core dendrimer in 23% and 5% yields, respectively. Because of the hydrophobic dendritic moiety, these [n]rotaxane-core dendrimers are soluble in a wide range of organic solvents, despite the polycationic natures of their bipyri-

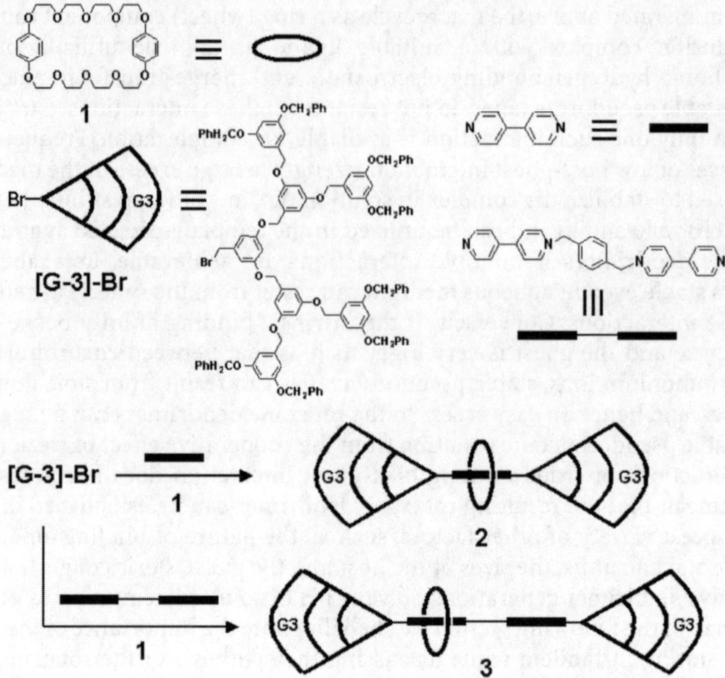

Fig. 1. Rotaxane dendrimers from crown ether type macrocycle ring, bipyridinium derivatives rod component and Fréchet-type dendron stoppers

dinium-based backbones. In all cases, they can be purified by column chromatography employing relatively low polar eluants. In **3**, shuttling of the macrocycle between the two different bipyridinium sites on the rod is observed. The rate of shuttling increases with the polarity of the medium: on going from chloroform to acetone the rate const increases from 200 to 33,000 times per second. The dramatic solvent dependence of the rate of this process is a result of conformational changes induced by varying the solvent polarity.

Vögtle and coworkers have synthesized a wide variety of mechanically interlocked molecules incorporating isophthalamide macrocycles as a "wheel" (ring) component. Recently, they developed a new procedure for the templated synthesis of rotaxanes based on the complexation of intermediate anionic stopper building block by neutral macrocyclic isophthalamides and their subsequent reaction with electrophilic stoppers to produce rotaxanes with phenyl ether "axles" [32]. They applied this method to the construction of several different types of rotaxane dendrimers [33–35]. A series of rotaxane dendrimers (type I-A) bearing dendron stoppers were synthesized [33]. Reaction of modified Fréchet's dendrons **4** containing a phenol functionality at the focal point with dendritic bromides **5** in the presence of macrocyclic isophthalamide **6** and K_2CO_3 produces [2]rotaxane-core dendrimers **7** having dendritic stoppers of different sizes (Fig. 2). The phenolate stoppers complexed to the macrocycle can act as "wheeled" nucleophiles and react through the "wheel" with centerpieces such as

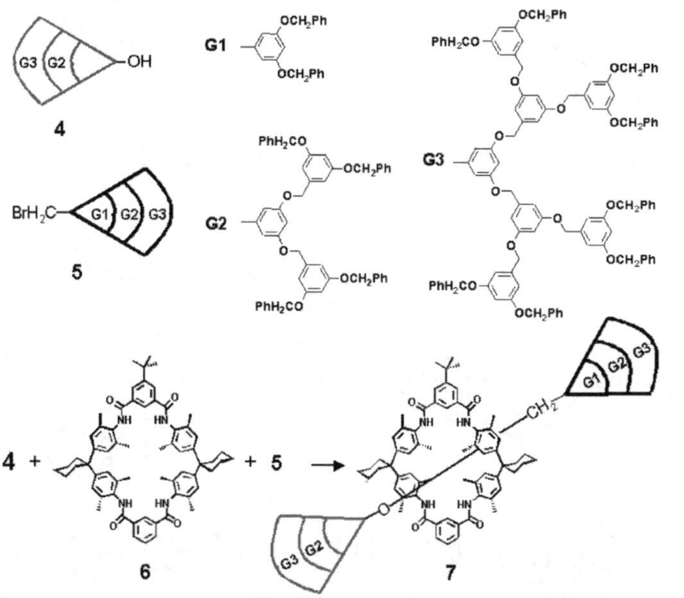

Fig. 2. Rotaxane dendrimers from isophthalamide [2]rotaxane-core and Fréchet-type dendron stoppers

dendritic bromides to yield the [2]rotaxane-core dendrimers. Thus, the macrocycle not only acts as a receptor for anions, but also orients the guest appropriately for threading the "axle" through the "wheel". All of these dendritic rotaxanes are indefinitely stable at room temperature, but deslipping of the "wheel" occurs at elevated temperatures. The deslipping experiment allowed them to estimate the "dynamic spatial demand" of the dendritic stoppers and rank them accordingly. For example, the third generation Fréchet dendritic stopper is bulkier than the *tert*-butyl-substituted trityl stopper, which is in turn bulkier that the second generation dendritic stopper.

Cucurbit[6]uril (CB[6]) or cucurbit[7]uril (CB[7]) form stable inclusion complexes with 1,6-di(pyridinium)hexane (log K=4.40 at 25 °C for CB[6]) [21]. Taking advantage of the fact, Kim and coworkers [36] synthesized Type I-A rotaxane dendrimers **10** incorporating CB[6] or CB[7] as a "bead" and Fréchet dendrons as stoppers (Fig. 3). Reaction of [2]pseudorotaxane **9-A**, which comprises a CB[6] bead threaded on 1,6-di(bipyridinium)hexane, with three equiv of Fréchet dendrons (G-1–G-3) **8** in DMSO affords [2]rotaxane-core dendrimers **10-A** in 75–88% yield. Similarly, [2]Rotaxane-core dendrimers **10-B** incorporating CB[7] can also be synthesized utilizing CB[7]-based [2]pseudorotaxane (**9-B**) as a starting material. The molecular bead CB[6] in **10-A** resides on the hexamethylene site of the "rod" component regardless of the generation of the dendritic stoppers whereas the larger molecular bead CB[7] in **10-B** occupies partially the hexamethylene site and partially the bipyridinium site as judged by ^{1}H NMR spectroscopy. The occupancy of the two different sites varies depending on the generation of the dendritic stoppers. On going from G-1 to G-3, the occupation of

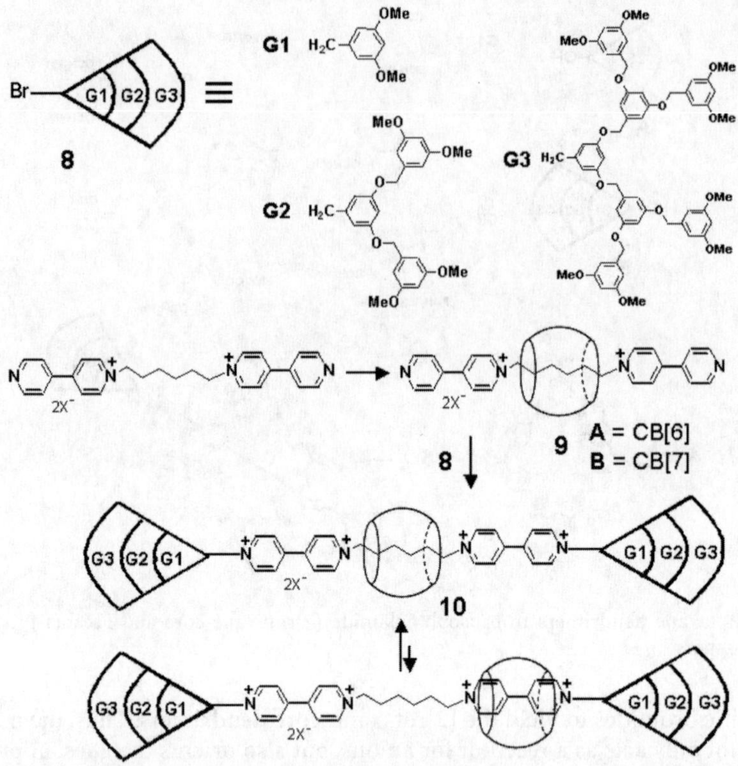

Fig. 3. Rotaxane dendrimers incorporating cucurbituril threaded on 1,6-di(bipyridinium)hexane with Fréchet dendrons as stoppers

the hexamethylene site steadily decreases from 94% to 33%. This behavior may be due to the change in polarity of the local medium affected by the dendritic stoppers. Both bead (CB[6] or CB[7]) and dendritic stoppers of **8** are also expected to influence the photochemical and electrochemical properties of the redox-active bipyridinium unit at the core of the rotaxane dendrimers.

2.2
Rotaxane Dendrimers with Dendron Units Attached to the Ring: Type I-B

Dendrimers containing a macrocycle at the core may serve as a ring component to form a rotaxane or pseudorotaxane, which are classified as Type 1-B pseudorotaxane dendrimers. Diederich et al. reported a series of dendrimers containing a cyclophane at the core, which were named *dendrophanes* (*dendr*imer-cyc*lophanes*) [37–41]. The cyclophane cavity is accessible to appropriate substrates even at high dendritic generation. In aqueous solution, the dendrophanes **11G-*n*** bind naphthalene-2,7-diol or 6-(*p*-Toluidino)naphthalene-2-sulfonate (TNS) (Fig. 4) [37, 38] to form 1:1 host guest complexes **12G-*n*,** which may be considered Type 1-B pseudorotaxane dendrimers. The binding constants for the dendrophanes are

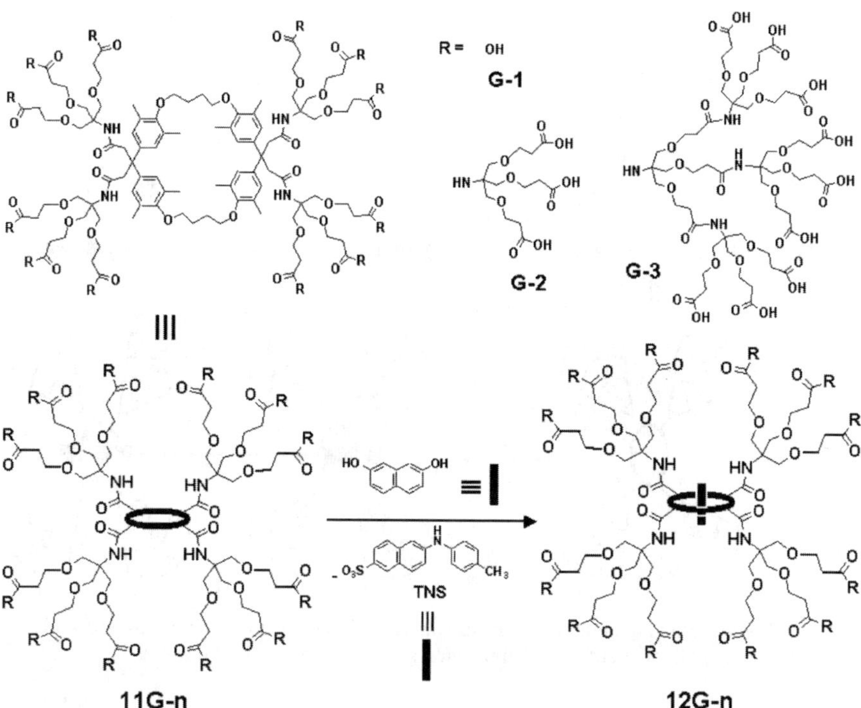

Fig. 4. Dendrophanes at the core, bind naphthalene-2,7-diol or 6-(*p*-toluidino)naphthalene-2-sulfonate to form [2]pseudorotaxane dendrimers

similar to those for non-dendritic cyclophane. However, the complexation rates for the formers are significantly lower than that for the latter. A titration study using TNS as a fluorescent probe shows a blue-shift of the emission maximum with increasing dendritic generation indicating that the microenvironment around the cavity becomes more apolar with increasing degree of branching.

Extending this work, Diederich et al. used dendrophanes **13** as building blocks for the assembly of new supramolecular architectures in aqueous solution (Fig. 5) [41]. Type I-B [3]pseudorotaxane dendrimer **15** was obtained by 2:1 complexation of the dendrophanes and molecular rods **14** consisting of rigid oligo(phenylacetylene) spacers with terminal steroid units. NMR study confirms that the steroid unit of the rod is included in the cyclophane cavities and the rigid rod effectively bridges the two dendrophanes involved in complexation. Apolar bonding interactions and hydrophobic desolvation provide the driving force for formation of these complexes. In addition, ion-pair interactions also play a role due to the anionic nature of the dendritic end groups and cationic nature of the rods. Diederich et al. further investigated the effects of the rod length and dendron generation on the complex formation. With the shorter rod **14a** (*n*=1) both **13G-0** and **13G-1** form 2:1 complexes, but **13G-2** forms only a 1:1 complex. For **13G-0** the threading of two cyclophane rings through **14a** occurs with nearly identical binding constants, whereas for **13G-1** the second threading is thermo-

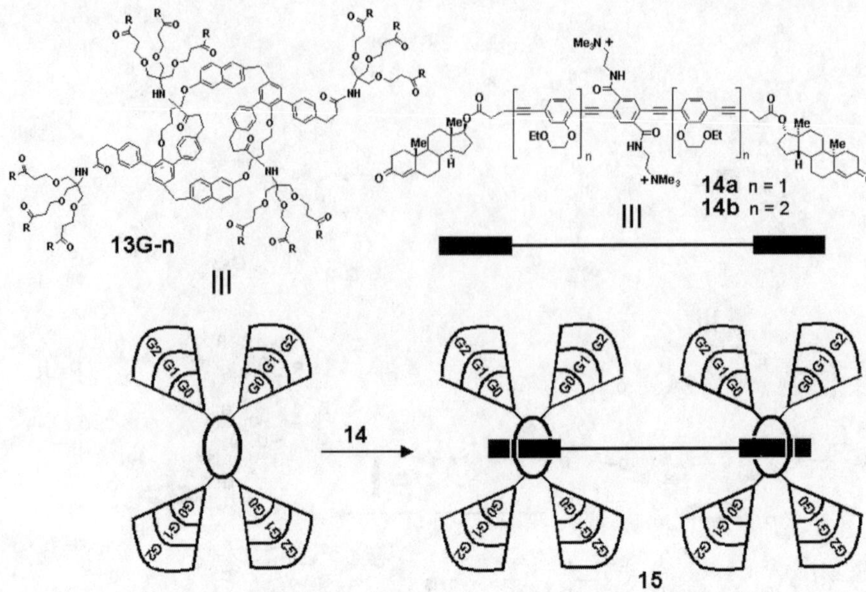

Fig. 5. [3]Pseudorotaxane dendrimers from complexation of the dendrophanes and molecular rod with rigid oligo(phenylacetylene) spacers and terminal steroid units

dynamically less favored. On the other hand, with the longer rod **14b** (n=2), both **13G-1** and **13G-2** are threaded onto both terminal steroids leading to a 2:1 complex. However, the first and second binding constants of **13G-1** to **14b** are almost the same whereas in the case of **13G-2**, the second threading is much less favorable than the first one probably due to the repulsion between the negatively charged dendritic shells. Given that two rods **14a** and **14b** are rigid and possess steroid-steroid distances of 41 and 55 Å, respectively, the extensions of the dendrophane shells in these [3]pseudorotaxane dendrimers can be estimated.

Newkome et al. also reported self-assembly of Type 1-B pseudorotaxane dendrimers using water-soluble dendritic β-CD **16G-n** (n=1, 2) of first and second generation as a ring component [42]. Addition of dendritic β-CD **16G-1** to a moderately basic aqueous solution of phenolphthalein results in the formation of Type I-B [2]pseudorotaxane dendrimer **17** (Fig. 6), as indicated by disappearance of the deep purple color of phenolphthalein. The driving forces for the complex formation include hydrophobic effects, van der Waals interactions and hydrogen bonding. Upon addition of adamantaneamine, which is known to bind very strongly to β-CD to form a 1:1 inclusion complex (stability constant $\sim 10^4\,M^{-1}$), to the solution, the deep purple color of phenolphthalein is regenerated, which illustrates that adamantaneamine displaces the indicator from [2]pseudorotaxane dendrimer **17** to produce new [2]pseudorotaxane dendrimer **18**. From these results, they concluded that the dendritic CDs **16G-1** and **16G-2** retain their binding sites and can incorporate, on the basis of molecular recognition, hydrophobic guests in basic aqueous media. In other words, the dendritic shell does not prevent recognition in the binding cavity. However, there are no

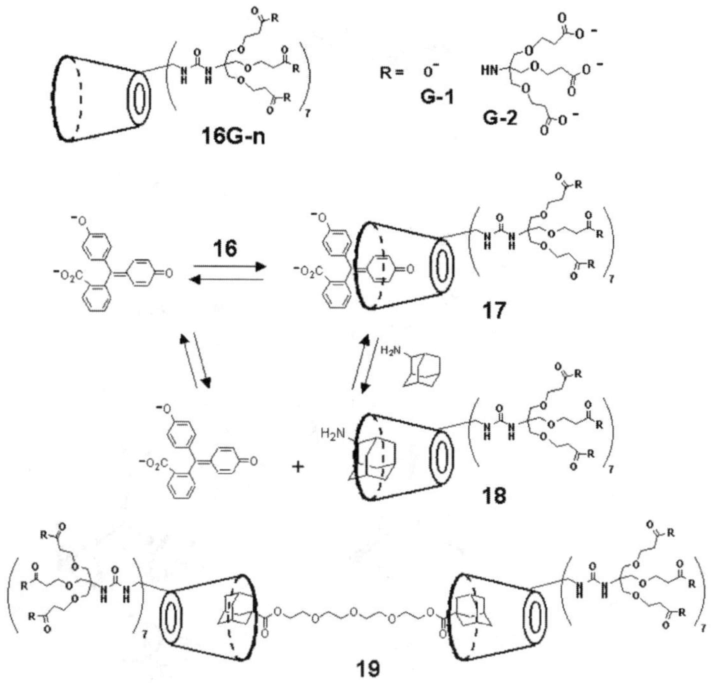

Fig. 6. [2]Pseudorotaxane and [3]pseudorotaxane dendrimers incorporating water-soluble dendritic β-CD

quantitative binding studies and the effect of the dendritic shell on binding strength or host-guest exchange kinetics. [3]Pseudorotaxane dendrimer **19** has also been assembled by forming a 2:1 complex between the dendritic β-CD **16G-2** and the bis(adamantane ester) of tetraethylene glycol (Fig. 6). In this case, no effect of the dendritic branching on the assembly process was reported.

Gibson et al. reported the self-assembly of Type I-B [4]pseudorotaxane dendrimers by threading dendrons with a macrocycle at the focal point onto a triply branched molecular rod [43–45]. The tritopic ligand **21**, which contains an ammonium unit on each arm, is completely insoluble in $CDCl_3$, but becomes soluble upon addition of Fréchet-type dendrons with a crown ether at the focal point (**20G-n**, $n=1$, 2, and 3) to form [4]pseudorotaxane dendrimers **22G-n** (Fig. 7). ^{1}H NMR spectroscopy and MALDI-TOF mass spectrometry confirmed the formation of 1:3 complexes between **21** and **20G-n**. The time required to reach equilibrium depends on the generation number of the dendrons: ~36 h for **20G-1**, ~48 h for **20G-2**, and ~72 h for **20G-3**. The slow formation of [4]pseudorotaxane dendrimer for **20G-3** is presumably due to the poor solubility of **21** in $CDCl_3$ and the steric hindrance experienced by neighboring dendron units in the 1:3 complex. Because these pseudorotaxanes lack one blocking group along the rod component, the threading and dethreading equilibrium is subject to environmental conditions such as the polarity of solvents and the generation of dendrons. A molecular modeling study of the [4]pseudorotaxanes indicates

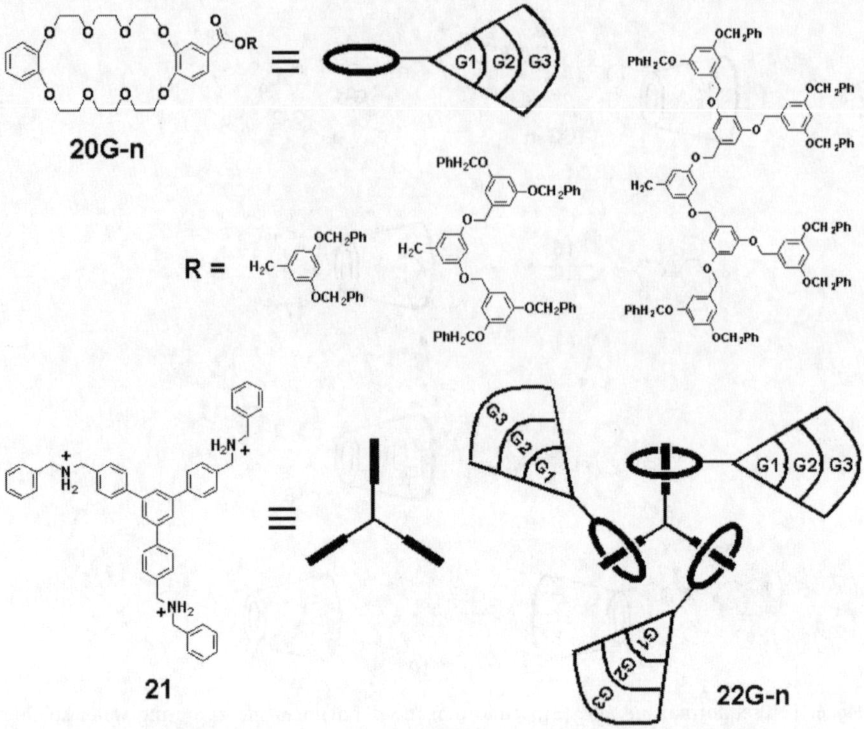

Fig. 7. [4]Pseudorotaxane dendrimers from dendrons with a macrocycle at the focal point and a triply branched molecular rod

that these dendritic structures range from 8–12 nm in diameter. These results demonstrate the feasibility of using this self-assembly protocol for the production of larger and more complex, nano-scale dendritic structures.

2.3
Rotaxane Dendrimers with Dendron Units Attached to Both the Ring and Rod: Type I-C

Only one example of Type I-C rotaxane dendrimers, which contain dendron units attached to both wheel and axle, is available in the literature. Vögtle et al. [34] synthesized such dendro[2]rotaxanes by attaching a dendron to the sulfonamide groups in the wheel and axle components. Reaction of rotaxane **23** with Fréchet-type dendritic bromides **5G-n** produces the dendro[2]rotaxanes **24G-n** (Fig. 8). Both **23** and **24G-n** are cycloenantiomerically chiral [2]rotaxanes [34]. The separation of the racemic mixture of these dendrimers was achieved by HPLC on chiral stationary phases. The cycloenantiomerically chiral dendro[2]rotaxanes containing G-0 and G-1 dendrons can be resolved by chiral HPLC, but the one with G-2 dendrons is eluted as a single peak. The chiroptical properties of the dendro[2]rotaxanes are also dependent on the generation number of the dendrons. Similarly, reaction of a [2]catenane containing a sul-

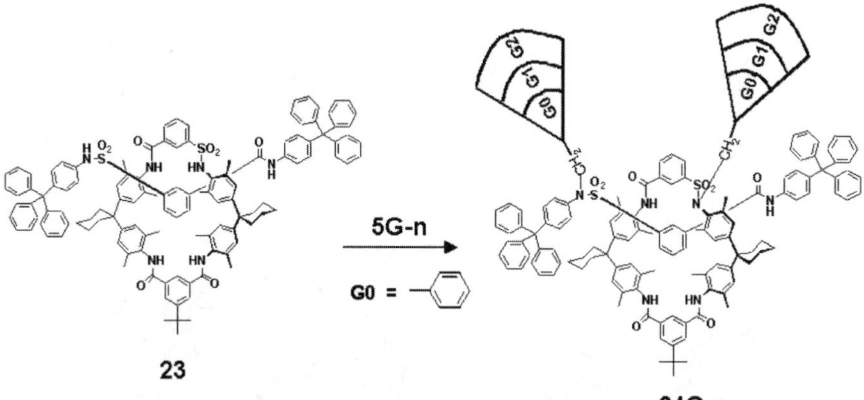

Fig. 8. A rotaxane dendrimer with dendron units attached to both wheel and axle

fonamide group in both of its macrocycles with Fréchet-type dendritic bromides produces dendro[2]catenanes, which are topologically chiral [2]catenanes [34].

3
Dendrimers with Pseudorotaxane-Decorated Periphery: Type II Rotaxane Dendrimers

As described earlier, Type II pseudorotaxane dendrimers have pseudorotaxane-like features at the *periphery* of dendrimers. Depending on whether the terminal units of the dendrimers serve as rod components or ring components, they can be further classified as Type II-A and Type II-B pseudorotaxane dendrimers, respectively.

3.1
(Pseudo)rotaxane-Terminated Dendrimers with Covalently-Attached Rod Components at the Periphery: Type II-A

Several examples this type of pseudorotaxane-terminated dendrimers are available in the literature, most of which exploit the host-guest interactions between β-CD and aromatic guests. Since ferrocene is a good substrate for inclusion complexation by β-CD (K_a=~1230 M^{-1}), ferrocenyl-functionalized polypropylimine (PPI) dendrimers **25** containing up to 16 (G-3) ferrocene units on their surfaces were synthesized [46] and their host-guest chemistry with β-CD was investigated using ^{1}H-NMR spectroscopy and electrochemistry (Fig. 9) [47]. Although the solubility of the dendrimers in aqueous media is very low, it is greatly enhanced in the presence of β-CD. This is rationalized by the formation of β-CD-ferrocene inclusion complexes on the surface of the dendrimers **25** to generate the pseudorotaxane-terminated dendrimers **26**. Although the complexation equilibria in these systems are very complicated, all the ferrocene

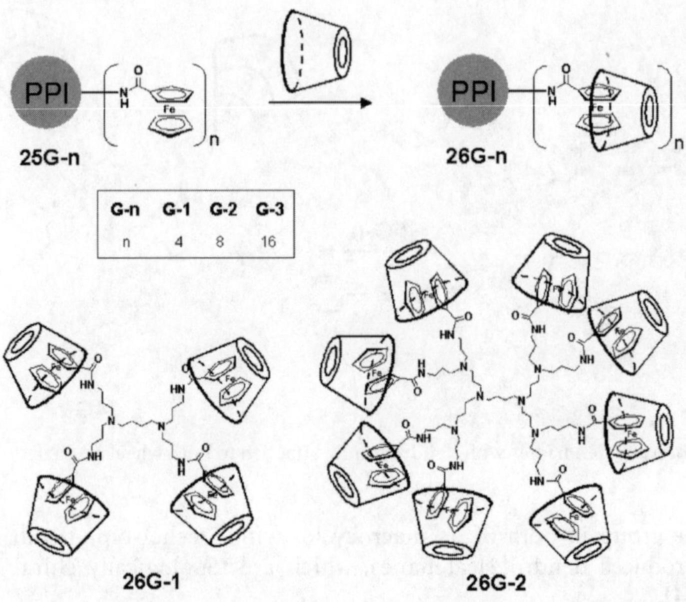

Fig. 9. Pseudorotaxane-terminated dendrimers: ferrocenyl-functionalized polypropylimine dendrimers terminated with β-CD

residues in dendrimers **25G-1** and **25G-2** are accessible for inclusion complexation by β-CD to generate pseudorotaxane-terminated dendrimers **26G-1** and **26G-2**. However, not all the ferrocene residues in dendrimers **25G-3** are accessible for inclusion complexation by β-CD presumably due to the substantial steric congestion on the surface of dendrimer for the higher generation dendrimer. The addition of sodium 2-naphthalenesulfonate to a solution containing dendrimer **26G-2** results in regeneration of **25G-2** as the 2-naphthalenesulfonate anion, a better substrate for inclusion complexation by β-CD, displaces the ferrocene subunits from the CD cavities. Here, the dendrimers serve as a template to organize β-CD in the periphery of the dendrimers, giving rise to large supramolecular assemblies, which is classified as pseudorotaxane-terminated dendrimers. Furthermore, the redox-active ferrocene end groups offer a reversible electrochemical mechanism to break up these supramolecular species as the binding affinity constant of the β-CD/ferrocene complex is strongly diminished by ferrocene oxidation.

As an extension of this work, Kaifer et al. prepared a series of PPI dendrimers (G-1–G-4) functionalized with cobaltocenium at the periphery [48] and studied their electrochemical behavior and binding interactions with β-CD. While the positively charged cobaltocenium-terminated dendrimer **27G-1** is not complexed by β-CD in aqueous media, electrochemical reduction of the dendrimer in the presence of excess β-CD triggers the formation of a multisite inclusion complex with this host to provide pseudorotaxane-terminated dendrimer **28G-1** (Fig. 10). Similar electrochemical and binding behavior toward β-CD was observed for the G-2 and G-3 dendrimers. However, the resulting multisite inclu-

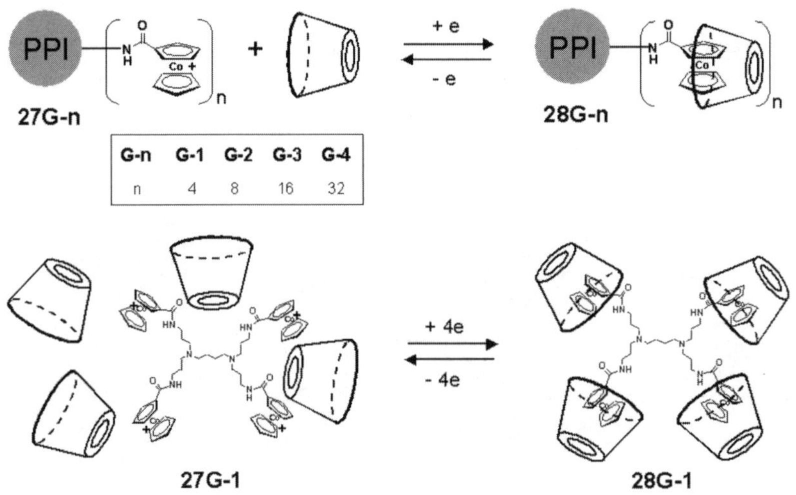

Fig. 10. Pseudorotaxane-terminated dendrimers generated by electrochemical reduction

sion complexes were not fully characterized; e.g. exact host-guest binding stoichiometry was not established. Nevertheless, this work illustrates the formation of large supramolecular assemblies triggered by "electrochemical activation" of the guest.

Meijer et al. reported adamantyl-terminated PPI dendrimers **29G-n** (*n*=1–5) containing from 4 (**G-1**) to 64 (**G-5**) adamantane units on their surfaces [49]. Since adamantane forms a very stable inclusion complex (association constant 10^5–10^6) with β-CD in aqueous solution, Reinhoudt and coworkers [50] investigated the interactions between β-CD and the adamantyl-terminated PPI dendrimers (Fig. 11). Here again, these adamantyl-terminated PPI dendrimers are insoluble in water but readily dissolve in the presence of β-CD through strong non-covalent interactions between the β-CD and adamantyl groups to generate the pseudorotaxane-terminated dendrimers **30G-n**. The formation of the pseudorotaxane-terminated dendrimer is most effective at pH=2, because at this pH there is complete protonation of the tertiary amino groups present in the dendritic cores, which causes the dendrimers to adopt a stretched conformation. The pseudorotaxane-terminated dendrimers remain in solution at pH <7, but precipitate under basic conditions, except for **G-1**, which remains in solution. The binding stoichiometry between CD and the dendrimers was determined by ^{1}H-NMR. All the terminal adamatyl units are complexed by β-CD except for **G-5** wherein about 40 β-CD instead of 64 were complexed, presumably due to steric congestion. The ability of the pseudorotaxane-terminated dendrimers (**G-1** to **G-5**) as a host was studied by fluorescence spectroscopy using the fluorescent probe 8-anilinonaphthalene-1-sulfonate (ANS) as a guest for the supramolecular dendritic host system. The shielding of ANS from the quenching water molecules by the dendrimers becomes more efficient with increasing dendrimer generation.

These β-CD/adamantyl pseudorotaxane-terminated dendrimers can be used as nanoreactors in the preparation of gold and platinum nanoparticles in water

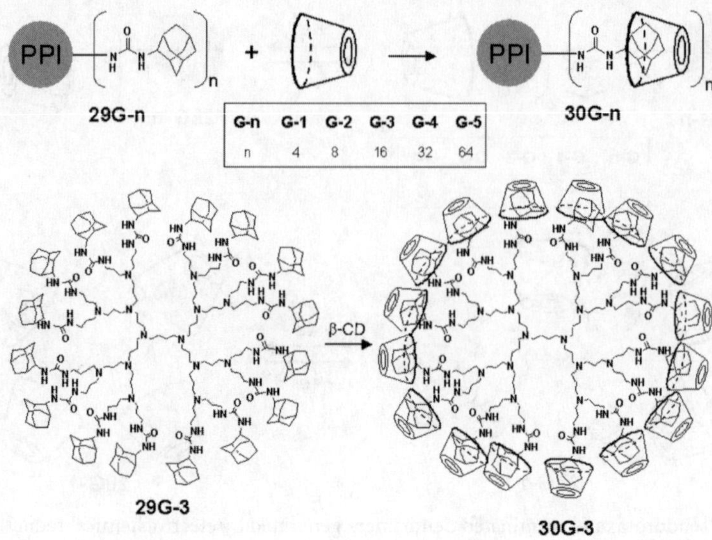

Fig. 11. Pseudorotaxane-terminated dendrimers from the adamantyl-terminated polypropyl-imine dendrimers and β-CD

[51]. These particles are formed by the reduction of aurate or platinate anions in the presence of **30G-4** (32 β-CD) and **30G-5** (40 β-CD). The lower generation assemblies (up to **30G-3**) do not provide stable nanoparticles. TEM images of the freshly prepared solutions confirmed the formation of nm-sized gold and platinum colloids stabilized by the **30G-4** (32 β-CD) and **30G-5** (40 β-CD) assemblies. The sizes of the Pt colloids (d=2.0±0.5 and 2.1±0.5 nm for **30G-4** (32 β-CD) and **30G-5** (40 β-CD), respectively) are slightly larger than for the Au colloids (d=1.6±0.7 and 1.7±0.9 nm for **30G-4** (32 β-CD) and **30G-5** (40 β-CD), respectively). The dispersity of the Au and Pt particles is significantly lower than that of most of the earlier reported dendrimer-stabilized nanoparticles. For both metals, the difference in particle size of the colloids stabilized by **30G-4** (32 β-CD) and **30G-5**(40 β-CD) is not significant. These Au and Pt nanoparticles are stable for 12 days at room temperature in daylight, and for a week at 4 °C in the dark. The dense shell of adamantyl-β-CD complexes provides a kinetic barrier for nanoparticle escape thus increasing their lifetime.

Reinhoudt et al. also reported water-soluble pseudorotaxane-terminated dendrimers possessing a radio-active metal core for radiotherapeutical applications [52]. First, Fréchet-type dendritic wedge [G-*n*]-amide-thiols **31G-*n*** are synthesized via a convergent growth strategy, i.e., starting from the adamantane periphery and working inward toward what is to become the amide-thiol ligand part. Two different convergent routes were used to produce the desired pseudorotaxane-terminated dendrimers (Fig. 12). In the first route, two dendritic wedge [G-*n*]-amide-thiols containing adamantane units are complexed with ReO(PPh$_3$)$_2$Cl$_3$ in organic solvents to produce adamantane-terminated dendrimers **32G-*n*** which are then complexed with β-CD to form pseudorotaxane-terminated dendrimers **34G-*n*.** In the second route, two dendritic wedge [G-*n*]-

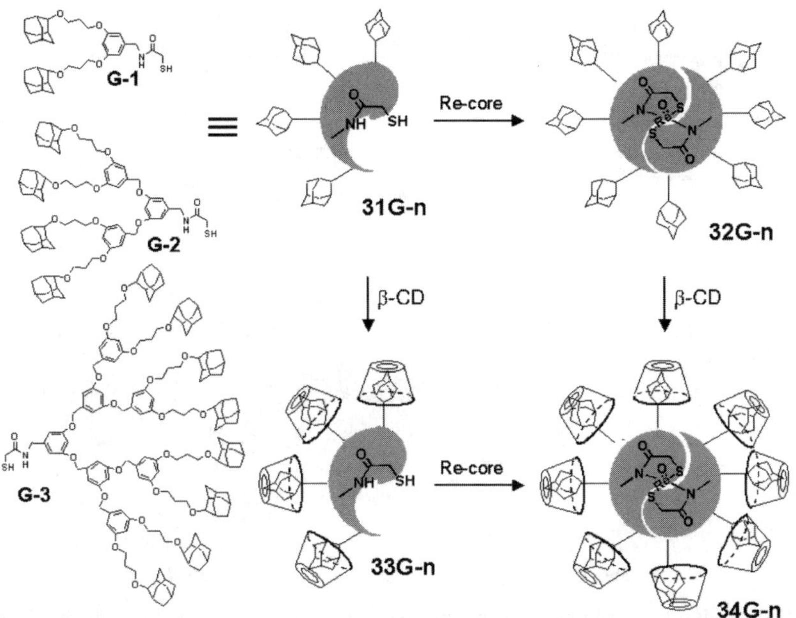

Fig. 12. Convergent routes to pseudorotaxane-terminated dendrimers with a metal complex at the cores and β-CD at the periphery

amide-thiols containing adamantane units are complexed with β-CD to provide the dendritic pseudorotaxanes **33G-*n***, followed by complexation with ReO(PPh$_3$)$_2$Cl$_3$ in water to produce the pseudorotaxane-terminated dendrimers **34G-*n***. The second route produces the desired products in far better yields than the first. Unfortunately, both methods fail to give [**G-3**] pseudorotaxane-terminated dendrimer presumably because the focal point is too sterically hindered for the ligand to undergo Re complex formation. Most importantly, complexation of the terminal adamantly groups by β-CD makes these dendrimers soluble in water (solubility, 9.6, 0.4 and 0.2 mM for **34G-0**, **34G-1** and **34G-2**, respectively) and facilitate the formation of the Re complex.

Taking advantage of the fact that protonated diaminobutane forms very stable complexes with cucurbituril (CB[6]) with a formation constant much greater than 10^5 Kim et al. reported CB[6]-based pseudorotaxane-terminated dendrimers [53]. They synthesized diaminobutane-functionalized PPI dendrimers **35 (G1-G5)** and corresponding pseudorotaxane-terminated dendrimers **36** (Fig. 13) by threading CB[6] onto the terminal diaminobutane units. NMR spectroscopy reveals all the termini to be occupied by CB[6] beads even in **G-5**, which contains 64 beads threaded at the periphery of the dendrimers with a molecular weight of ~94,500. However, they failed to obtain the exact mass of the dendrimers by ESI mass spectrometry except **G-1** due to their high number of charges and molecular weights. Upon threading molecular "beads", the large terminal pseudorotaxane units form a rigid shell at the exterior of the dendrimers as demonstrated by molecular dynamic simulation and ^{1}H NMR spin-lattice (T_1)

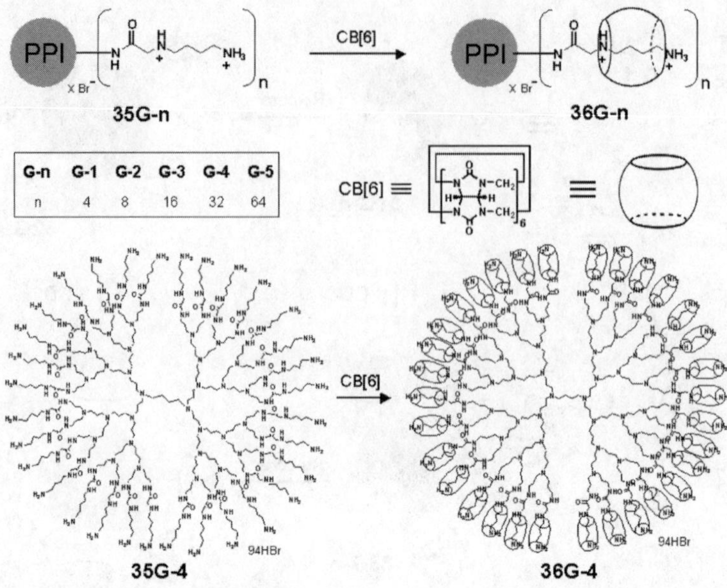

Fig. 13. Pseudorotaxane-terminated dendrimers from diaminobutane-functionalized polypropylimine dendrimers and CB[6]

relaxation measurements. According to a molecular dynamics study, the overall size of the **G-3** dendrimer increases from 40~41 Å to 44~49 Å upon threading of CB[6]. An atomic force microscopy image of the **G-5** dendrimer on a mica surface reveals a flattened sphere with a diameter of ~15 nm and a height of ~1 nm. Upon treatment with a base all or a part of the molecular beads of these dendrimers are dethreaded. This reversible threading and dethreading of molecular beads at the termini may provide them with a mechanism for reversible encapsulation and release of guest molecules, which may find useful applications including drug delivery.

Nakamura and Kim carried out DNA binding studies of pseudorotaxanes comprising CB[6] and polyamines [54], which showed CB[6] to modulate the DNA binding ability of polyamines. Encouraged by these results, Park and Kim investigated the potential utility of these novel dendrimers as gene carriers because the polycationic pseudorotaxane-terminated dendrimers **36** contain a large number of surface amines, which allow them to bind polyanionic DNA. Gene transfer mediated by **36** was investigated with two kinds of cell lines (293 cells and Vero 76 cells) [55]. Transfection efficiencies (TE) were dependent on the cell type, 300 times higher in 293 cells than in Vero 76 cells for **36G-5**. It is well known that 293 cells are very susceptible to transfection. The transfection efficiency of **36** increases with increasing dendrimer generation. The best transfection efficiency is only one order of magnitude lower than that of polyethylenimine (PEI), which is one of the most efficient gene carriers reported up till now. These results indicate that this type of pseudorotaxane-terminated dendrimer may be useful as a gene delivery vector.

3.2
(Pseudo)rotaxane-Terminated Dendrimers with Covalently-Attached Ring Components at the Periphery: Type II-B

Although a number of examples of Type II-A pseudorotaxane dendrimers are known, only one example of Type II-B pseudorotaxane dendrimer has been reported. Gibson and co-workers synthesized pseudorotaxane-terminated dendrimers utilizing dendrimers functionalized with crown ether at the periphery [56]. Crown ether-terminated PPI dendrimers **38G-*n*** (*n*=1, 2, and 3), in which the basic polyamine interior is fully protonated, form complexes with dibenzylammonium ion **37** in organic solvent to produce pseudorotaxane-terminated dendrimers **39G-*n*** (Fig. 14). The formation of the pseudorotaxane-terminated dendrimers was studied by ^{1}H-NMR spectroscopy. The binding is cooperative with a Hill coefficient of 2.1 for all generations. This positive cooperativity may be due to the favorable π-π interactions between the aromatic rings of neighboring bound guest molecules. Another possibility is that the polarity of the dendritic host increases upon complexation, creating an environment more receptive to the next charged guest. The reversibility of the process has been demonstrated. The addition of DABCO to a solution containing protonated **38G-3** and **37** immediately leads to deprotonation and subsequent decomplexation of guest. The same releasing mechanism has been used for designing molecular machines by Stoddart et al. [13].

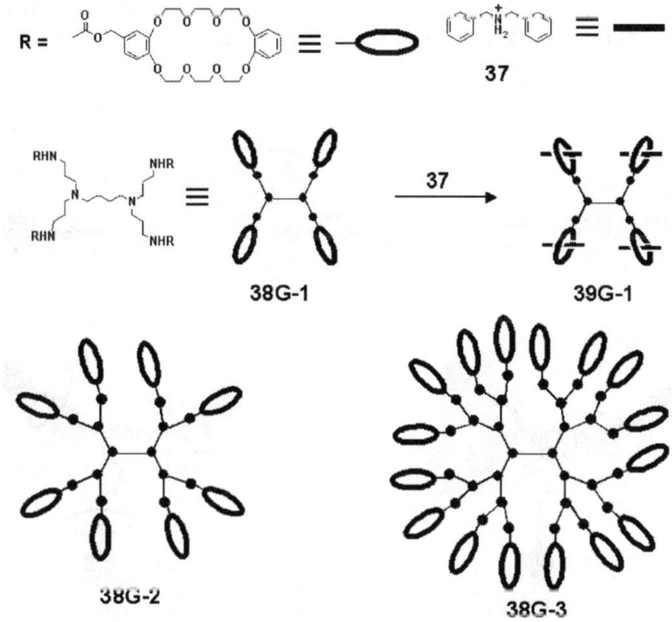

Fig. 14. Rotaxane dendrimers from crown ether-terminated polypropylimine dendrimer complexation with dibenzylammonium ion

4
Dendritic Polyrotaxanes: Type III Rotaxane Dendrimers

As described earlier, we classify dendritic polyrotaxanes in which rotaxane building units grow like a dendrimer, as Type III rotaxane dendrimers. Depending on whether ring components are located *on the branches* or *at the branching points*, Type III rotaxane dendrimers are further classified as III-A and III-B, respectively.

4.1
Dendritic Polyrotaxanes Incorporating Ring Components on the Branches: Type III-A

The first study towards building dendritic polyrotaxanes was reported in 1995 by Stoddart et al. [57]. A branched [4]rotaxane, which can be considered as a G-1 dendritic rotaxane, was successfully synthesized by self-assembly (by a slippage mechanism) between the triply-branched tris(bipyridinium)derivative 40 and bis-*p*-phenylene-34-crown-10 (BPP34C10) 1 (Fig. 15). Treatment of 1 (2 equiv) with 40 at 50 °C in acetonitrile for 10 days affords [4]rotaxane 41 (6%) along with [2]rotaxane (46%) and [3]rotaxane (26%). When 15 equiv of 1 is used under identical conditions, the [4]rotaxane is obtained in 22% yield together with a [2]rotaxane (19%) and a [3]rotaxane (41%). Interestingly, even when a large excess of macrocycle 1 is employed, the major product is not the fully-occupied [4]rotaxane 41, but rather the [3]rotaxane. The reason for this result may be a consequence of the steric crowding around the central core, which inhibits the slipping on of the third macrocycle. In principle, this approach can be extended to higher generations, but the synthesis of higher generation dendritic rotaxanes has not been reported.

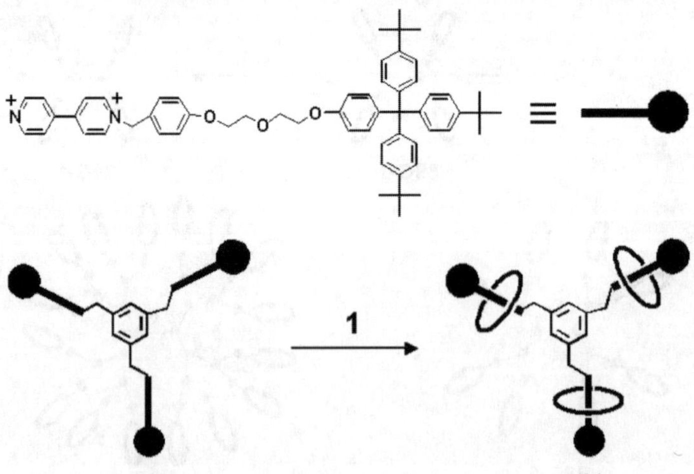

Fig. 15. A branched [4]rotaxane from self-assembly of the triply-branched tris(bipyridinium)derivative and bis-*p*-phenylene-34-crown-10

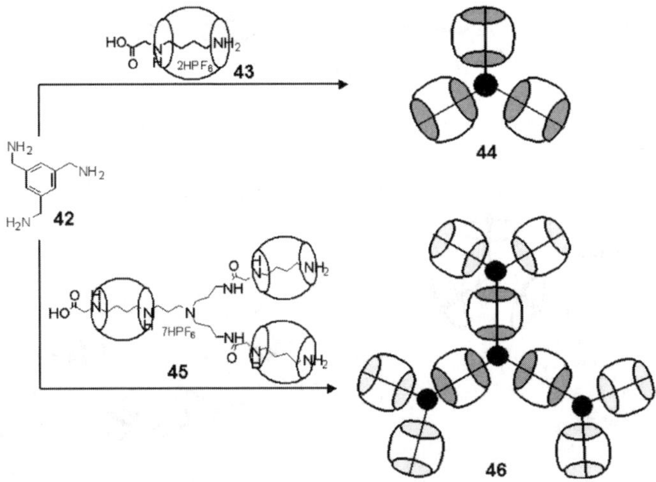

Fig. 16. Coupling triply-branched amine and pseudorotaxane furnishes pseudorotaxane containing CB[6] beads threaded on the dendritic framework

Kim et al. synthesized similar dendritic pseudorotaxanes as shown in Fig. 16 [58]. Reaction of triply-branched amine **42** with pseudorotaxane **43** incorporating CB[6] threaded on a string with a carboxylic acid terminus in DMF in the presence of EDC produces branched [4]pseudorotaxane **44**, which is a G-1 dendritic pseudorotaxane. Coupling between **42** and pseudorotaxane dendritic wedge **45** under similar conditions yields G-2 dendritic pseudorotaxane **46**, which contains 9 beads threaded on a dendritic framework ([10]pseudorotaxane). These dendritic pseudorotaxanes have been characterized by NMR spectroscopy, but further characterization remains to be done.

More recently, Kim et al. synthesized dendritic [n]pseudorotaxane based on the stable charge-transfer complex formation inside cucurbit[8]uril (CB[8]) (Fig. 17) [59]. Reaction of triply branched molecule **47** containing an electron deficient bipyridinium unit on each branch, and three equiv of CB[8] forms branched [4]pseudorotaxane **48** which has been characterized by NMR and ESI mass spectrometry. Addition of three equivalents of electron-rich dihydroxynaphthalene **49** produces branched [4]rotaxane **50**, which is stabilized by charge-transfer interactions between the bipyridinium unit and dihydroxynaphthalene inside CB[8]. No dethreading of CB[8] is observed in solution. Reaction of [4]pseudorotaxane **48** with three equiv of triply branched molecule **51** having an electron donor unit on one arm and CB[6] threaded on a diaminobutane unit on each of two remaining arms produced dendritic [10]pseudorotaxane **52** which may be considered to be a second generation dendritic pseudorotaxane.

A polyrotaxane with a dendrimer-like structure is known [60]. Based on the observation that β-CD and sodium deoxycholate (NaDC) **54** forms a 2:1 host guest complex in water, Tato et al. constructed hyperbranched polyrotaxanes **55** by slowly reacting triply branched receptor **53** containing β-CD and NaDC

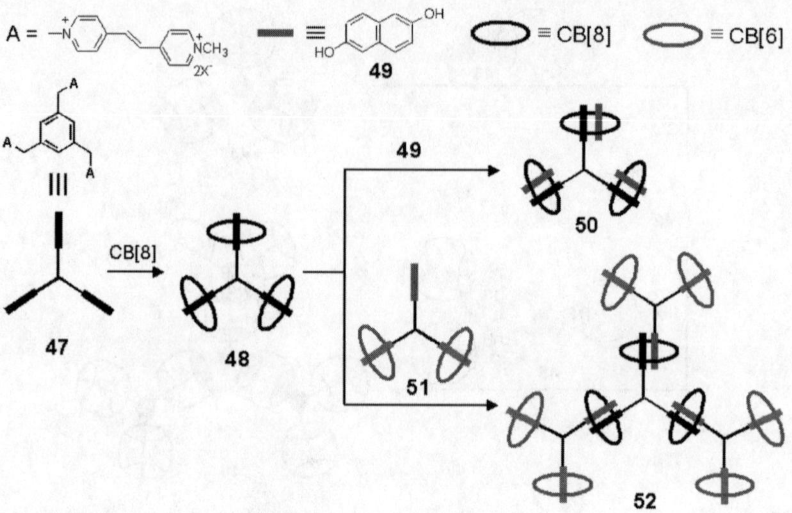

Fig. 17. Dendritic [4]pseudorotaxanes and [10]pseudorotaxanes based on the stable charge-transfer complex formation in CB[8] cavity

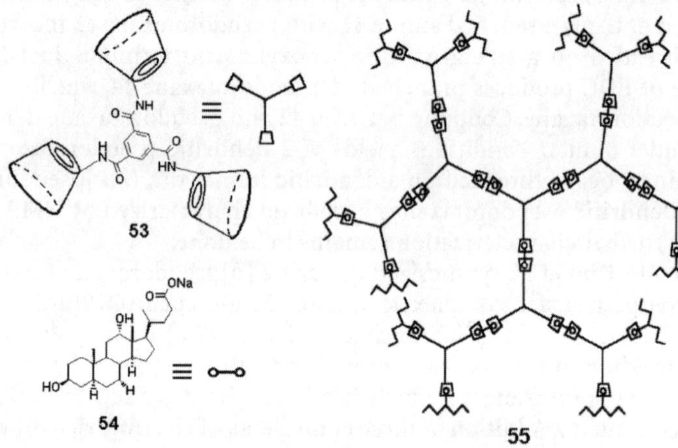

Fig. 18. Hyperbranched polyrotaxanes incorporating β-CD

54 in a 2:1 molar ratio (Fig. 18). Investigation of the precipitate formed from the solution after 2–3 months by scanning electron microscopy and confocal laser scanning microscopy revealed fractal structures of the self-assembled supramolecular species. Several different types of fractal structures are observed which may arise from different complexation stoichiometries for each β-CD trimer unit.

4.2
Dendritic Polyrotaxanes Incorporating Ring Components at Branching Points: Type III-B

Novel dendrimers in which mechanical bonds replace covalent bonds for *branching connections* have been reported recently. Vögtle et al. [35] proposed two strategies, a convergent and a divergent approach, towards dendritic rotaxanes with mechanical branching units (Fig. 19). In the divergent approach (Fig. 19a), two of wheel **A**, each containing a reactive functional group, react with a suitable axle in the presence of macrocycle **B** to form the rotaxane of generation zero **C**. To obtain a stable rotaxane **C**, the outer two macrocycles must be large enough to prevent dethreading. In the second reaction, the available coordination sites of the two outer wheels are used to trap two more "macrocycle-stopped" axles. This leads to the rotaxane assembly of generation one **D**. Continuing this strategy, the formation of higher generations should be possible. If non-cyclic, sterically demanding stoppers are used in the last reaction step, the dendritic oligorotaxane of second generation **E** is formed and then further growth towards higher generations is no longer possible. In the convergent strategy shown in Fig. 19b, the synthesis starts with a protected wheel to produce rotaxane **F**. After removing the protective group, rotaxane **F** reacts in the presence of wheel **G** to form the rotaxane assembly of first generation **H**. If the central wheel carries a functional group, it can react further to form the assembly of second generation **J**.

Using the divergent approach, Vögtle et al. [35] attempted to synthesize G1 dendrimer of this type but failed. Reaction of the phenolic 'stopper wheel' **56** with 1,4-bis(bromomethyl)benzene **57** as an axle in the presence of wheel **58** produces rotaxane **59** in 57% yield (Fig. 20a). Unfortunately, attempts to synthesize higher generation rotaxanes in the next step failed. They therefore switched to the convergent strategy. The phenolic 'stopper rotaxane' **60**, synthesized by the 'anion template method', was used as the building unit **F** in Fig. 19b to form rotaxane-branched dendrimers (Fig. 20b). Rotaxane **60** reacts with **57** in the presence of **58** in chloroform with a base to produce the first generation rotaxane-branched dendrimer **61** in 39% yield. In principle, higher generation dendritic rotaxanes can be synthesized using this strategy but the next coupling

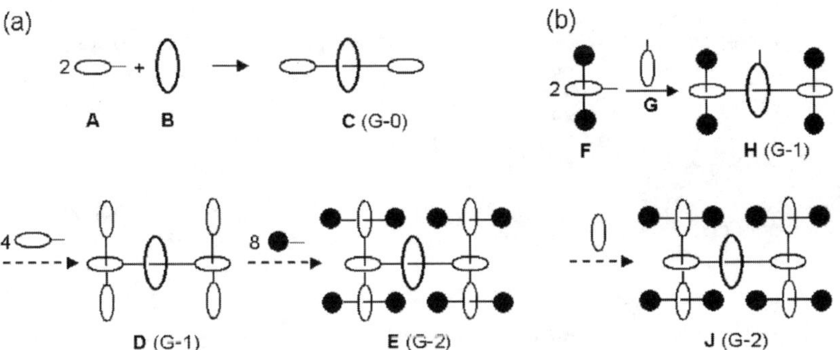

Fig. 19. (a) Divergent and (b) convergent synthesis of dendritic polyrotaxanes with mechanical branching units

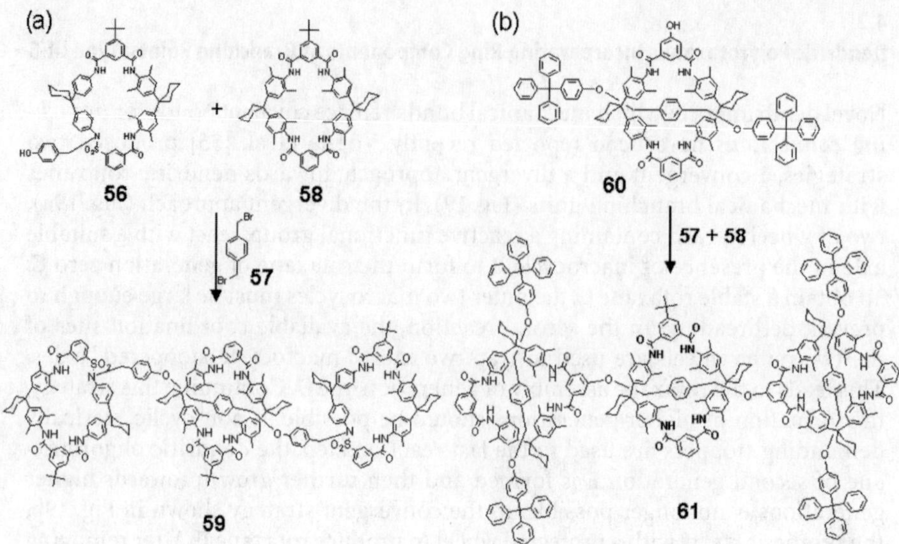

Fig. 20. Toward dendritic polyrotaxanes with mechanical branching units

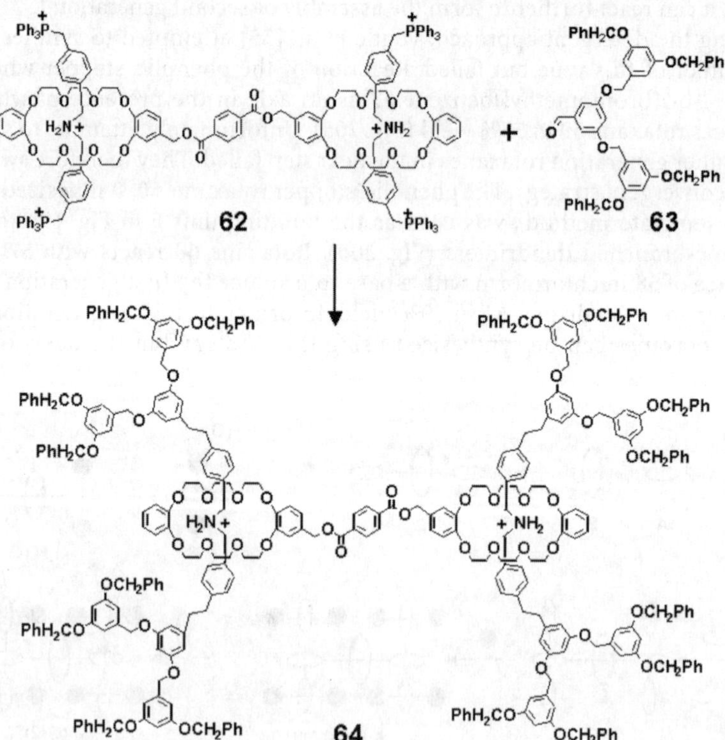

Fig. 21. Dendritic polyrotaxanes with mechanical branching units containing covalently linked bis-dendrons and a core unit fused to polyether macrocycles

reactions would be more difficult to achieve due to large steric hindrance between the bulky rotaxane components.

Very recently, Stoddart et al. reported a dendrimer possessing rotaxane-like mechanical branching [61], which contains two identical covalently linked bis-dendrons and a core unit fused to two polyether macrocycles that encircle the two bis-dendrons (Fig. 21). They used the approach of "threading-followed-by-stoppering" and then "stopper exchange". The bis[2]rotaxane **62** carrying four benzylic triphenylphosphonium stoppers, was prepared in 37% yield by complex formation (*threading*) between a pseudorotaxane bis-macrocycles and the bis-(bromomethyl)-substituted dibenzylammonium derivative followed by *stoppering* using excess of triphenylphosphine. The stopper exchange of the bis[2]-rotaxane **62** was conducted by a Wittig reaction. Reaction of **62** with the dendron-aldehyde **63** followed by reduction of the resulting olefinic bond to give the bis[2]rotaxane-branched dendrimer **64** in 75% yield. In principle, this strategy can be extended to dendrimers with mechanical branching points at more than one generation but the actual synthesis may not be easily achieved.

5
Summary and Outlook

In this review, we tried to cover all the supramolecular species that may be classified as rotaxane dendrimers. We classified them by their structures – where in dendrimer rotaxane-like features are introduced. Several different types of macrocycles have been employed as a ring component in the templated synthesis of rotaxane dendrimers. While the synthesis of Type I and II rotaxanes dendrimers is relatively straightforward, that of well-defined Type III rotaxane dendrimers, particularly those of second and higher generations, is still challenging.

In the synthesis of rotaxane dendrimers, several aspects should be carefully considered. To achieve a rotaxane dendrimer of well-defined structure and of high generation, use of a pseudorotaxane unit with high stability is a prerequisite that entails a careful choice of the macrocycle and the guest. The synthesis of rotaxane dendrimers using low-stability pseudorotaxane building blocks often leads to a number of complex problems. The rotaxane dendrimers in such cases are either unstable in solution or isolated in low yields. Sometimes a mixture of dendrimers of different generations may result, posing considerable difficulty in their separation and characterization. Especially, the complex stoichiometries in pseudorotaxane-terminated dendrimers (Type II-A) formed from hosts and guest-functionalized dendrimers are influenced by the intrinsic binding affinity between host and guest. For example, for a ferrocene-terminated poly(propylimine) dendrimer, it is not possible to cap all the terminal ferrocene groups by β-CD, unless a large excess of β-CD is used, due to relatively weak binding affinity of the host and the guest. However, well-defined assemblies of pseudorotaxane-terminated dendrimers are achievable with as many as 64 CB[6] beads for a G-5 dendrimer through the high binding affinity of CB[6] for the diaminoalkane strings on the periphery.

In addition to this, the subtle importance of several other factors warrants due consideration. For example, in making a choice of the proper host attention should be paid to the space limitations that many arms of the dendrimer can impose. Steric congestions can severely limit the formation of higher generation of rotaxane dendrimers. For dendrimers to be stable in solution, additional factors such as solvent, the pH of solution, temperature, etc. turn out to be evenly important.

The properties of these rotaxane dendrimers are quite different from those of the individual rotaxanes or dendrimers and often a blend of both. In view of the versatile characteristics that a dendron or dendrimer can manifest, several new properties can be imparted to the rotaxanes. For example, the solubility of rotaxanes in organic solvents as well as in water can be significantly improved when large dendrimer units are appended enhancing the prospects of their use as molecular machines. The dendritic units can also influence the photo/electrochemical properties of the rotaxanes. Employing photo-receptive dendron units, photochemically driven molecular machines may be developed, where the dendrons act as antenna for photo-harvesting [62].

Also, from the dendrimer point of view, the introduction of mechanical bonds to dendrimers has an enormous potential to alter the properties of dendrimers in a controlled way. For example, in the synthesis of a Type II rotaxane dendrimers, the wheel components are introduced to the terminal groups of the dendrimers. This can improve the solubility of dendrimer in organic and/or aqueous media due to the formation of complexes soluble in such solvents.

In particular, rotaxane dendrimers capable of reversible binding of ring and rod components, such as Type II, pseudorotaxane-terminated dendrimers, can be reversibly controlled by external stimuli, such as the solvent composition, temperature, and pH, to change their structure and properties. This has profound implications in diverse applications, for instance in the controlled drug release. A trapped guest molecule within a closed dendrimeric host system can be unleashed in a controlled manner by manipulating these external factors. In the type III-B rotaxane dendrimers, external stimuli can result in perturbations of the interlocked mechanical bonds. This behavior can be gainfully exploited to construct controlled molecular machines.

There has been quite a revolution in rotaxane dendrimers with the discovery of many interesting molecules. Nevertheless, it is only a modest beginning and many challenges pertaining to their synthesis and applications lie ahead. For example, efficient synthesis of well-defined Type III rotaxane dendrimers, particularly those of second and higher generations, remains to be achieved. Although a number of applications such as molecular nanoreactors, drug delivery, and gene delivery have been proposed or studied, many other applications are still to be explored.

Acknowledgments. Our work described here has been supported by the Creative Research Initiatives Program and International Joint R&D Projects of the Korean Ministry of Science and Technology, and in part by the Bain Korea 21 Program of the Korean Ministry of Education. We also thank Dr. S. Samal for helpful discussions in preparing the manuscript.

6
References

1. Newkome GR, Moorefield CN, Vögtle F (1996) Dendritic molecules: concepts, synthesis, perspectives. VCH, Weinheim
2. Tomalia DA, Naylor AM, Goddard WA III (1990) Angew Chem Int Ed Engl 29:138
3. Feuerbacher N, Vögtle F (1998) Top Curr Chem 197:1
4. Fischer M, Vögtle F (1999) Angew Chem Int Ed 38:884
5. Bosman AW, Janssen HM, Meijer EW (1999) Chem Rev 99:1665
6. Grayson SM, Fréchet JMJ (2001) Chem Rev 101:3819
7. Balzani V, Gómez-López M, Stoddart JF (1998) Acc Chem Res 31:405
8. Balzani V, Credi A, Raymo FM, Stoddart JF (2000) Angew Chem Int Ed 39:3348
9. Schalley CA, Beizai K, Vögtle F (2001) Acc Chem Res 34:465
10. Whang D, Kim K (1997) J Am Chem Soc 119:451
11. Lee E, Heo J, Kim K (2000) Angew Chem Int Ed 39:2699
12. Asakawa M, Ashton PR, Ballardini R, Balzani V, Bělohradský M, Gandolfi MT, Kocian O, Prodi L, Raymo FM, Stoddart JF, Venturi M (1997) J Am Chem Soc 119: 302
13. Ashton PR, Ballardini R, Balzani V, Baxter I, Credi A, Fyfe MCT, Gandolfi MT, Gómez-López M, Martínez-Díaz M-V, Piersanti A, Spencer N, Stoddart JF, Venturi M, White AJP, Williams DJ (1998) J Am Chem Soc 120: 11,932
14. (a) Diederich F (1988). Angew Chem Int Ed Engl 27: 362; (b) Diederich F (1991) Cyclophanes. The Royal Society of Chemistry, Cambridge; (c) Vögtle F (1993) Cyclophane chemistry. Wiley, New York
15. Vögtle F, Dünnwald T, Schmidt T (1996) Acc Chem Res 29: 451
16. Hübner GM, Gläser J, Seel C, Vögtle F (1999) Angew Chem Int Ed 38: 383
17. (a) Atwood JL, Davies JED, MacNicol DD, Vögtle F (1996) (eds) Comprehensive supramolecular chemistry, vol 3. Pergamon, Oxford; (b) Connors KA (1997) Chem Rev 97:1325; (c) Rekharsky MV, Inoue Y (1998) Chem Rev 98:1875
18. Nepogodiev SA, Stoddart JF (1998) Chem Rev 98:1959
19. Mock WL (1996) Cucurbituril. In: Vögtle F (ed) Comprehensive supramolecular chemistry vol 2. Pergamon, Oxford, p 477
20. Mock WL (1995) Top Curr Chem 175:1
21. (a) Buschmann H-J, Meschke C, Schollmeyer E (1998) An Quím Int Ed 94:241; (b) Lee JW, Kim K, Kim K (2001) Chem Commun 1042
22. Kim K (2002) Chem Soc Rev 31:96
23. (a) Kim J, Jung I-S, Kim S-Y, Lee E, Kang J-K, Sakamoto S, Yamaguchi K, Kim K (2000) J Am Chem Soc 122:540; (b) Day A, Arnold AP, Blanch RJ, Snushall B (2001) J Org Chem 66:8094
24. Kim H-J, Heo J, Jeon WS, Lee E, Kim J, Sakamoto S, Yamaguchi K, Kim K (2001) Angew Chem Int Ed 40:1526
25. Kim S-Y, Jung I-S, Lee E, Kim J, Sakamoto S, Yamaguchi K, Kim K (2001) Angew Chem Int Ed 40:2119
26. Jon SY, Ko YH, Park S-H, Kim H-J, Kim K (2001) Chem Commun 1938
27. Kim H-J, Jeon WS, Ko YH, Kim K (2002) Proc Natl Acad Sci USA 99:5007
28. Marquez C, Nau WM (2001) Angew Chem Int Ed 40:4387
29. Day AI, Blanch RJ, Arnold AP, Lorenzo S, Lewis GR, Dance I (2002) Angew Chem Int Ed 41:275
30. Blanch RJ, Sleeman AJ, White TJ, Arnold AP, Day AI (2002) Nano Lett 2:147
31. Amabilino DB, Ashton PR, Balzani V, Brown CL, Credi A, Fréchet JMJ, Leon JW, Raymo FM, Spencer N, Stoddart JF, Venturi M (1996) J Am Chem Soc 118:12,012
32. Seel C, Vögtle F (2000) Chem Eur J 6:21
33. Hübner GM, Nachtsheim G, Li QY, Seel C, Vögtle F (2000) Angew Chem Int Ed 39:1269
34. Reuter C, Pawlitzki G, Wörsdörfer U, Plevoets M, Mohry A, Kubota T, Okamoto Y, Vögtle F (2000) Eur J Org Chem 3059
35. Osswald F, Vogel E, Safarowsky O, Schwanke F, Vögtle F (2001) Adv Synth Catal 343:303
36. Lee JW, Hwang IH, Ko YH, Jeon YJ, Choi SW, Kim K unpublished results

37. Mattei S, Seiler P, Diederich F, Gramlich V (1995) Helv Chim Acta 78:1904
38. Mattei S, Wallimann P, Kenda B, Amrein W, Diederich F (1997) Helv Chim Acta 80:2391
39. Wallimann P, Seiler P, Diederich F (1996) Helv Chim Acta 79:779
40. Wallimann P, Mattei S, Seiler P, Diederich F (1997) Helv Chim Acta 80:2368
41. Kenda B, Diederich F (1998) Angew Chem Int Ed Engl 37:3154
42. Newkome GR, Godínez LA, Moorefield CN (1998) Chem Commun 1821
43. Yamaguchi N, Hamilton LM, Gibson HW (1998) Angew Chem Int Ed Engl 37:3275
44. Gibson HW, Hamilton L, Yamaguchi N (2000) Polym Adv Technol 11:791
45. Gibson HW, Yamaguchi N, Hamilton L, Jones JW (2002) J Am Chem Soc 124:4653
46. Cuadrado I, Morán M, Casado C M, Alonso B, Lobete F, Garcia B, Ibisate M, Losada J (1996) Organometallics 15:5278
47. Castro R, Cuadrado I, Alonso B, Casado CM, Morán M, Kaifer AE (1997) J Am Chem Soc 119:5760
48. González B, Casado CM, Alonso B, Cuadrado I, Morán M, Wang Y, Kaifer AE (1998) Chem Commun 2569
49. Baars M W P L, Karlsson A, Sorokin V, de Waal B F M, Meijer EW (2000) Angew Chem Int Ed 39:4262
50. Michels JJ, Baars MWPL, Meijer EW, Huskens J, Reinhoudt DN (2000) J Chem Soc Perkin Trans 2 1914
51. Michels JJ, Huskens J, Reinhoudt DN (2002) J Chem Soc Perkin Trans 2 102
52. van Bommel KJC, Metselaar GA, Verboom W, Reinhoudt DN (2001) J Org Chem 66:5405
53. Lee JW, Ko YH, Park S-H, Yamaguchi K, Kim K (2001) Angew Chem Int Ed 40:746
54. Isobe H, Tomita N, Lee JW, Kim H-J, Kim K, Nakamura E (2000) Angew Chem Int Ed 39:4257
55. Lim Y-b, Kim T, Lee JW, Kim S-m, Kim H-J, Kim K, Park J-S (2002) Bioconjugate Chem 13:1181
56. (a) Gibson HW, Bosman AW, Bryant WS, Jones JW, Jannsen RAJ, Meijer EW (2001) Polym Mater Sci Eng 84:66; (b) Jones JW, Bryant WS, Bosman AW, Janssen RAJ, Meijer EW, Gibson HW (2003) J Org Chem 68:2385
57. Amabilino DB, Ashton PR, Bělohradský M, Raymo FM, Stoddart JF (1995) J Chem Soc Chem Commun 751
58. Lee JW, Kim K unpublished results
59. Kim S-Y, Lee JW, Kim K unpublished results
60. Alvarez-Parrilla E, Cabrer PR, Al-Soufi W, Meijide F, Núñez ER, Tato JV (2000) Angew Chem Int Ed 39:2856
61. Elizarov AM, Chiu S-H, Glink PT, Stoddart JF (2002) Org Lett 4:679
62. Jang DL, Aida T (1997) Nature 388:454

Note added in proof

After submission of this review article, two more rotaxane dendrimers have been reported: a Type I-B rotaxane dendrimer (Dykes GM, Smith DK, Seeley GJ (2002) Angew Chem Int Ed 41:3254 and a Type I-C rotaxane dendrimer (Elizarov AM, Chang T, Chiu S-H, Stoddart JF (2002) Org Lett 4:3565).

Top Curr Chem (2003) 228: 141–158
DOI 10.1007/b11009

Effect of Dendrimers on the Crystallization of Calcium Carbonate in Aqueous Solution

Kensuke Naka

Department of Polymer Chemistry, Graduate School of Engineering, Kyoto University, Yoshida, Sakyo-ku, Kyoto, 606-8501, Japan
E-mail: ken@chujo.synchem.kyoto-u.ac.jp

Construction of organic-inorganic hybrid materials with controlled mineralization analogous to those produced by nature is now a current interest for both organic and inorganic chemists to understand the mechanism of natural biomineralization processes as well as to seek industrial and technological applications. Model systems, in which low-molecular-weight, linear polymeric organic materials have been used to study the effect of molecular properties such as charge and functionality on inorganic crystallization, are providing insights into the possible mechanisms operating in biology. Due to unique and well-defined secondary structures of the dendrimers, the starburst dendrimers should be a good candidate for studying inorganic crystallization. This review provides a general survey of recent research on crystal nucleation and growth of calcium carbonate by a carboxylic acid derivative of hyperbranched polyphenylene polymer, poly(amidoamine) (PAMAM) dendrimers with carboxylate groups at the external surface, and poly(propyleneimine) dendrimers modified with long aliphatic chains.

Keywords. Calcium carbonate, Mineralization, Hyperbranched polymer, Supramolecular assembly, Anionic dendrimer

1
Introduction

In nature, biological organisms produce polymer-inorganic hybrid materials such as bone, teeth, diatoms, and shells. These hybrids have superior mechanical properties as compared to synthetic hybrids. For example, the abalone shell, a composite of calcium carbonate with a few percent of the organic component (Fig. 1), is 3000 times more fracture resistant than a single crystal of the pure mineral [1]. The core of the organic template is composed of a layer of β-chitin layered between "silk-like" glycine- and alanine-rich proteins. The outer surfaces of the template are coated with hydrophilic acidic macromolecules. Natural inorganic-organic hybrid materials are formed through mineralization of inorganic materials on self-assembled organic materials. Biopolymers and low-molecular-weight organic molecules are organized to nanostructures and used as flame-works for specifically oriented shaped inorganic crystals such as calcium carbonate, hydroxyapatite, iron oxide, and silica. These processes use aqueous solution at temperature below 100 °C. Moreover, no toxic intermediates are produced in these systems. Therefore, the processes and materials that control such crystal growth are of great interest to materials scientists who seek to make composite materials and crystalline forms analogous to those produced by nature.

In these mineralized tissues, crystal morphology, size, and orientation are determined by local conditions and, in particular, the presence of "matrix" proteins or other macromolecules [2]. Morphological control can also be accomplished by adsorption of soluble additives onto specific faces of growing crystals, altering the relative growth rates of the different crystallographic faces and leading to different crystal habits. These processes take place usually at an organic-inorganic interface, the organic portion providing the initial structural information for the inorganic part to nucleate on and grow outwards in the desired manner. Most researcher's interest lies in understanding how organized inorganic materials with complex morphological form can be produced by biomineralization processes, and how such complexity can be reproducibly synthesized in biomimetic systems. Due to the complexity of the natural biomineralization systems, mineralization research has been studied on model organic interfaces. Several excellent review papers have already been devoted to biomineralization, usually focusing mainly on inorganic materials by biomimetic processes [3–7]. Recently, several organic and polymer research scientists have been interested in the research area of biomineralization [8, 9]. Various types of

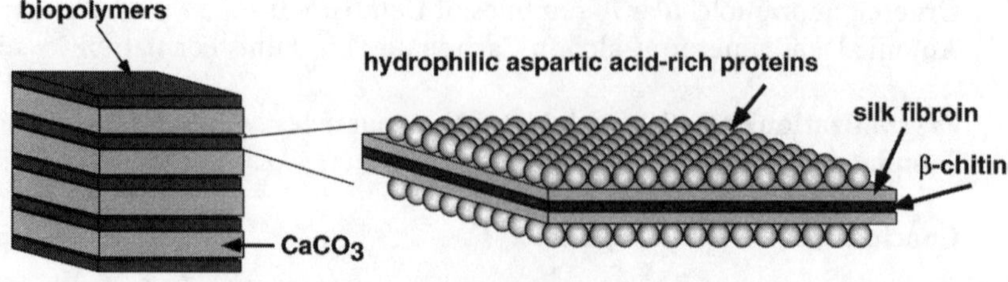

Fig. 1. Schematic illustration of nacre of abalone shell

organic matrices as a structural template for crystallization of inorganic materials should be easily designed and synthesized. Current significant development of so-called "supramolecular chemistry" also provided a highly ordered template for biomineralization.

Calcium carbonate makes up an attractive model mineral for studies in the laboratory, since its crystals are easily characterized and the morphology of $CaCO_3$ has been the subject of control in biomineralization processes. The precipitation of calcium carbonate in aqueous solution is also of great interest for industrial and technological applications. The particular interest in this system is due to the polymorphism of calcium carbonate, which has three anhydrous crystalline forms, i.e., vaterite, aragonite, and calcite in order of decreasing solubility and increasing stability. The three polymorphs have markedly different physicochemical characteristics, and it is often found that less stable forms are stabilized kinetically. Vaterite transforms into the thermodynamically most stable calcite via a solvent-mediated process [10]. Organizations of calcium carbonate crystals in biological systems in the three polymorphs, calcite, vaterite, and aragonite, are well known illustrations of the biomineralization processes. Calcite and aragonite are widespread in marine organisms, and vaterite, mono-hydrocalcite, and amorphous calcium carbonate are formed and stabilized by some organisms [11]. The manner in which organisms control polymorph formation is not well understood.

2
Overview of Calcium Carbonate Crystallization by Synthetic Substrates

An in vitro study of biomineralization provides useful information for design of organic templates. Falini and co-workers assembled in vitro a complex containing the major matrix components present in a mollusk shell, namely β-chitin, silk-fibroin-like protein, and water-soluble acidic macromolecules [12]. When this assemblage was placed in a saturated solution of calcium carbonate, multicrystalline spherulites formed within the complex. They extracted aspartic-acid-rich glycoproteins from an aragonitic mollusk shell layer or a calcitic layer. These were aragonite if the added macromolecules were from an aragonitic shell layer, or calcite if they were derived from a calcitic shell layer. In the absence of the acidic glycoproteins, no mineral formed within the complex. Since the high-dimensional structures of these proteins were unknown, detail mechanism of the effect of nucleation at molecular level remains unknown. Even though, these in vitro study motivate us to design artificial templates for controlled nucleation of minerals.

Model systems, in which low-molecular-weight organic additives are used to study the effect of molecular properties such as charge and functionality on inorganic crystallization, are providing insights into the possible mechanisms operating in biology. These additives were chosen to mimic the active protein ligands. The influence of organic molecules on the nucleation and crystal growth of calcium carbonate has been studied by a number of authors [13–17]. Because the proteins that have been found to be associated with biominerals are usually highly acidic macromolecules, simple water-soluble polyelectrolytes, such as the sodium salts of poly(aspartic acid) and poly(glutamic acid), were examined for the model of biomineralization in aqueous solution. Studies of inorganic crystallization in the presence of soluble polymers, modeled on biogenic proteins, have shown that selectivity for certain crystal faces appears to be highly dependent on the secondary structure of the macromolecules. For example, poly-L-asparatate, with a predominantly β-sheet conformation, produces more argonite than poly-L-glutamate, which has a random conformation [18]. Crystallization of $CaCO_3$ in the presence of various synthetic non-peptide polymers has been investigated as a model of biomineralization [8, 9].

3
Dendrimers

For synthetic linear polymers, it has been difficult to assign unambiguously structure-function relationship in the context of their activity in crystallization assays, since they mostly occur in random-coil conformation. Dendrimers are monodisperse macromolecules with a regular and highly branched three-dimensional architecture. As shown in Fig. 2, the starburst structures are disk-like shapes in the early generations, whereas the surface branch cell becomes substantially more rigid and the structures are spheres [19]. Due to unique and well-defined secondary structures of the dendrimers, the starburst dendrimers should be a good candidate for studying inorganic crystallization. The interaction of the negative surface of dendrimers with metal ions has been extensively examined [20]. Dendrimers were used as templates for formation of metal nanoparticles [20–22]. It was found that stable gold nanoparticles with diameters less than 1 nm are prepared in the presence of the later generation dendrimers, and the dendrimer concentrations required for obtaining stable gold nanoparticles are extremely low compared to those of other linear polymers. The following sections show recent research on crystal nucleation and growth of calcium carbonate by a carboxylic acid derivative of hyperbranched polyphenylene polymer, poly(amidoamine) (PAMAM) dendrimers with carboxylate groups at the external surface, and poly(propyleneimine) dendrimers modified with long aliphatic chains.

3.1
Hyperbranched Polymers

A carboxylate derivative of a fully aromatic, water-soluble, hyperbranched polyphenylene is considered as a "unimolecular micelle" due to its ability to complex and solubilize non-polar guest molecules [23]. The carboxylic acid derivative of hyperbranched polyphenylene polymer (HBP) (M_w=5750–7077, M_n=3810–3910) consists of 40–60 phenyl units that branch outward from a central point forming a roughly spherical molecule with carboxylates on the outer surface. The free acid form of HBP was suspended in distilled water and dissolved by adding a minimum quantity of NaOH. The solution was adjusted to pH 6.2 with aqueous HCl. Calcium carbonate crystals were growth from supersaturated calcium hydrogencarbonate solution at room temperature. HBP gave

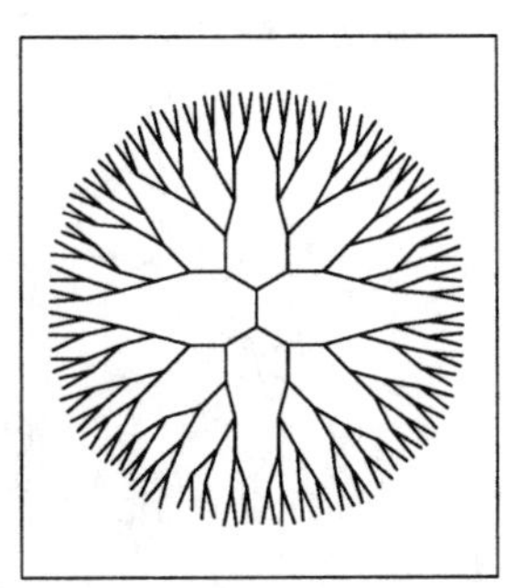

Fig. 2. Poly(amidoamine) (PAMAM) dendrimers with carboxylate groups at the external surface

**Carboxylic acid derivative of hyperbranched
polyphenylene polymer (HBP)**

oriented nucleation owing to partial segregation of the macromolecule at the air/water interface [13]. HBP exhibited distinct surface activity at the air/water interface. The polymer has a roughly spherical shape and is relatively rigid. When segregated at the air/water interface, the molecules appeared to promote calcite nucleation in a similar manner to previous crystallization experiments carried out under compressed monolayers of stearic acid [24].

3.2
Anionic Dendrimers

Poly(amidoamine) (PAMAM) dendrimers with carboxylate groups at the external surface termed half-generation or G=n.5 dendrimers have been proposed as mimics of anionic micelles or proteins [25]. Later than 4-generation of the PAMAM dendrimers are nearly spherical in shape according to molecular simulations [26]. The half-generations PAMAM dendrimers using various photophysical probes indicate a structural surface transition in the dendrimer appearance on going from generation 3.5 to 4.5 [27]. The morphology of the early generation (0.5–3.5) dendrimers is an open structure, and the internal tertiary amines and amide groups are available for bonding. The later generations have more spheroidal and close-packed surface structures. Recently, crystallization of $CaCO_3$ in the presence of anionic PAMAM dendrimers has been extensively studied [28–30].

3.2.1
Crystallization of CaCO₃ with the Anionic PAMAM Dendrimers by the Double Jet Method

The precipitation of $CaCO_3$ in the presence and the absence of the PAMAM dendrimer (G=1.5) was carried out in a double jet reactor [31] to prevent heterogeneous nucleation at the glass walls as shown in Fig. 3. The two reactants ($CaCl_2$ and Na_2CO_3) are injected via capillaries into the reaction vessel under vigorous stirring to prevent heterogeneous nucleation at the glass wall. The two capillary

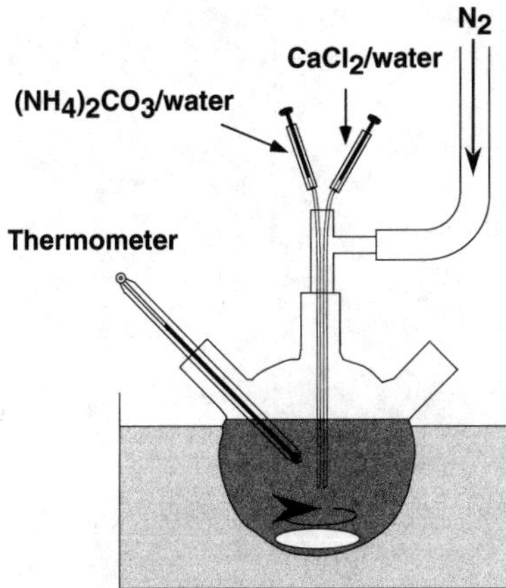

Fig. 3. The experimental set-up of a double jet reactor for the precipitation of calcium carbonate

ends are joined together so that a high local reactant concentration and thus extreme supersaturation is achieved at the moment when the two reactants leave the capillaries, which provides an immediate nucleation of $CaCO_3$. The nuclei are then immediately transported to regions of lower $CaCO_3$ concentration and can grow further. The $CaCO_3$ crystal formation occurring after excess addition of reactants was easily observed as a sudden increase in the turbidity of the solution. The main idea behind this technique, which was set up for the controlled precipitation of silver halides in the photographic industry, is to maintain a rapid nucleation of a constant particle number in the beginning of the experiment to enable growth of monodispersed particles. A crystallization of $CaCO_3$ in the presence of the PAMAM dendrimers with carboxylate groups at the external surface resulted in the formation of spherical vaterite crystals (Fig. 4a), whereas rhombohedral calcite crystal was formed without the additive (Fig. 4b) [28, 29]. In the presence of the dendrimers, further washing of the vaterite crystal with water did not change the crystal morphology.

The yields of the crystalline products in the presence of the G1.5 and G3.5 dendrimers were 61 and 30%, respectively. Due to the complexation of the dendrimer with calcium ions in aqueous solutions, the saturated concentration of calcium ions in the presence of the G1.5 and G3.5 PAMAM dendrimers were 1.3 and 2.8 times higher than that in the absence of the additive. This result indicates that the higher generation of the dendrimer acts as an inhibitor for crystal formation. The inner cores of the PAMAM dendrimers are hydrophilic and potentially open to small hydrophilic molecules, since interior nitrogen moieties serve as complexation sites. The amount of calcium ions on the anionic starburst dendrimers is considerably higher for the higher generations than for the lower generations.

The precipitation of $CaCO_3$ in the presence of a Na-salt of poly(acrylic acid) (PAA) (M_n=5100) was also carried out under the same condition described above. In the presence of the Na-salt of poly(acrylic acid) (PAA), the formation

Fig. 4a,b. Scanning electron micrographs of the crystalline products: **a** in the presence of PAMA dendrimer (G=1.5); **b** in the absence of PAMA dendrimer (G=1.5) (reproduced from [28]

of crystalline $CaCO_3$ was prevented. The crystalline $CaCO_3$ was hardly collected after incubation at 25 °C under N_2 for four days. This indicates that PAA acts as an inhibitor for crystal formation.

Vaterite transforms easily and irreversibly into thermodynamically more stable forms when in contact with water. The complete phase transformation into the thermodynamically stable calcite occurs within 80 h, usually much faster under the conditions described above [31]. It is well known that vaterite transforms into stable calcite via a solvent-mediated process [32]. Under the precipitation condition, the extreme supersaturation is achieved at the moment and provides an immediate nucleation of $CaCO_3$, which is not affect by the organic additives. If the Ca-O bonds of the polymer ligand are easily dissociated by water, the polymer is thought not to be occluded in the $CaCO_3$ crystal. Then the solvent-mediated vaterite-calcite transformation might be performed. Strong Ca-O bonds are thus required to control the polymorph of $CaCO_3$. In the presence of the dendrimers, the vaterite crystal with incubation in water for one week did not change the crystal morphology, which indicates that the vaterite surface was stabilized by the carboxylate-terminated PAMAM dendrimer to prevent phase transformation to calcite in an aqueous solution.

The complexes of anionic starburst dendrimers with calcium ions are considerably stronger for higher generations than for lower generations. The particle sizes of the spherical vaterite crystals obtained in the presence of the PAMAM dendrimers were depended on the generation numbers and concentration of the PAMAM dendrimers [29]. As the generation number of the PAMAM dendrimer increased from G1.5 to G3.5, the particle size of the spherical vaterite was decreased from 5.5±1.1 to 2.3±0.7 μm. Further increase in the generation number to G4.5, the particle size was not changed (2.3±0.8 μm). At the lower concentration of the G1.5 PAMAM dendrimer corresponding to 0.13 mmol/l of –COONa, the particle size of the vaterite crystals was 5.6±1.4 μm. As the concentration of –COONa increased from 0.26 to 8.33 mmol/l, the particle sizes of the spherical vaterite were reduced from 5.5±1.1 to 2.5±0.6 μm. In the case of the G3.5 PAMAM dendrimer, the concentration of –COONa increased from 0.13 to 0.26 mmol/l, the particle size also decreased from 5.8±1.8 to 2.3±1.8 μm. With further increase of the concentration to 8.33 mmol/l, the particle size decreased to 1.5±0.6 μm. The particle sizes by the lower generation (G1.5) were larger than that by the higher generation (G3.5) at any concentration of the PAMAM dendrimers.

The spherical vaterite crystals were aggregates of vaterite nanoparticles with diameters of 10 to 30 nm [31]. Formation of the vaterite particles involves two processes, i.e., the nucleation of vaterite nanocrystals and the aggregation of the nanocrystals. The nanoparticle surface might be covered with the PAMAM dendrimer. Complexation ability of the higher generation of the PAMAM dendrimer is stronger than that of the lower generation of the PAMAM dendrimer [25]. Aggregation of the nanoparticles in the presence of the higher generation of the G3.5 PAMAM dendrimer might be prevented compared to that in the presence of the lower generation of the G1.5 PAMAM dendrimer.

Vaterite is thermodynamically most unstable in the three crystal structures. Vaterite, however, is expected to be used in various purposes, because it has some features such as high specific surface area, high solubility, high dispersion, and small specific gravity compared with the other two crystal systems. Spherical vaterite crystals have already been reported in the presence of divalent cations [33], a surfactant [bis(2-ethylhexyl)sodium sulfate (AOT)] [32], poly(styrenesulfonate) [34], poly(vinylalcohol) [13], and double-hydrophilic block copolymers [31]. The control of the particle size of spherical vaterite should be important for application as pigments, fillers and dentifrice.

3.2.2
Crystallization of CaCO₃ with the Anionic PAMAM Dendrimers by Carbonate Diffusion Method

Crystallization of $CaCO_3$ is highly dependent on nucleation condition. The precipitation of $CaCO_3$ in the absence or the presence of the G4.5 PAMAM dendrimer was carried out by a "carbonate diffusion method" similar to the method described by Addadi et al. [35]. A solution of the dendrimer with calcium chloride in 200 ml of distilled water was adjusted to pH 8.5 with aqueous NH_3, and then placed in a closed desiccator containing crushed ammonium carbonate (Fig. 5). Carbon dioxide was introduced to the solution via vapor diffusion. The critical point of the appearance in the turbidity of the solution was observed at around 5 min. These solutions were kept at 30 °C under N_2 for one day. The crys-

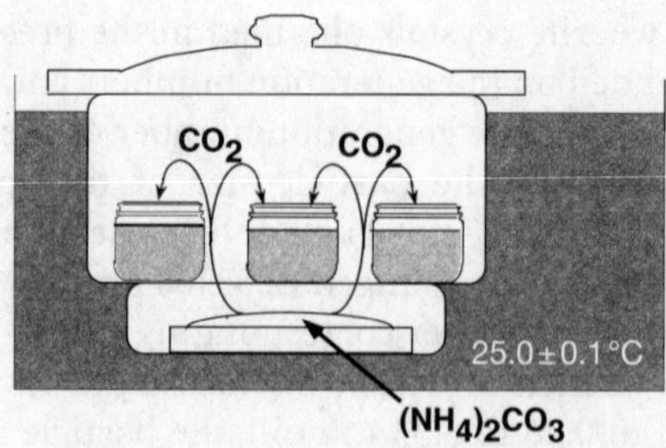

Fig. 5. The experimental set-up of a carbonate diffusion method for the precipitation of calcium carbonate

talline $CaCO_3$ was collected and washed with water several times to remove contaminated dendrimers that were not involved in the crystal. Formation of the crystalline $CaCO_3$ at different feed ratio of the G4.5 PAMAM dendrimer to calcium ions was studied with the constant concentration of calcium ions at 0.1 mol/l. The results are summarized in Table 1. At the low concentration of the G4.5 PAMAM dendrimer (runs 2 and 3), calcite was predominantly formed. However, as the concentration of –COONa increased to 5.3 mmol/l, vaterite formation strongly appeared. Further increase of the concentration of –COONa to 10.6 mmol/l also formed vaterite. Although calcite was predominantly formed at the concentration of the G4.5 PAMAM dendrimer corresponding to 2.65 mmol/l of –COONa (run 3), the crystal phase of the obtained $CaCO_3$ at the higher concentration of the G4.5 PAMAM dendrimer at 5.3 mmol/l consisted entirely of vaterite (98% by IR). These results indicate that the presence of the dendrimer affected polymorphs of $CaCO_3$ crystallization.

The precipitation of $CaCO_3$ in the absence of any additives was carried out under the same nucleation condition (run 1 in Table 1). The crystal phase of the obtained $CaCO_3$ was a mixture of calcite and vaterite by IR. The vaterite content

Table 1. The precipitation of $CaCO_3$ in the absence and the presence of the G4.5 PAMAM dendrimer (adapted from [30]

Run	[–COONa] (mmol/l)	[Ca^{2+}] (mol/l)	[–COONa]/ [Ca^{2+}]	Polymorphs[a]	Vaterite content[a] (%)
1	0	0.1	0	Calcite+Vaterite	56
2	0.53	0.1	0.0053	Calcite>>Vaterite	15
3	2.65	0.1	0.027	Calcite	0
4	5.3	0.1	0.053	Vaterite>>>Calcite	98
5	10.6	0.1	0.11	Vaterite>>>Calcite	93
6	0	0.05	0	Calcite+Vaterite	70
7	0.53	0.05	0.011	Calcite	0
8	1.33	0.05	0.027	Calcite>>>Vaterite	13
9	2.65	0.05	0.053	Vaterite>>>Calcite	89
10	5.3	0.05	0.11	Vaterite>>>Calcite	93

[a] Polymorphism and the fraction of vaterite were characterized by FTIR.

was slightly higher than the calcite content. However, calcite was predominantly formed at the low concentration of the G4.5 PAMAM dendrimer corresponding to 0.53 and 2.65 mmol/l of –COONa (runs 2 and 3). Due to the complexation of the PAMAM dendrimer with calcium ions in aqueous solutions, crystallization of calcium carbonate in the presence of the dendrimer is inhibited compared with that in the absence of the additive with constant concentration of calcium ions [29]. The interior nitrogen moieties of the dendrimers serve as complexation sites. According to the literature, relatively high supersaturations in high pH values favor the precipitation of vaterite [36]. The presence of the dendrimer might decreased the concentration of free calcium ions which were not bound to the dendrimer, resulting in the higher calcite contents compared with that in the absence of the dendrimer.

SEM observations showed that the most crystals obtained in the absence and the low concentration of the G4.5 PAMAM dendrimer were rhombohedral. In the high concentration of the G4.5 PAMAM dendrimer, the vaterite products were spherical. Each shape of $CaCO_3$ is a typical morphology for each polymorph. The particle sizes of the spherical vaterite particles obtained in the presence of the PAMAM dendrimer were depended on the concentration of the dendrimer. As the concentration of –COONa increased from 5.3 to 10.6 mmol/l, the particle sizes of the spherical vaterite particles were reduced from 8.7±1.0 to 5.2±3.0 µm.

The results of the precipitation of $CaCO_3$ in the absence and the presence of the G4.5 PAMAM dendrimer at a lower concentration of calcium ion (0.05 mol/l) are also summarized in Table 1. Although the vaterite content was higher than the calcite content in the absence of the dendrimer, calcite was predominantly formed at the concentration of the G4.5 PAMAM dendrimer corresponding to 0.53 and 1.33 mmol/l of –COONa. As the concentration of –COONa increased to 2.65 mmol/l, vaterite formation was strongly induced. At the higher concentration of calcium ions (0.1 mol/l), the critical point of the morphology change from calcite to vaterite was observed as the concentration of –COONa increased from 2.65 to 5.3 mmol/l. These results indicate that the feed ratio of –COONa and Ca^{2+} is a key factor for inducing vaterite formation.

The precipitations of $CaCO_3$ in the presence of the G1.5, G3.5, and G4.5 PAMAM dendrimers were carried out with constant –COONa unit and calcium ions of 0.1 mol/l. Although vaterite was predominantly formed by the G4.5 dendrimer, relatively high amount of calcite was observed in the case of the G3.5 and G1.5 dendrimers (Table 2). These results suggest that the G4.5 dendrimer effec-

Table 2. The precipitation of $CaCO_3$ in the presence of the G1.5, G3.5, and G4.5 PAMAM dendrimers (Adapted from [30])

Run	Generation	[-COONa] (mmol/l)	[Ca^{2+}] (mol/l)	[Ca^{2+}]/ [-COONa]	Polymorphs[a]	Vaterite content[a] (%)
1	1.5	5.3	0.1	0.053	Vaterite>Calcite	77
2	3.5	5.3	0.1	0.053	Vaterite+Calcite	48
3	4.5	5.3	0.1	0.053	Vaterite>>>Calcite	98

[a] Polymorphism and the fraction of vaterite were characterized by FTIR.

tively induces vaterite formation compared with the earlier generation of the dendrimers. When precipitation of $CaCO_3$ in the presence of the G1.5, G3.5, and G4.5 PAMAM dendrimers was carried out by the double jet method as described above, in all cases stable vaterite particles were obtained, in contrast to the present results. Under the previous precipitation condition, the extreme supersaturation is achieved at the moment and provides an immediate nucleation of $CaCO_3$, which is not affected by the organic additives. The nuclei are then immediately transported to regions of lower $CaCO_3$ concentration and can grow further. The spherical vaterite crystals were stabilized by the anionic PAMAM dendrimers in aqueous solution for more than seven days. Vaterite transforms easily and irreversibly into thermodynamically more stable forms when in contact with water. These results suggested that the surface of the vaterite particles in the previous case was stabilized by the dendrimers to avoid water contact. On the other hand, the vaterite particles obtained by the carbonate diffusion method were transformed to calcite when the solution was incubated for four days. Although the dendrimer content in the vaterite particle was 7.27 wt% as determined by elemental analysis, these results indicate that the vaterite surface was not effectively stabilized by the PAMAM dendrimer. Under the present precipitation condition, the dendrimers can play a role in initiating the nucleation of vaterite. Alternatively, the vaterite growth could be explained by a kinetic inhibition of the calcite nuclei by the dendrimers.

The protonation behavior of the three generations of the PAMAM dendrimers in 1.0 mol/l KCl is presented in Fig. 6. For the early generation dendrimers, the rather pronounced difference in dissociation constant between carboxylic groups and amines yields two distinct steps in their titration curves. Only around pH 4 is the carboxylic groups involved in protonation [37]. The titration curves of the G3.5 and G4.5 dendrimers shifted downward in the high pH region, in which the present crystallization took place, compared with that of the G1.5 dendrimer, suggesting that the deprotonation of the ammonium groups of the later generation is easier than that of the early generation by electrostatic repulsions between neighboring amine groups due to their close-packed structures. This indicates that the tertiary amine groups of the later generation can act as effective bonding sites for cations compared with those of the early generation in the high pH region. Although the amide groups also coordinate with calcium ions, the presence of the internal ammonium groups in the early generation might inhibits the complexation with cations. In the same –COONa concentration, the amount of calcium ions entrapped in the PAMAM dendrimer was higher in the G3.5 dendrimer than that in the G1.5 dendrimer. The titration curve of the G4.5 dendrimer shows no distinct step corresponding to the protonation of the carboxylic groups, suggesting electrostatic repulsions between neighboring carboxylic groups due to the close-packed surface structure.

The titration curves of the dendrimers in the presence of calcium ions are also shown in Fig. 6. In the presence of calcium ions, the titration curve of the G1.5 dendrimer shifted downward, suggesting chelate formation. Both titration curves of the G3.5 dendrimers in the absence and presence of calcium ions are the same, especially in the high pH region. The protonation of the G4.5 dendrimer, however, was promoted in the presence of calcium ions. These results

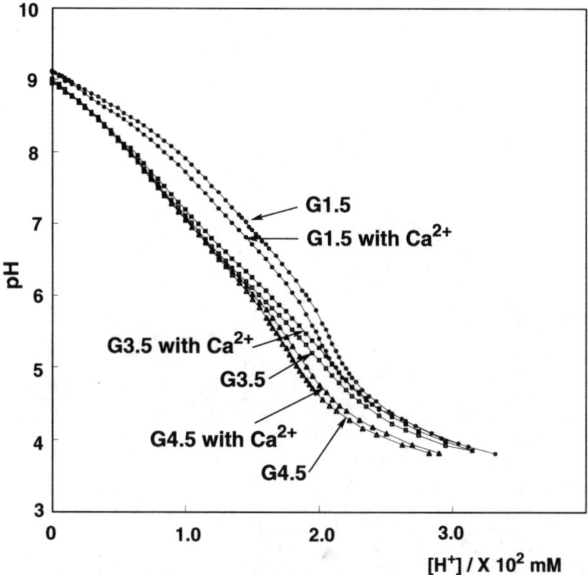

Fig. 6. Potentiometric titrations of the G1.5, G3.5, and G4.5 PAMAM dendrimers in the absence (*open symbols*) and the presence (*close symbols*) of calcium ions in 1.0 mol/l KCl solution with 0.01 mol/l HCl solution. The –COONa unit of the PAMAM dendrimers was constant at 3.0×10^{-2} mmol (adapted from [30]

indicate that coordination between the divalent cations and the coordination sites of the G4.5 dendrimer was weaker than that with the univalent cations due to reduced flexibility of the structure compared with the early generations. The present mineralization results showed a transition in polymorphous change of $CaCO_3$ occurring between G3.5 and G4.5. Although the calcite content increased with increasing the generation from 1.5 to 3.5, the G4.5 dendrimer effectively induced vaterite formation. Complexation properties of the G4.5 dendrimers due to the stereochemical factor would be a major role for the vaterite mineralization.

3.3
Ordered Supramolecular Assemblies of Dendrimers

The use of ordered supramolecular assemblies, such as micelles, monolayers, vesicles, inverted micelles, and lyotropic liquid crystalline systems, allows for the controlled nucleation of inorganic materials on molecular templates with well-defined structure and surface chemistry. Poly(propyleneimine) dendrimers modified with long aliphatic chains are a new class of amphiphiles which display a variety of aggregation states due to their conformational flexibility [38]. In the presence of octadecylamine, poly(propyleneimine) dendrimers modified with long alkyl chains self-assemble to form remarkably rigid and well-defined aggregates. When the aggregate dispersion was injected into a supersaturated

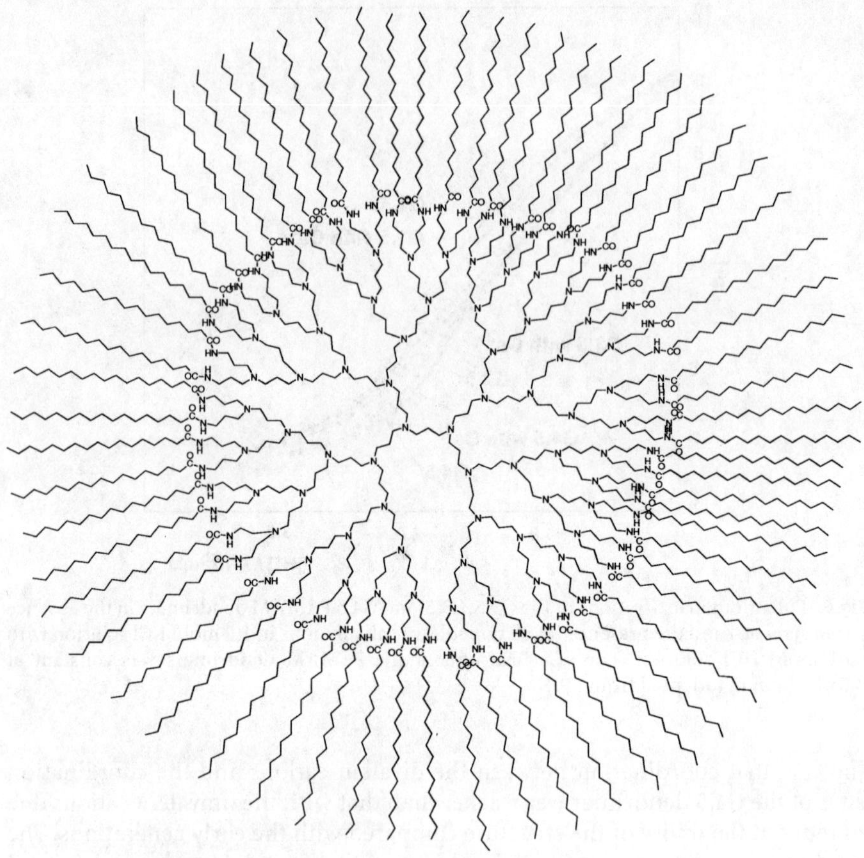

solution of calcium carbonate, the aggregates stabilized the normally unstable amorphous calcium carbonate [39]. Over the course of four days a discrete population of rhombohedral calcite crystals was also identified, alongside large isolated amorphous calcium carbonate particles. This process might affect the gradual release of the aggregates. Although amorphous calcium carbonate has been observed as a short-lived intermediate in the presence of various scale prevention agents, only biological systems allow the coexistence of amorphous and crystalline calcium carbonate for extended periods of time. Stabilization of the amorphous phase against transformation to the thermodynamically stable form results in the formation of a unique inorganic-organic hybrid material.

The geometry and surface chemistry of the dendrimer assemblies can be varied through the addition of surfactants. These dendrimer/surfactant aggregates can be tuned to template the formation of the different phases of calcium carbonate [40]. In combination with hexadecyltrimethylammonium bromide (CTAB), small spherical aggregates were formed that induce the formation of vaterite. Over a period of five days, the vaterite was transformed into calcite. The use of the negatively charged surfactant, sodium dodecylsulfonate (SDS), resulted in growth retardation of dendrimer/surfactant aggregates into giant spherical with diameters of 1–20 µm. Eventually these particles become overgrowth by rhombohedral calcite.

3.4
Anionic Dendrimer for Calcium Carbonate Thin Films Formation

An in vitro precipitation of calcium carbonate in the form of a thin film has been of considerable interest. Ceramic films are also in demand as active and passive components in microelectronic circuits as capacitors, memories, and insulating or passivating layers. Conventional ceramic processing, which involves high-temperature sintering, cannot be used for ceramic coatings on plastics. With biomimetic processing, high-quality, oriented, and patterned ceramic films can be deposited on plastics and other materials at temperature below 100 °C. Biomimetic film-formation can produce materials of both industrial and scientific interest. The solution nucleation and growth of thin films of iron hydroxides onto sulfonated polystyrene substrates and sulfonated self-assembling monolayers attached to oxidized silicon was achieved [41]. Biomimetic processing of inorganic thin films except $CaCO_3$ has already been reviewed [42, 43].

Several factors could influence mineral nucleation and crystal growth on a polymer film, such as the degree of saturation of a supersaturated solution and the surface charges on a polymer film. It is very crucial for successful biomimetic synthesis that only heterogeneous nucleation is promoted on the polymer film and that homogeneous nucleation is suppressed in the mother liquid. One key factor for successfully mimicking biomineralization to coat a dense ceramic film on organic polymer substrates is to increase the density of charges or polarity on the substrate surface. Poly(acrylic acid) (PAA) proved to be an effective additive, promoting calcium carbonate heterogeneous nucleation on chitosan-film surfaces and suppressing homogeneous nucleation in solution [44]. In the presence of poly(acrylic acid), calcium carbonate crystals heterogeneously nucleated and grew only on the chitosan-film surface and covered the whole film. In the absence of PAA, homogeneous nucleation occurred in the supersaturated calcium carbonate solution. Gower and Tirrell reported that a highly birefringent $CaCO_3$ film formed in streaks and patches on the glass substrate by addition of poly(asparatate) [45]. Kato and Amamiya have independently introduced a series of $CaCO_3$ thin film formation on organic substrates [46]. Thin film states of $CaCO_3$ crystals have been obtained as organic/inorganic composites with chitosan that acts as a solid matrix in the presence of poly(acrylic acid) or poly(glutamic acid) as a soluble additive. The template/inhibition strategy for the synthesis of $CaCO_3$ thin film has been also employed independently by Groves's group [47, 48]. They reported the synthesis of macroscopic and continuous calcium carbonate thin films at a porphyrin template/subphase interface by employing poly(acrylic acid) as a soluble inhibitor to mimic the cooperative promotion-inhibition in biogenic thin film production.

Anionic poly(amidoamine) (PAMAM) dendrimer was selected as a model of the soluble acidic-rich proteins to prepare $CaCO_3$ film on a poly(ethylenimine) film [49]. The $CaCO_3$/poly(ethylenimine) composite film was obtained in the presence of anionic PAMAM dendrimer (G=3.5), whereas the formation of composite film was not observed without PAMAM dendrimer or with PAMAM dendrimer of lower generation (G=1.5). PAMAM dendrimer with calcium ion was adsorbed on the poly(ethylenimine) surface through an insoluble polymer complex. In the absence of PAMAM dendrimer, poly(ethylenimine) was dissolved in an aqueous phase. The complexes of anionic dendrimers with calcium ions are considerably stronger for the higher generations than the lower generations. The amount of calcium ions entrapped in the internal coordination site of PAMAM dendrimer with higher generation might be higher than that with lower generation. Higher generation of PAMAM dendrimer (G=3.5) inhibits the crystallization of $CaCO_3$ in the solution. The adsorption of PAMAM dendrimer on poly(ethylenimine) film caused high local concentration of calcium ion.

4
Crystallization of CaCO₃ with Metal Nanoparticles as Spherical Templates

The discovery that thiols of long chain organic molecules bind strongly and specifically to metal (usually mercury, silver, or gold) surfaces forming ordered monolayers, the self-assembled monolayers (SAM), has yielded an interesting class of molecular assemblies that permit the production of highly specific interfaces spread over relatively large areas. The SAM surfaces can be modified through the organic functionality at the end of the molecule, usually the ω functionality on a long-chain thiol. The chemistry of the thiol SAM on the metal surfaces has been extended to metal colloids. Tremel and his co-workers used *p*-mercaptophenol-protected gold colloids as templates for the growth of inorganic crystallization [50]. Going from flat surfaces to colloids in solution introduces some interesting new aspects. First, the crystallization is carried out heterogeneously (at an interface) in a homogeneous solution. Second, the templating substrate is changed from two-dimensional to the curved two-dimensional. When the crystallization experiments were carried out in the presence of *p*-mercaptophenol-protected gold colloids by the carbonate diffusion method, the initial deep-pink solutions of protected gold colloids turned completely colorless at the end of the crystallization. This observation indicates that the precipitation of $CaCO_3$ in some way resulted in an entrapment of the colloids. Comparing flat templates, they observed significant differences in the nature of the crystallization products. An important issue in using a small, round template is that the crystals mutually frustrate the growth of one another in the direction tangential to the growing spherical.

Spherical vaterite crystals were obtained with 4-mercaptobenzoic acid protected gold nanoparticles as the nucleation template by the carbonate diffusion method [51]. The crystallization of calcium carbonate in the absence of the 4-MBA capped gold nanoparticles resulted in calcite crystals. This indicates that the polymorphs of $CaCO_3$ were controlled by the acid-terminated gold nanoparticles. This result indicates that the rigid carboxylic acid structures can play a role in initiating the nucleation of vaterite as in the case of the G4.5 PAMAM dendrimer described above.

5
Conclusions

Over the past decade, considerable work has been carried out on mineralization of calcium carbonate in the presence of various low-molecular-weight organic molecules and polymers as templates and additives. These interesting efforts have led to fundamental developments in areas relating to the biomineralization process. However, there still remain many unknowns as to how natural organisms produce inorganic-organic hybrid materials. Since most synthetic polymers are not monodisperse molecules with random-coil conformations, the use of dendrimers allows one to assign structure-function relationships in crystallization assays. Although the template mechanism of the dendrimers still seems to be of a complex nature because simultaneous $CaCO_3$ nucleation and interaction with the polymer can be expected, the stereochemical factor is important for the affinity of the templates to manipulate the polymorph of $CaCO_3$. I expect that the continuous cooperation of organic and polymer chemists with inorganic and biochemists, which is desirable to clarify the biomineralization process, will lead to the next industrial revolution.

6
References

1. Smith BL (1999) Nature 399:761
2. Addadi L, Weiner S (1985) Proc Natl Acad Sci USA 82:4110
3. Mann S (1997) J Chem Soc Dalton Trans 3953
4. Mann S (1988) Nature 332:119
5. Mann S (1993) Nature 365:499
6. Mann S, Ozin G (1996) Nature 382:313
7. Stupp SI, Braun PV (1997) Science 277:1242
8. Naka K, Chujo Y (2001) Chem Mater 13:3245
9. Estroff LA, Hamilton AD (2001) Chem Mater 13:3227
10. Kralj D, Brecevic L, Nielsen AE (1990) J Cryst Growth 104:793
11. Lowenstam HA (1981) Science 211:1126
12. Falini G, Albeck S, Weiner S, Addai L (1996) Science 271:67
13. Didymus JM, Oliver P, Mann S, Devries AL, Hauschka PV, Westbroek P (1993) J Chem Soc Faraday Trans 89:2891
14. Mann S, Didymus JM, Sanderson NP, Heywood BR, Samper EJA (1990) J Chem Soc Faraday Trans 86:1873
15. Geffroy C, Foissy A, Persello J, Cabane B (1999) J Colloid Interface Sci 211:45
16. Ogino T, Tsunashima N, Suzuki T, Sakaguchi M, Sawada K (1988) Nippon Kagaku Kaishi 6:899
17. Keum DK, Kim K-M, Naka K, Chujo Y (2002) J Mater Chem 12:2449
18. Levi Y, Albeck S, Brack A, Weiner S, Addadi L (1998) Chem Eur J 4:389
19. Ottaviani MF, Bossmann S, Turro NJ, Tomalia DA (1994) J Am Chem Soc 116:661
20. Balogh L, Tomalia DA (1998) J Am Chem Soc 120:7355
21. Esumi K, Suzuki A, Aihara N, Usui K, Torigoe K (1998) Langmuir 14:3157
22. Esumi K, Suzuki A, Yamahira A, Torigoe K (2000) Langmuir 16:2604
23. Kim YH, Webster OW (1990) J Am Chem Soc 112:4592
24. Rajam S, Heywood BR, Walker JBA, Mann S, Davey RJ, Birchall JD (1991) J Chem Soc Faraday Trans 87:727
25. Tomalia DA, Naylor AM, Goddard WA III (1990) Angew Chem Int Ed Engl 29:138

26. Naylor AM, Goddard WA III, Kiefer GE, Tomalia DA (1989) J Am Chem Soc 111:2341
27. Caminati G, Turro NJ, Tomalia DA (1990) J Am Chem Soc 112:8515
28. Naka K, Tanaka Y, Chujo Y, Ito Y(1999) Chem Commun 1931
29. Naka K, Tanaka Y, Chujo Y (2002) Langmuir 18:3655
30. Naka K, Kobayashi A, Chujo Y (2002) Bull Chem Soc Jpn 75:2541
31. Cölfen H, Antonietti M (1998) Langmuir 14:582
32. Lopezmacipe A, Gomezmorales J, Rodriguezclemente R (1996) J Cryst Growth 166:1015
33. Brecevic L (1996) J Chem Soc Faraday Trans 92:1017
34. Kawaguchi H, Hirai H, Sakai K, Nakajima T, Ebisawa Y, Koyama K (1992) Colloid Polym Sci 270:1176
35. Addadi L, Moradian J, Shay E, Maroudas NG, Weiner S (1987) Proc Natl Acad Sci USA 84:2732
36. Spanos N, Koutsoukos PG (1982) J Phys Chem B 102:6679
37. van Duijvenbode RC, Rajanayagam A, Koper GJM, Baars MWPL, de Waal BFM, Meijer EW, Borkovec M (2000) Macromolecules 33:46
38. Stevelmans S, van Hest CJM, Jansen JFGA, van Boxtel DAFJ, de Brabander-van den Berg EM, Meijer EW (1996) J Am Chem Soc 118 7398
39. Donners JJJM, Heywood BR, Meijer EW, Notle RJM, Roman C, Schenning APHJ, Sommerdijk NAJM (2000) Chem Commun 1937
40. Donners JJJM, Heywood BR, Meijer EW, Notle RJM, Schenning APHJ, Sommerdijk NAJM (2002) Chem Eur J 8:2561
41. Tarasevich BJ, Rieke PC, Liu J (1996) Chem Mater 8:292
42. Bunker BC, Rieke PC, Tarasevich BJ, Campbell AA, Fryxell GE, Graff GL, Song L, Liu J, Virden JW, McVay GL (1994) Science 264:48
43. Aksay IA, Trau M, Mann S, Honma I, Yao N, Zhou L, Fenter P, Eisenberger PM, Gruner SM (1996) Science 273:892
44. Zhang S, Gonsalves KE (1998) Langmuir 14:6761
45. Gower LA, Tirrell DA (1998) J Crystal Growth 191:153
46. Kato T, Amamiya T (1999) Chem Lett 199
47. Lahiri J, Xu G, Dabbs DM, Yao N, Aksay IA, Groves JT (1997) J Am Chem Soc 119:5449
48. Xu G, Yao N, Aksay IA, Groves JT (1998) J Am Chem Soc 120:11,977
49. Tanaka Y, Nemoto T, Naka K, Chujo Y (2000) Polym Bull 45:447
50. Küther J, Seshadri R, Nelles G, Assenmacher W, Butt H-J, Mader W, Tremel W (1999) Chem Mater 11:1317
51. Lee I, Han SW, Choi HJ, Kim K (2001) Adv Mater 13:1617

Top Curr Chem (2003) 228: 159–191
DOI 10.1007/b11010

Luminescent Dendrimers. Recent Advances

Vincenzo Balzani · Paola Ceroni · Mauro Maestri · Christophe Saudan
Veronica Vicinelli

Dipartimento di Chimica "G. Ciamician", Università di Bologna, Via Selmi 2,
40126 Bologna, Italy. *E-mail: vbalzani@ciam.unibo.it*

Luminescent dendrimers are currently attracting much attention since coupling luminescence and dendrimer research topics can lead to valuable new functions. Indeed, luminescence is a valuable tool to monitor both basic properties and possible applications (sensors, displays, lasers), and dendrimers are macromolecular compounds exhibiting a well-defined chemical structure with the possibility of containing selected chemical units in predetermined sites and of encapsulating ions or neutral molecules in their internal dynamic cavities. In this paper we will review recent advances in this field focusing our attention on their properties in fluid solution related to light harvesting, changing the "color" of light, sensing with signal amplification, quenching and sensitization processes, shielding effects, elucidation of dendritic structures and superstructures, and investigation of dendrimer rotation in solution.

Keywords. Dendrimers, Luminescence, Photochemistry, Light-harvesting, Sensors

1
Introduction

Luminescence can be defined as the emission of light (intended in the broader sense of ultraviolet, visible, or near infrared radiation) by electronic excited states of atoms or molecules. Luminescence is an important phenomenon from a basic viewpoint (e.g., for monitoring excited state behavior) [1] as well as for applications (lasers, displays, sensors, etc.) [2, 3].

Dendrimers are complex but well-defined chemical compounds, with a tree-like structure, a high degree of order, and the possibility of containing selected chemical units in predetermined sites of their structure [4]. Dendrimer chemistry is a rapidly expanding field for both basic and applicative reasons [5]. From a topological viewpoint, dendrimers contain three different regions: core, branches, and surface. Luminescent units can be incorporated in different regions of a dendritic structure and can also be noncovalently hosted in the cavities of a dendrimer or associated at the dendrimer surface as schematically shown in Fig. 1 [6].

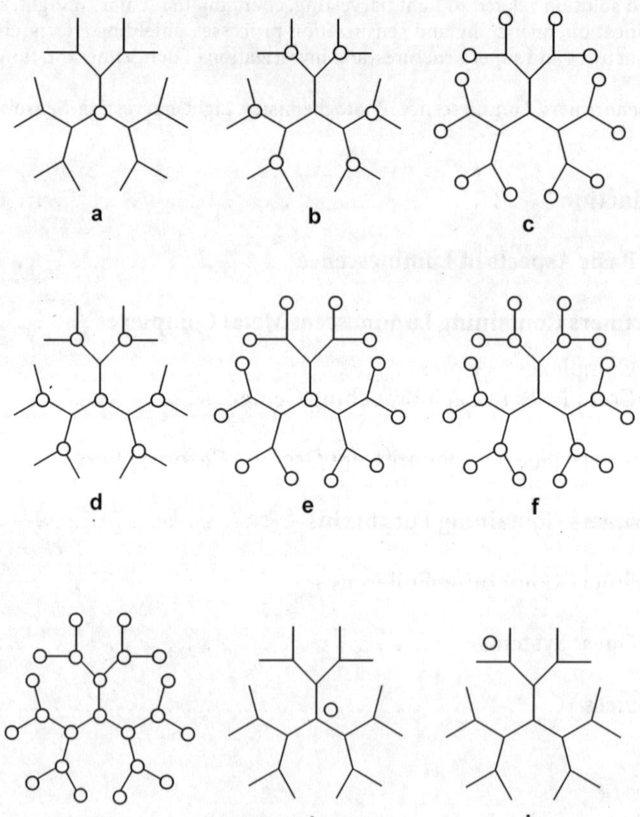

Fig. 1a–i. Schematic illustration of the possible location of photoactive units, represented by *circles*, covalently linked (types **a–g**) or associated (types **h–i**) to a dendrimer

Coupling luminescence with dendrimer chemistry can lead to systems capable of performing very interesting functions [7–10]. In this paper we will review recent advances in the field of luminescent dendrimers in fluid solution related to (i) light harvesting, (ii) changing the "color" of light, (iii) sensing with signal amplification, (iv) quenching and sensitization processes, (v) shielding effects, (vi) elucidation of dendritic structures and superstructures, and (vii) investigation of dendrimer rotation in solution. For space reasons, we will illustrate each selected topic with a few examples, and we will only mention topics which are currently the object of extensive studies, such as dendrimer fluorescence at the single-molecule level.

2
Some Basic Aspects of Luminescence

Before reviewing recent advances in the field of luminescent dendrimers, it is worthwhile recalling a few elemental principles of electronic spectroscopy. Interested readers are referred to several books and reviews for detailed discussions [2, 11, 12].

Figure 2 shows a schematic energy level diagram for a generic molecule. In most cases the ground state of a molecule is a singlet state (S_0), and the excited states are either singlets (S_1, S_2, etc.) or triplets (T_1, T_2, etc.). In principle, transitions between states with the same spin value are allowed, whereas those between states of different spin are forbidden. Therefore, the electronic absorption bands observed in the UV–visible spectrum of molecules usually correspond to $S_0 \rightarrow S_n$ transitions. Excited states are unstable species that undergo fast deactivation by intrinsic (first-order kinetics) processes. When a molecule is excited to upper singlet excited states, it usually undergoes a fast and 100% efficient radiationless deactivation (internal conversion, ic) to the lowest excited singlet, S_1. Such an excited state undergoes deactivation via three competing

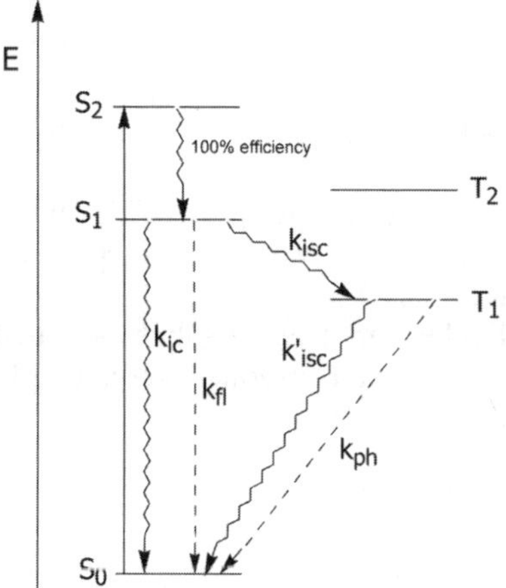

Fig. 2. Schematic energy level diagram for a generic molecule

processes: nonradiative decay to the ground state (internal conversion, rate constant k_{ic}); radiative decay to the ground state (fluorescence, k_{fl}); conversion to the lowest triplet state T_1 (intersystem crossing, k_{isc}). In turn, T_1 can undergo deactivation via nonradiative (intersystem crossing, k'_{isc}) or radiative (phosphorescence, k_{ph}) decay to the ground state S_0. When the molecule contains heavy atoms, the formally forbidden intersystem crossing and phosphorescence processes become faster. The lifetime (τ) of an excited state, that is, the time needed to reduce the excited state concentration by 2.718, is given by the reciprocal of the summation of the deactivation rate constants:

$$\tau(S_1) = 1 / (k_{ic} + k_{fl} + k_{isc}) \tag{1}$$

$$\tau(T_1) = 1 / (k'_{isc} + k_{ph}) \tag{2}$$

The orders of magnitude of $\tau(S_1)$ and $\tau(T_1)$ are approximately 10^{-9}–10^{-7} s and 10^{-3}–10^0 s, respectively.

The quantum yield of fluorescence (ratio between the number of photons emitted by S_1 and the number of absorbed photons) and phosphorescence (ratio between the number of photons emitted by T_1 and the number of absorbed photons) can range between 0 and 1 and are given by the following expressions:

$$\Phi_{fl} = k_{fl} / (k_{ic} + k_{fl} + k_{isc}) \tag{3}$$

$$\Phi_{ph} = k_{ph} / [(k'_{isc} + k_{ph})(k_{ic} + k_{fl} + k_{isc})] \tag{4}$$

Deactivation of an excited state can occur not only by the abovementioned intrinsic (first-order) decay channels, but also by interaction with other species (called "quenchers") following second-order kinetics. The two most important types of interactions are those leading to energy [Eq. (5)] or electron transfer [Eqs. (6) and (7)] (*A and *B stand for excited molecules) [1]:

$$^*A + B \rightarrow A + {}^*B \tag{5}$$

$$^*A + B \rightarrow A^+ + B^- \tag{6}$$

$$^*A + B \rightarrow A^- + B^+ \tag{7}$$

In both cases, the luminescence of the species A is quenched, but in the case of energy transfer the luminescence of species A is replaced by the luminescence of species B (sensitization process).

Besides energy and electron transfers between distinct molecules, such processes can take place between components contained in the same supramolecular structure [Eqs. (8–10)] [13]:

$$^*A{-}B \rightarrow A{-}{}^*B \tag{8}$$

$$^*A{-}B \rightarrow A^+{-}B^- \tag{9}$$

$$^*A{-}B \rightarrow A^-{-}B^+ \tag{10}$$

The occurrence of energy transfer requires electronic interactions and therefore its rate decreases with increasing distance. Depending on the interaction mechanism, the distance dependence may follow a $1/r^6$ (resonance (Förster) mechanism) or e^{-r} (exchange (Dexter) mechanisms) [1]. In both cases, energy transfer is favored by overlap between the emission spectrum of the donor and the absorption spectrum of the acceptor.

Excited state electron transfer also needs electronic interaction between the two partners and obeys the same rules as electron transfer between ground state molecules (Marcus equation and related quantum mechanical elaborations [14]), taking into account that the excited state energy can be used, to a first approximation, as an extra free energy contribution for the occurrence of both oxidation and reduction processes [8].

Because of their proximity, the various functional groups of a dendrimer may easily interact with one another. Interaction can also occur between dendrimer units and molecules hosted in the dendritic cavities or associated to the dendrimer surface.

3
Dendrimers Containing Luminescent Metal Complexes

3.1
Metal Complexes as Cores

Dendrimer 1^{2+} is a classical example of a dendrimer containing a luminescent metal complex core. In this dendrimer the 2,2′-bipyridine (bpy) ligands of the $[Ru(bpy)_3]^{2+}$-type core carry branches containing 1,2-dimethoxybenzene- and 2-naphthyl-type chromophoric units [15].

In the $[Ru(bpy)_3]^{2+}$ metal complex, the lowest excited state is a metal-to-ligand charge-transfer triplet, ^{3}MLCT, which, because of the presence of the heavy metal atom, is populated with unitary efficiency from the upper lying excited state. The ^{3}MLCT excited state emits a relatively strong phosphorescence (λ_{max}=610 nm). In fact, all three types of chromophoric groups present in the dendrimer, namely, $[Ru(bpy)_3]^{2+}$, dimethoxybenzene, and naphthalene, are potentially luminescent species. In the dendrimer, however, the fluorescence of the dimethoxybenzene- and naphthyl-type units is almost completely quenched in acetonitrile solution, with concomitant sensitization of the $[Ru(bpy)_3]^{2+}$ phosphorescence. These results show that very efficient energy transfer processes take place converting the very short lived (nanosecond time scale) UV fluorescence of the aromatic units of the wedges to the relatively long lived (microsecond time scale) orange phosphorescence of the metal-based dendritic core. This dendrimer is therefore an excellent example of a light-harvesting antenna system as well as of a species capable of changing the color of the incident light. It should also be noted that in aerated solution the phosphorescence intensity of the dendritic core is more than twice as intense as that of the $[Ru(bpy)_3]^{2+}$ parent compound because the dendrimer branches protect the Ru(bpy)-based core from dioxygen quenching [16].

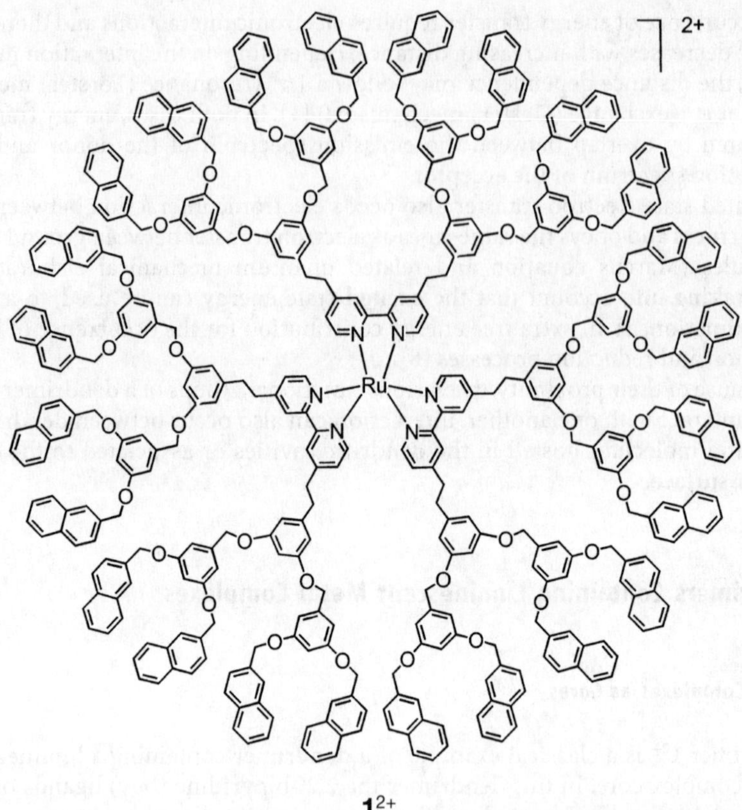

$\mathbf{1}^{2+}$

The $[Ru(bpy)_3]^{2+}$ core has more recently been used to construct first-generation dendrimers $\mathbf{2}^{2+}$ containing coumarin-450 chromophoric groups.

In acetonitrile solution, energy transfer from the excited coumarin units to the $[Ru(bpy)_3]^{2+}$ core, whose excited state is again protected from dioxygen quenching, takes place with almost unitary efficiency, so that the absorbance of near UV light leads again to the characteristic orange emission of the core [17]. When two first- or second-generation 1,3,5-phenylene-based dendrons are appended to the 3 and 8 positions of the phenanthroline ligand of the $[Ru(bpy)_2(phen)]^{2+}$ complex (phen=1,10-phenanthroline), excitation in the phenylene dendrons is followed by a very efficient (>98%) energy transfer to the Ru(II)-based moiety [18].

Self-assembly of functionalized carboxylate-core dendrons around Er^{3+}, Tb^{3+}, or Eu^{3+} ions leads to the formation of dendrimers [19]. Experiments carried out in toluene solution showed that UV excitation of the chromophoric groups contained in the branches caused the sensitized emission of the lanthanide ion, presumably by an energy transfer Förster mechanism. The much lower sensitization effect found for Eu^{3+} compared with Tb^{3+} was ascribed to a weaker spectral overlap, but it could be related to the fact that Eu^{3+} can quench the donor excited state by electron transfer [20].

$$\mathbf{2}^{2+}$$

3.2
Metal Complexes in Each Branching Center

Polypyridine ligands have been extensively used to build up luminescent polynuclear complexes with dendritic structures [6, 21, 22]. Dendrimers containing Ru(II) and, less frequently, Os(II) polypyridine complexes have been carefully investigated from the photophysical viewpoint. As mentioned above, the lowest excited state in these complexes is a metal-to-ligand charge-transfer triplet, ^{3}MLCT. In dendrimers, the small but not negligible electronic interaction between nearby units is sufficient to cause a very fast energy transfer that leads to the quenching of the potentially luminescent units with higher energy ^{3}MLCT levels and the sensitization of the luminescence of the units with lower energy ^{3}MLCT levels. Recent studies on dinuclear model compounds have suggested that energy transfer between nearby units occurs within 200 fs, probably from nonthermalized excited states [23].

The energy of the ^{3}MLCT excited state of each unit depends on the metal and ligands in a predictable way and a modular synthetic strategy [6, 24] allows a high degree of synthetic control in terms of the nature and position of metal centers, bridging ligands, and terminal ligands. Such synthetic control translates into control of the direction of energy flow within the dendritic array [6, 9]. On increasing nuclearity, however, a unidirectional gradient (center-to-periphery or vice versa) for energy transfer cannot be obtained with only two types of metals [Ru(II) and Os(II)] and ligands (bpy and 2,3-dpp).

An extension of this kind of antennae is a first-generation heterometallic dendrimer with appended organic chromophores like pyrenyl units [25, 26]. In the tetranuclear species consisting of an Os(II)-based core surrounded by three Ru(II)-based moieties and six pyrenyl units in the periphery, 100% efficient energy transfer is observed to the Os(II) core regardless of the light-absorbing unit.

4
Dendrimers Based on Fluorescent Organic Chromophores

A classical example of dendrimer containing fluorescent organic chromophores is **3**. In this dendrimer, an energy gradient from the peripheral units to the perylene core allows energy transfer to occur with a high rate (1.9×10^{11} s^{-1}) and large efficiency (approximately 98%) [27].

3

Dendrimers with a polyphenyl core around a central biphenyl unit decorated at the rim with peryleneimide chromophores have been investigated both in bulk and at the single-molecule level in order to understand their time and space-resolved behavior [28]. The results obtained have shown that the conformational distribution plays an important role in the dynamics of the photophysical processes. Energy transfer in a series of shape-persistent polyphenylene dendrimers substituted with peryleneimide and terryleneimide chromophoric units (4–7) has been investigated in toluene solution [29].

Energy hopping among the peryleneimide chromophores, revealed by anisotropy decay times [30], occurs with a rate constant of 4.6×10^9 s^{-1}. When three peryleneimide and one terryleneimide chromophores are attached to the dendrimer rim, energy transfer from the former to the latter units takes place with

4 $R^1 = D, R^2 = R^3 = R^4 = H$

5 $R^1 = R^2 = R^3 = D, R^4 = H$

6 $R^1 = R^2 = R^3 = R^4 = D$

7 $R^1 = R^2 = R^3 = D, R^4 = A$

D

A

efficiency >95%. All the observed energy transfer processes can be interpreted on the basis of the Förster mechanism. Polyphenylene dendrimers with a perylene diimide as a luminescent core have also been investigated [31]. In a dendrimer consisting of a terrylenediimide core and four appended peryleneimide units the antenna effect has been studied at the single-molecule level [32].

Very interesting antenna systems have been constructed by functionalizing the chain ends of a poly(arylether) convergent dendritic backbone with coumarin-2 (λ_{em}=435 nm) and its focal point with coumarin-343 (λ_{em}=490 nm) [33]: dendrimer **8** represents the fourth generation.

8

In these dye-functionalized dendrimers, light absorbed by the numerous peripheral coumarin-2 units is funneled to the coumarin-343 core with remarkably high efficiency (toluene solution: 98% for the first three generations; 93% for compound **8**). Given the large transition moments and the good overlap between donor emission and acceptor absorption, energy transfer takes place by Förster mechanism [34].

Oligo(*p*-phenylenevinylene) (OPV) units are increasingly used to obtain photoactive dendrimers and polymers [35]. In poly(propylene amine) dendrimers (POPAM) substituted at the periphery with OPV units, interchromophoric interactions are sufficiently strong in 2-methyltetrahydrofuran to induce delocalization of the excitation over more than one chromophoric group [36], which is a phenomenon observed in natural light-harvesting complexes [37]. In OPV-terminated dendritic wedges functionalized with C_{60} in the focal point, the excited OPV moieties transfer energy to the fullerene core by Förster-type singlet–singlet energy transfer in dichloromethane solution [38].

A luminescent unit extensively used to functionalize dendrimers is the so-called dansyl (5-dimethylamino-1-naphthalenesulphonamido group). Dendrimers (up to the third generation, compound **9**) containing a single dansyl unit attached "off center" [39] show that this fluorescent unit, which is very sensitive to environment polarity, is progressively shielded from interaction with water molecules as the dendrimer generation increases.

The emission maximum moves to the blue (from 547 nm for the first generation to 535 nm for the third generation); the emission quantum yields and lifetimes become larger (Φ=0.044 and 0.086, τ=4.90 and 7.94 ns for the first and the third generation, respectively). This behavior demonstrates that the dendritic environment surrounding the dansyl moiety is less polar than water. Anisotropy decays are monoexponential within the dendritic family, indicating no independent movement of the dansyl residue within the dendritic framework, and the corresponding lifetimes monotonically increase with dendrimer generation,

9

showing a concomitant growth of dendrimer volume. Another proof of core isolation is given by the decreasing association constants between dansyl and β-cyclodextrin from the dansyl model compound up to the second-generation dendrimer and the evidence of no binding at all for the third-generation one. On the other hand, in the case of anti-dansyl antibody, steady-state fluorescence anisotropy and emission measurements demonstrate an association even with the third-generation dendrimer, but with a K_a value 30-fold lower than for a dansyl model compound. Therefore, while association between a nonselective host (β-cyclodextrin) and dansyl group can be completely prevented by dendritic arms, a selective host (anti-dansyl antibody) can complex the dansyl core of each dendrimer to some degree, although its molecular mass is much higher than for the former host. This result suggests that the dendrimer conformation can be altered to allow access to the core, if the binding affinity is great enough. Phenylacetylene dendrons have been connected to a (S)-1,1′-bi-2-naphthol core obtaining three generations of optically pure dendrimers (**10–12**) [40].

Luminescence experiments in dichloromethane solution indicated that the fluorescence of the phenylacetylene branches is quenched, whereas intense emission is observed from the binaphthol core. This antenna effect represents the first example of efficient (>99%) energy migration in an optically pure dendrimer. The fluorescence quantum yield increases slightly with increasing generation; the values of 0.30, 0.32, and 0.40 were obtained, respectively, for **10–12**.

(S)-**10**, R =

(S)-**11**, R =

(S)-**12**, R =

This increase was ascribed to diminished molecular flexibility on increasing dendrimer size.

The optically pure dendrimers containing a 1,1′-bi-2-naphthol core discussed above (**10–12**) can be used also as enantioselective fluorescent sensors of amino alcohols [41–43]. Both enantiomers of the chiral alcohols 2-amino-3-methyl-1-butanol, 2-amino-4-methyl-1-pentanol (in dichloromethane), and 2-amino-3-phenyl-1-propanol (in benzene/hexane) were found to quench the fluorescence emission of the S enantiomer of dendrimers **10–12**. The Stern–Volmer quenching constants increase with increasing dendrimer generation, so that higher generation dendrimers can be more sensitive fluorescent sensors, giving rise to a "signal amplification" effect. In the quenching of the S dendrimers, the Stern–Volmer quenching constants of each S alcohol, $K_{SV}(S)$, was always greater than that of the enantiomeric R alcohol, $K_{SV}(R)$. A ratio $K_{SV}(S)/K_{SV}(R)$ greater than 1 indicates that enantioselective quenching is taking place (the maximum $K_{SV}(S)/K_{SV}(R)$ ratio observed was 1.27). Moreover, in the case of **12**, the fluorescence lifetime of the dendrimer does not change on addition of the quencher. This indicates that quenching occurs by a static mechanism, presumably via a ground-state hydrogen bond formed between the hydroxyl groups of the dendrimer core and the amino alcohol.

Chiral dendrimers based on oligonaphthyl cores and Fréchet-type poly(aryl ether) dendrons have been investigated [44]. The absolute configuration of these dendrimers remains the same as that of their chiral cores. Both the nature of the core and the generation play a role in determining the fluorescence quantum yield.

Poly(aryl ether) branches of generation 1 to 3 have been appended to a photautomerizable quinoline core to investigate the effect of dendritic architecture on the excited state intramolecular proton transfer [45]. The changes observed in the absorption and emission spectra on increasing dendrimer generation indicate that the dendritic branches affect the planarity of the core and therefore the efficiency of the excited state intramolecular proton transfer and of the related fluorescence processes.

5
Dendrimers Containing Porphyrins

Porphyrins have extensively been used to construct light-harvesting antenna arrays based on covalent bonds [46] as well as on self assembly [47]. Much attention has been focused on light-harvesting antenna dendrimers comprising a porphyrin core and properly chosen dendrons. Morphology-dependent antenna properties have been reported [48] for a series of dendrimers consisting of a free-base porphyrin core bearing different numbers (1–4) of poly(benzylether) dendrons at the *meso* position of the central porphyrin (**13–17**).

In dichloromethane solutions, excitation of the chromophoric groups of the dendrons causes singlet–singlet energy transfer processes that lead to the excitation of the porphyrin core. It was found that the dendrimer **17**, which has a spherical morphology, exhibits a much higher energy transfer quantum yield (0.8) than the partially substituted species **13–16** (quantum yield <0.32). Fluo-

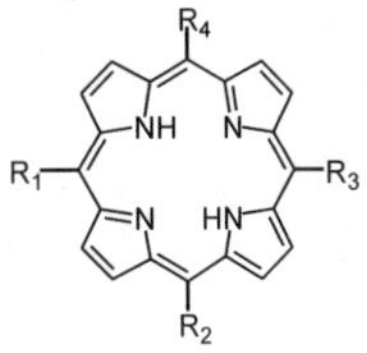

	R_1	R_2	R_3	R_4
13	L	tolyl	tolyl	tolyl
14	L	tolyl	L	tolyl
15	L	L	tolyl	tolyl
16	L	L	L	tolyl
17	L	L	L	L

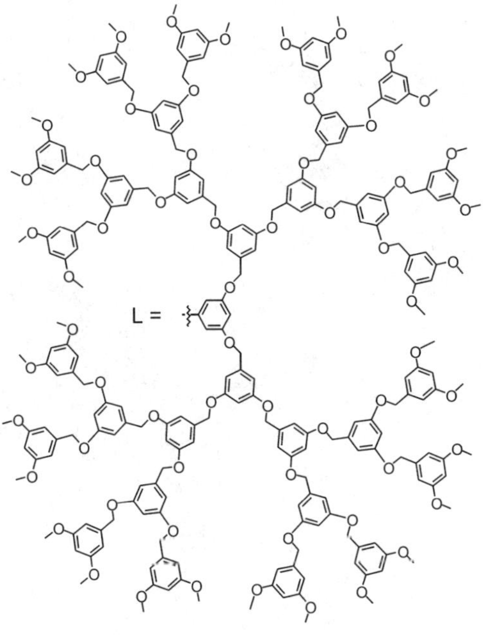

L =

rescence polarization studies on **17** showed that the excitation energy migrates very efficiently over the dendrons within the excited state lifetime, so that the four dendrons can be viewed as a single, large chromophore surrounding the energy trap. Temperature-dependent effects suggested that increased flexibility and conformational freedom were responsible for the decreased energy transfer efficiency on decreasing the number of dendrons. Only the highly crowded dendrimer **17** retained a constant level of energy transfer, even at high temperatures. It was also postulated that cooperativity between dendrons, which decreases with increasing conformational mobility, is necessary for efficient energy transfer [48]. Such behavior would mimic that of natural photosynthetic systems, where energy migration within "wheels" of chromophoric groups results in an efficient energy transfer to the reaction center [37]. More recently, the morphology effect was investigated by using much larger porphyrin dendrimers consisting of a free-base porphyrin core, P, with up to four appended dendrons, each containing seven zinc porphyrin units (compound **18**) [49].

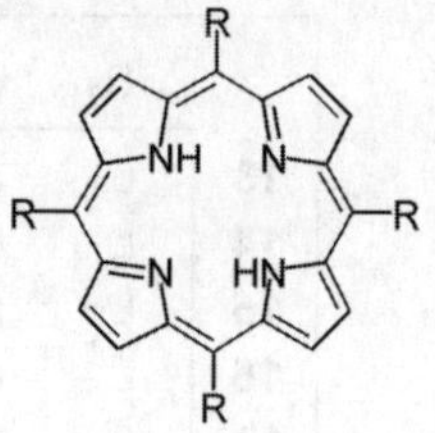

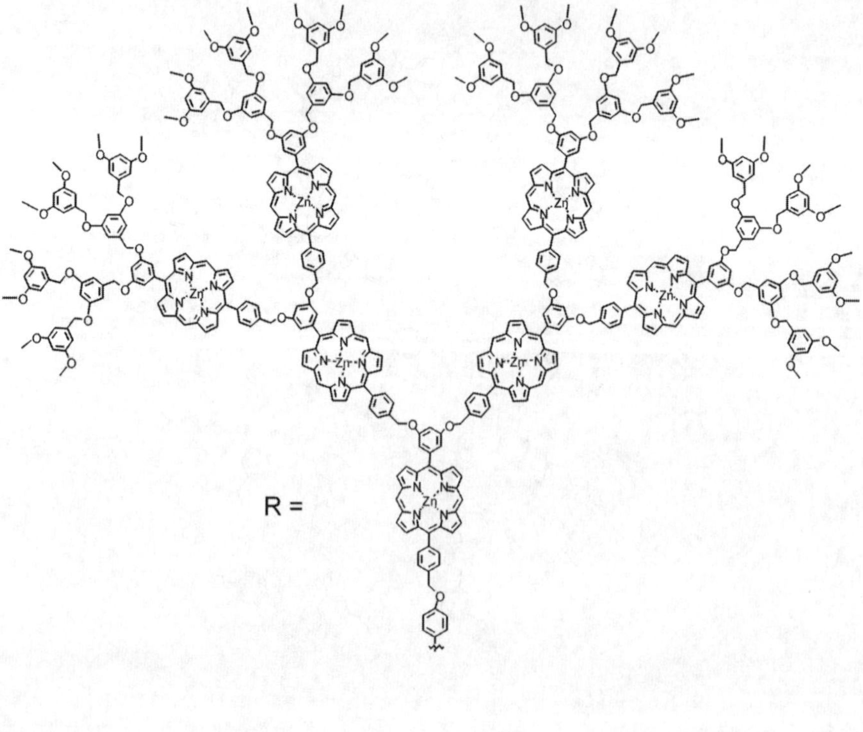

The presence of poly(benzylether) dendritic wedges at the periphery makes such dendrimers soluble in common organic solvents. The experiments performed in tetrahydrofuran showed that in the star-shaped dendrimer **18**, energy transfer from the excited singlet states of dendrons to the focal core takes place with rate constant of 1.0×10^9 s^{-1} and 71% efficiency, whereas in the conically shaped dendrimer containing a single dendron substituent the energy transfer rate constant was 10 times smaller and the efficiency was 19%. This result shows that morphology has indeed a noticeable effect on the energy transfer rate. Excitation of **18** at 544 nm with polarized light resulted in a highly depolarized fluorescence from the Zn porphyrin units (fluorescence anisotropy factor 0.03, to be compared with 0.19 of a monomeric reference compound), indicating an efficient energy migration among the Zn porphyrin units before energy transfer to the free-base core. In the case of the conically shaped compound containing a single dendron substituent, the fluorescence anisotropy factor was much higher (0.10). These results suggest a cooperation of the four dendrons of **18** in facilitating the energy migration among the Zn porphyrin units. Clearly, **18**, which incorporates 28 light-absorbing Zn porphyrin units into a dendritic scaffold with an energy-accepting core, mimics several aspects of the natural light-harvesting system. Fast dynamics of energy migration have also been observed for multiporphyrin functionalized poly(propylene amine) dendrimers [50].

When a porphyrin core is linked to four 1,3,5-phenylene-based dendrons (compounds **19–22**), energy transfer from the excited dendrons to the por-

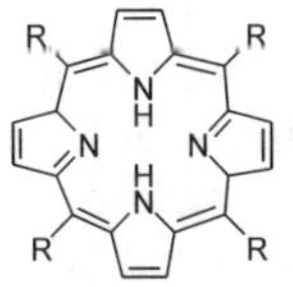

19 R = D^1

20 R = D^2

21 R = D^3

22 R = D^4

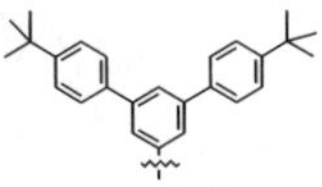

D^1

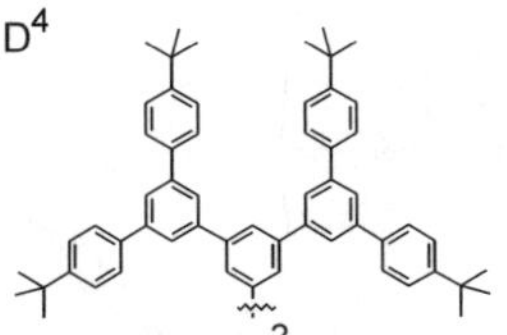

D^2

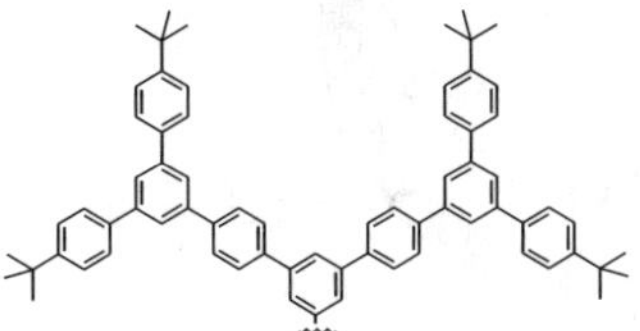

D^3

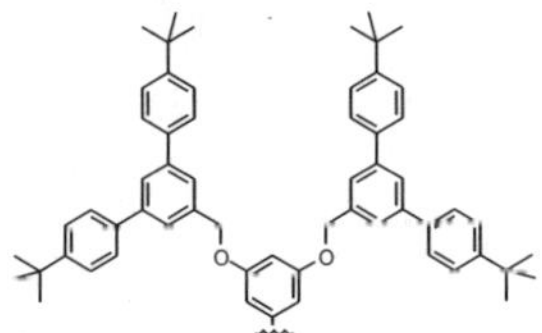

D^4

phyrin core takes place in dichloromethane solution and depends on the dendron structure [51, 52].

In particular, the energy transfer efficiencies are 0.66, 0.98, 0.74, and 0.42 for the dendrimers **19–22**, respectively. Such higher efficiency, compared with the aryl ether dendrimers described above, has been ascribed to energy transfer through the cross conjugation of 1,3,5-phenylene based dendrons and to a larger spectral overlap. Indeed, dendrimer **22** which contains ether bonds shows a lower energy transfer efficiency compared to dendrimer **20** that is completely cross conjugated, suggesting the presence of a through-bond (Dexter) contribution to the energy transfer processes in **20**.

Electrostatic assembly of negatively and positively charged dendritic porphyrins has been investigated by means of energy transfer measurements [53]. The compounds used were made of zinc and free-base porphyrin cores surrounded by second-generation Fréchet-type dendrons functionalized at the periphery with ammonium and carboxylate units, respectively. The energy transfer rate constant was found to be 3.0×10^9 s^{-1}, in satisfactory agreement (Förster mechanism) with the expected core-to-core distance in the assembly of the positively and negatively charged dendrimers.

A poly(L-lysine) dendrimer **23** which carries 16 free-base porphyrins in one hemisphere and 16 Zn porphyrins in the other has been synthesized and studied in dimethylformamide solution [54]. In such a dendrimer, energy transfer from the Zn porphyrins to the free-base units can occur with 43% efficiency. When the 32 free base and zinc porphyrins were placed in a scrambled fashion, the efficiency of energy transfer was estimated to be 83% [55]. Very efficient (98%) energy transfer from Zn to free-base porphyrins was also observed in a rigid, snowflake-shaped structure in which three Zn porphyrin units alternate with three free-base porphyrin units [56].

An interesting example of core shielding is represented by Pd porphyrins surrounded by polyglutamic dendrons up to the fourth generation (**24–26**) [57]. In

23

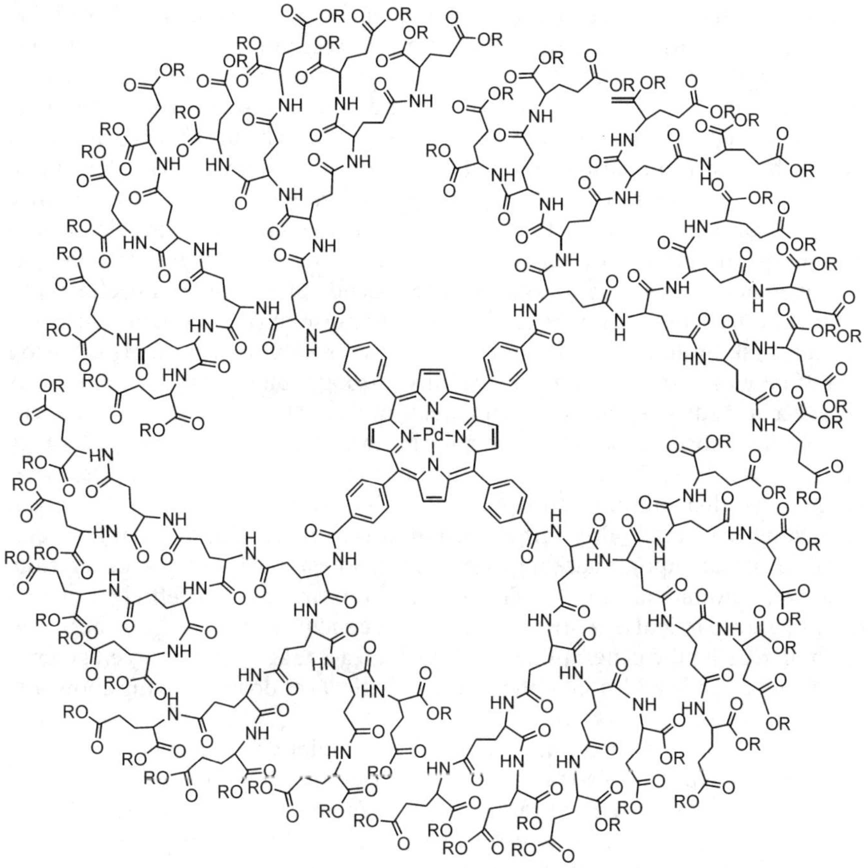

24 R = H

25 R = Et

26 R = Bz

deaerated solutions they show a strong phosphorescence at 692 nm and single exponential decays in the range of 400–600 µs. In the presence of dioxygen a quenching process has been observed. In particular, while in air-equilibrated dimethylformamide solutions the quenching rate constants are independent of the size and hydrophobicity of the dendrimer periphery, in air-equilibrated water solutions an increase in the dendrimer size causes a large decrease in the quenching rate constant. This peculiar behavior can easily be rationalized taking into account that in dipolar aprotic media, such as dimethylformamide, dendritic branches terminated by either carboxylic acid (**24**) or ester groups (**25, 26**) are neutral and adopt a rather unfolded conformation, so that dioxygen access is not hindered. On the other hand, in water solutions, hydrophobic interactions lead the dendritic shell to shrink around the core. Therefore, upon increasing dendrimer generation, the Pd porphyrin is encapsulated by a more densely packed cage, partly preventing dioxygen quenching. The same family of poly-glutamic dendrimers (up to the third generation) containing carboxylate groups at the periphery and a free-base porphyrin core, instead of the Pd porphyrin, shows strong changes of the absorption and emission properties of the por-phyrin core upon protonation in water solution [58]. The pK value correspond-ing to the first protonation gradually shifts towards alkaline pH values upon increasing generation. Indeed, a growing stabilization of the central positive charge is brought about by the increasing negative charge on the dendrimer periphery. Therefore, an electrostatic core shielding effect is active as long as dendritic branches are deprotonated. These dendrimers may find useful appli-cations as pH sensors in biological systems because of (i) a pK value close to 7 for the first protonation, (ii) unlimited water solubility, (iii) high emission quantum yield, and (iv) impermeability through biological membranes. A very recent study [59] compares the shielding ability towards dioxygen quench-ing of the phosphorescent Pd porphyrin core present in dendrimer **24** by three different types of dendrons, namely Fréchet poly(aryl ether)s, New-kome poly(ether amide)s, and polyglutamates. The changes of the dioxygen quenching constants are rather insignificant in dimethylformamide and tetrahydrofuran upon changing both the dendron type and the generation. However, in water solution the Fréchet dendrons are the most efficient in core shielding. In particular, a strong decline in quenching rate has been observed upon increasing the generation: a 24-fold decrease takes place between genera-tion 0 and 2, while addition of the next dendritic layer does not bring about any significant change.

The effect of core shielding of a porphyrin moiety by peripheral dendrons has been carefully investigated on two series of Zn–phthalocyanine-cored dendrimers with aryl-ether branches [60]. Generation 0, 1, and 2 (dendrimer **27**) species, terminated with ester groups, are soluble in organic solvents, while the species terminated with carboxylate units (e.g., **28**) are soluble in water.

Photo-induced electron transfer quenching of the fluorescent core of the ester-terminated dendrimers was investigated using anthraquinone as electron acceptor. The quenching occurs through a dynamic mechanism, and the rate of the electron transfer process decreases with growing dendrimer size. The fluorescence of the core of the water-soluble series was quenched with anion-ic (picrate anion), and cationic electron acceptors (5,10,15,20-tetrakis(1-methyl-4-pyridyl)porphyrin). With the picrate quencher, a sharp decrease of the bimolecular electron transfer rate constant is observed on increasing den drimer generation; **28** was virtually unquenched. This behavior was ascribed to the highly negative dendrimer surface that electrostatically repels the picrate anion. In contrast, using the cationic quencher, the electrostatic interac-tion with the dendrimer results in a very efficient quenching through a static mechanism.

Alternatively, the quencher can be directly connected to the dendrimer struc-ture. The photophysical behavior of generation 1, 2 (compound **29**), and 3 den-drimers containing a free-base porphyrin core and 12, 36, and 108 peripheral anthraquinone units, respectively, was studied in chloroform and dimethylac-

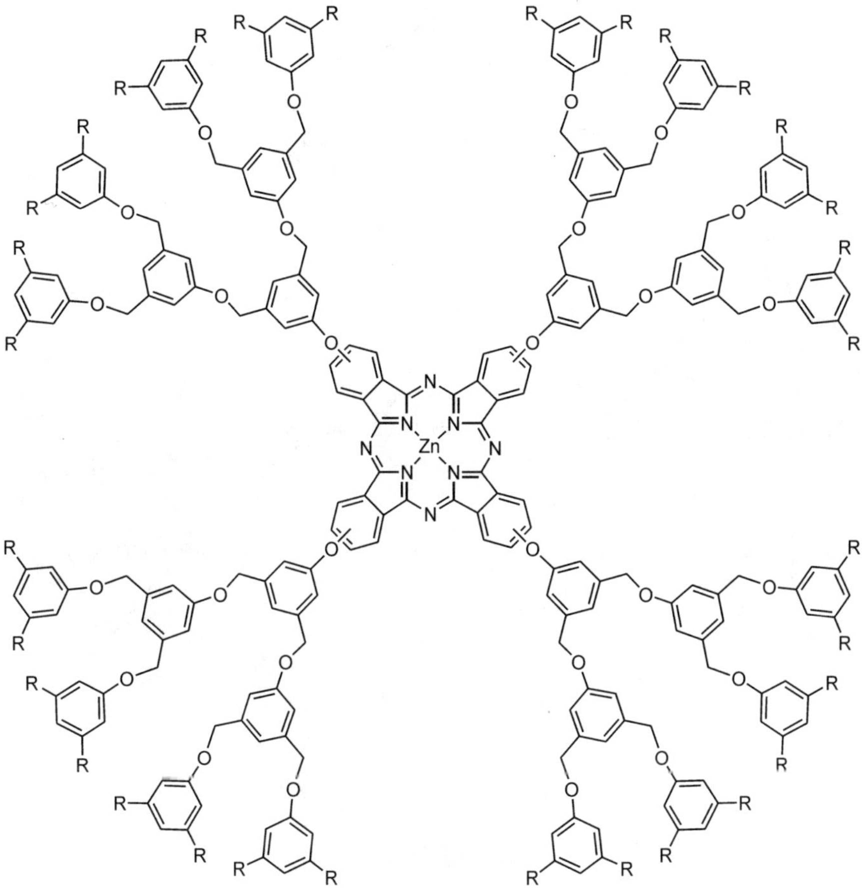

27 R = COOC$_5$H$_{11}$

28 R = COONa

etamide solution [61]. In all cases the core fluorescence is quenched by intramolecular electron transfer to the peripheral anthraquinone units; the most efficient quenching was observed for generation 1 and 2 species. Time-resolved fluorescence experiments revealed a complex decay behavior, requiring a three-exponential fit. This complexity was attributed to the flexibility of the dendritic framework, which gives rise to different conformers exhibiting different porphyrin–anthraquinone distances. The measured electron transfer rate constants indicate that the through-space (Förster) mechanism is operative in these quenching processes.

A comparison between two families of dendrimers containing poly(aryl ether) dendrons and either a Zn porphyrin (G*n*PZn) or a tetraphenylporphyrin (G*n*TPPH$_2$) core up to the fourth generation (**30** and **31**) shows that the core

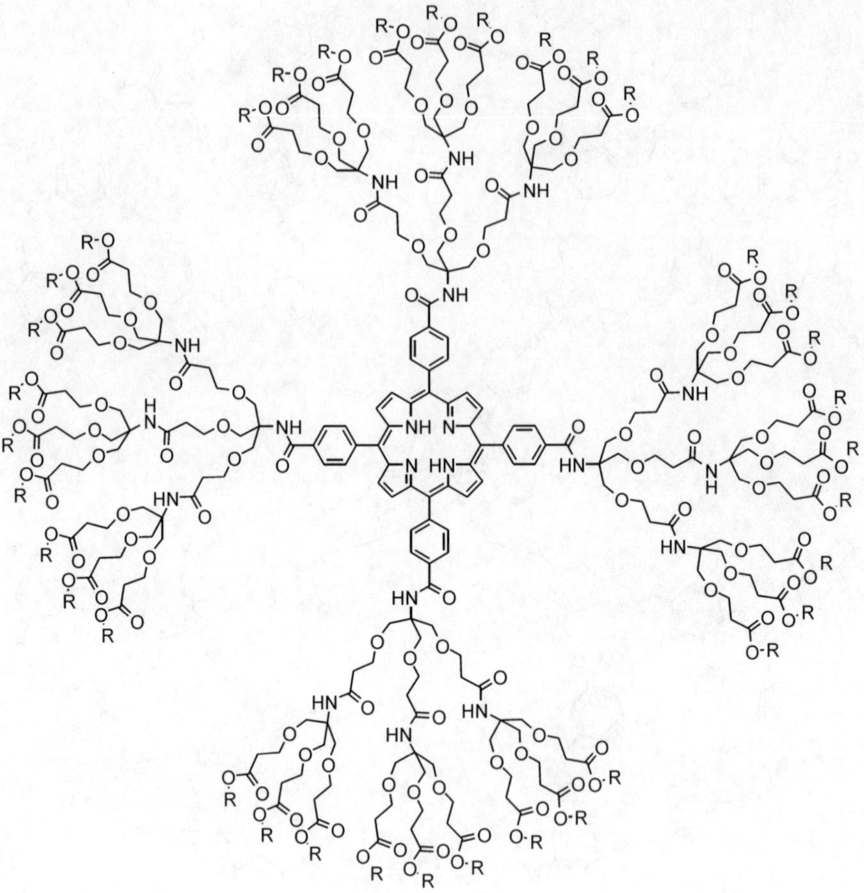

29 R =

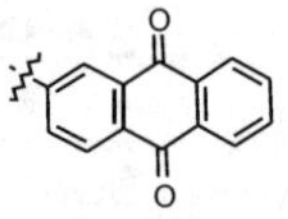

structure influences the hydrodynamic properties [62]. Fluorescence quantum yields and lifetimes in dimethylformamide and tetrahydrofuran solutions do not change upon increasing generation within a dendrimer family, suggesting little influence of the dendrons on the radiative and nonradiative processes of the porphyrin core. The anisotropy decays can be fitted by a single exponential function for all generations with a limiting value close to that of model porphyrin compounds. The intrinsic viscosity, obtained by fluorescence depolarization measurements, shows a maximum for the GnPZn family corresponding to the third generation, while that of the GnTPPH$_2$ family linearly increases with generation. The additional phenyl group in the GnTPPH$_2$ family increases the distance between dendrons, thus minimizing steric hindrance, compared to the GnPZn family, in which the dendrons are directly linked to the porphyrin core.

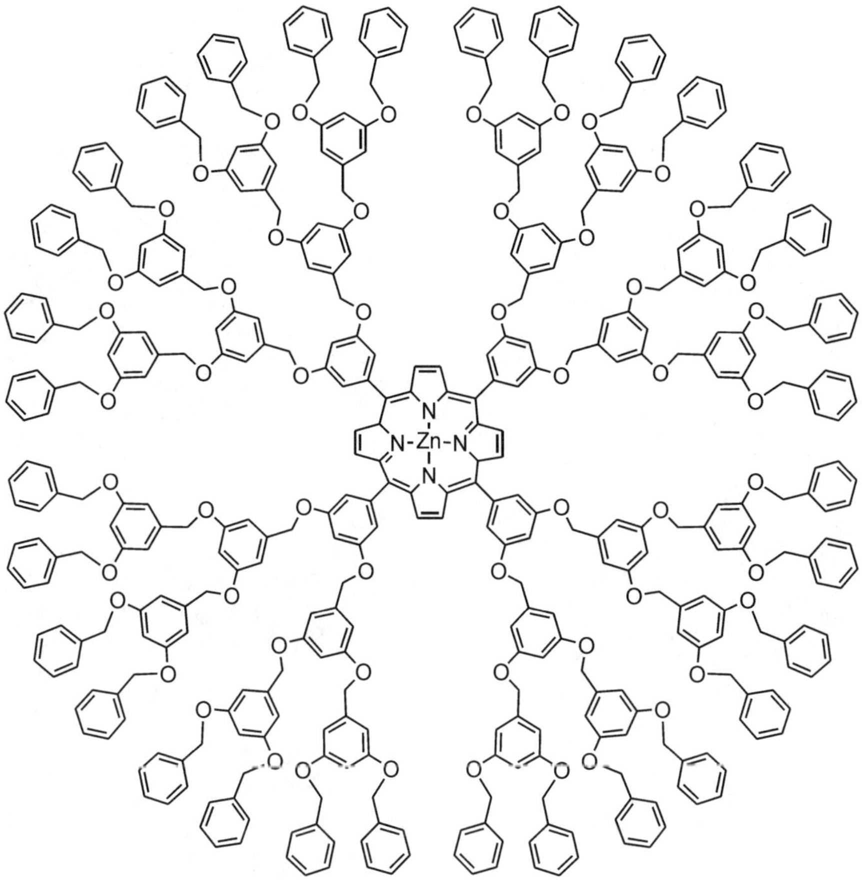

30

As a result, the branches in the $GnTPPH_2$ family are more flexible, thus causing a decrease in the hydrodynamic volume compared to the theoretical fully extended structure in the gas phase.

That dendrimers are unique when compared with other architectures is confirmed by an investigation on porphyrin core dendrimers and their isomeric linear analogues [63]. The isomers displayed dramatically different hydrodynamic properties, crystallinity, and solubility characteristics when compared to those of their dendritic analogues, and photophysical studies showed that energy transfer from the poly(benzylether) backbone to the core was more efficient in the dendrimer because of the shorter distance between the donor units and the acceptor core.

31

6
Dendrimers Containing Fullerene

Fullerene (C_{60}) is a very interesting molecule [64]. It has low-lying singlet and triplet excited states and it can be easily reduced in successive, reversible, one-electron processes. Its lowest singlet excited state exhibits fluorescence with a maximum around 720 nm, and its lowest triplet excited state is an excellent sensitizer for singlet oxygen formation. Investigations on dendrimers containing fullerene units are rapidly growing [65].

In a recent study, poly(aryl ether) dendritic branches terminated with triethyleneglycol chains were attached to C_{60} [66]: dendrimer **32** represents the fourth generation. The photophysical properties of these fullerodendrimers have been systematically investigated in three solvents, namely toluene, dichloromethane, and acetonitrile. On increasing dendrimer generation, it has been found that in each solvent (i) the maximum of the fullerene fluorescence band is red-shifted

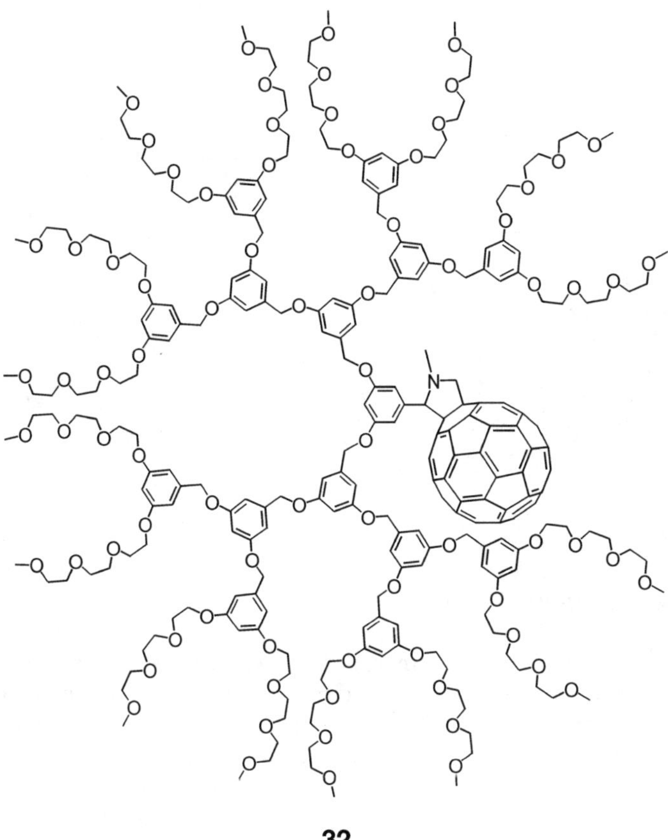

32

and (ii) the fluorescence lifetime increases slightly. These effects are attributed to changes in the dendrimer solvation environment. In air-equilibrated solutions, the triplet lifetime increases with increasing dendrimer generation in each solvent, showing that the dendrimer branches offer a shielding effect towards encounters with external molecules like dioxygen. In deaerated solution, the triplet lifetime decreases with increasing solvent polarity and increases slightly on increasing dendrimer generation.

In another recent study, the fullerene unit was connected via an acetylene bond with poly(aryl ether) dendritic branches from the second to the fourth generation (compound **33**) [67]. The photophysical properties and the quenching by energy transfer (quenchers: dioxygen and β-carotene) and electron transfer (quenchers: three different amines) were investigated. In this case, the lifetimes of the singlet and triplet excited states are essentially independent of dendrimer generation, but the rate constants of the intermolecular processes (triplet–triplet annihilation, triplet energy transfer, triplet electron transfer) decrease with increasing dendrimer generation. For the largest dendrimer, electron transfer takes place over a long distance because of the steric hindrance caused by the dendron groups and the resulting fullerene anion is stabilized.

33

7
Host–Guest Systems

An important property of dendrimers is the presence of internal cavities in which ions or neutral molecules can be hosted [68]. Such a property can potentially be exploited for a variety of purposes, which include catalysis and drug delivery.

Energy transfer from the numerous chromophoric units of a suitable dendrimer to an appropriate guest may be exploited to construct systems for changing the color of the incident light and for light harvesting. An advantage shown by such host–guest systems compared with dendrimers with a luminescent core is that the wavelength of the sensitized emission can be tuned by changing the guest hosted in the same dendrimer.

Dendrimers of the poly(propylene amine) family functionalized with fluorescent dansyl units at the periphery have been used as hosts for fluorescent dye molecules [69]. Each dendrimer nD, where the generation number n goes from 1 to 5, comprises 2^{n+1} (i.e., 64 for 5D) dansyl functions in the periphery and $(2^{n+1}-2)$ (i.e., 62 for 5D) tertiary amine units in the interior. Compound **34** represents the fourth generation dendrimer 4D. These dendrimers show intense absorption bands in the near UV spectral region (λ_{max}=252 and 339 nm; ε_{max}≈12,000 and 3,900 L mol^{-1} cm^{-1}, respectively, for each dansyl unit) and a strong fluorescence band in the visible region (λ_{max}=500 nm; Φ_{em}=0.46, τ=16 ns). In dichloromethane solution, the nD dendrimers extract eosin from aqueous solutions with the maximum number of eosin molecules hosted in the dendrimers increasing with increasing dendrimer generation, up to a maximum of 12 for the 5D dendrimer. The fluorescence of the peripheral dansyl units of the dendrimers is completely quenched via energy transfer (Förster mechanism) by the hosted eosin molecule, whose fluorescence (λ_{max}=555 nm) is, accordingly, sensitized. The behavior of fluorescein and rose bengal is qualitatively similar to that of eosin, whereas naphthofluorescein is not extracted.

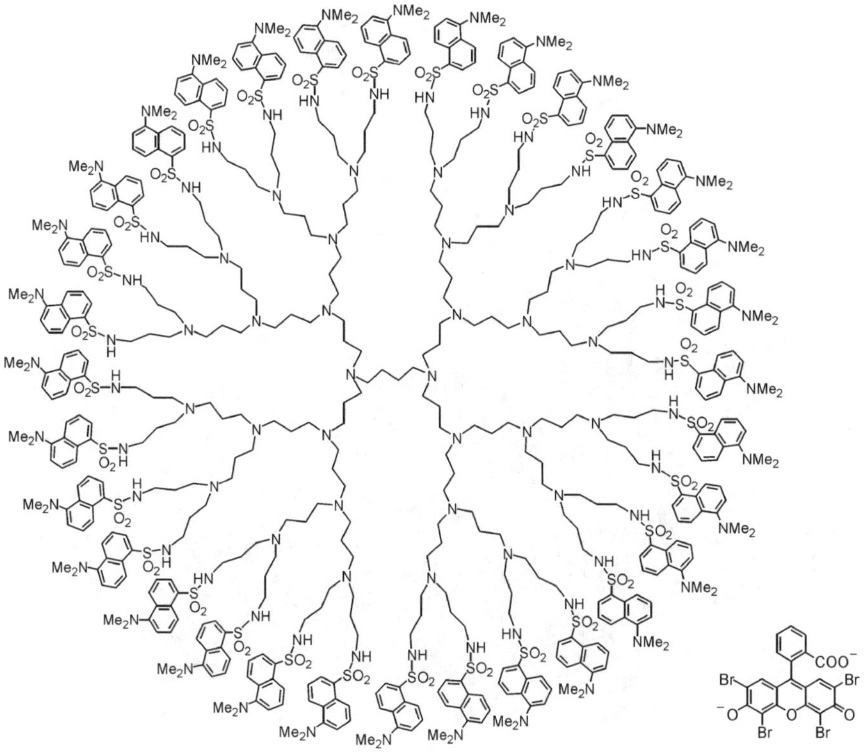

Quantitative analysis of the results obtained has shown that a single eosin guest is sufficient to completely quench the fluorescence of any excited dansyl unit of the hosting dendrimer. Fluorescence lifetime measurements indicated that the dye molecules can occupy two different sites (or two families of substantially different sites) in the interior of the dendritic structure.

Dendrimer **35**, consists of a hexaamine core surrounded by 8 dansyl-, 24 dimethoxybenzene-, and 32 naphthalene-type units [70]. In dichloromethane solution, **35** exhibits the characteristic absorption bands of the component units and a strong dansyl-type fluorescence. Energy transfer from the peripheral dimethoxybenzene and naphthalene units to the fluorescent dansyl units occur with >90% efficiency. When the dendrimer hosts a molecule of the fluorescent eosin dye (**35**⊃eosin), the dansyl fluorescence, in its turn, is quenched and sensitization of the fluorescence of the eosin guest can be observed. Quantitative measurements showed that the encapsulated eosin molecule collects electronic energy from all the 64 chromophoric units of the dendrimer with an efficiency >80%. Both intramolecular (i.e., within dendrimer) and intermolecular (i.e., dendrimer host→eosin guest) energy transfer processes occur very efficiently by a Förster-type mechanism because of the strong overlap between the emission and absorption spectra of the relevant donor/acceptor units.

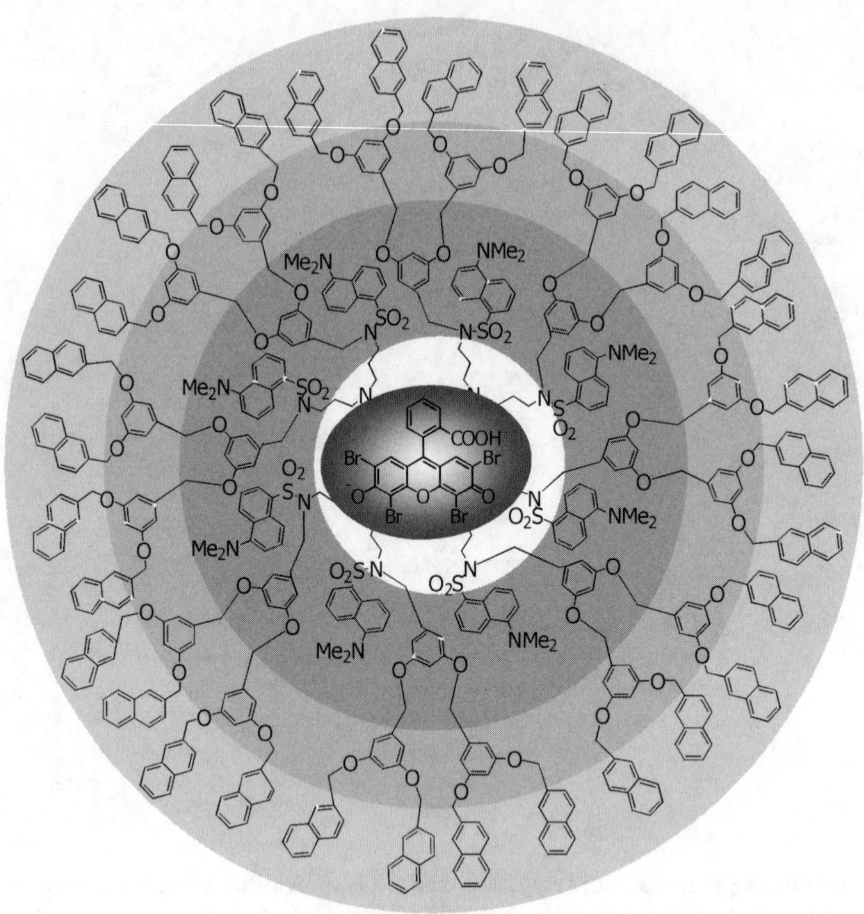

Dye molecules can also be hosted into poly(propylene amine) dendrimers peripherally modified with OPV units [71]. In these systems, energy transfer from the OPV fluorescent units (λ_{max}=492 nm) to the enclosed dye molecules is not efficient in solution (40% efficiency at maximal loading), but is very efficient in spin-coated films of dendrimer/dye assemblies.

Dendritic hosts can be used in aqueous solution to encapsulate water-soluble fluorescent probes. Changes in the photophysical properties of these encapsulated probes are useful to understand the properties of the microenvironment created by the dendritic interior. For example, adamantyl-terminated poly(propylene amine) dendrimers from the first to the fifth generation (**36** represents the third generation) can be dissolved in water at pH<7 in the presence of β-cyclodextrin because of encapsulation of the hydrophobic adamantyl residue inside the β-cyclodextrin cavity and the presence of protonated tertiary amine units inside the dendrimer [72]. Under these experimental conditions, 8-anili-

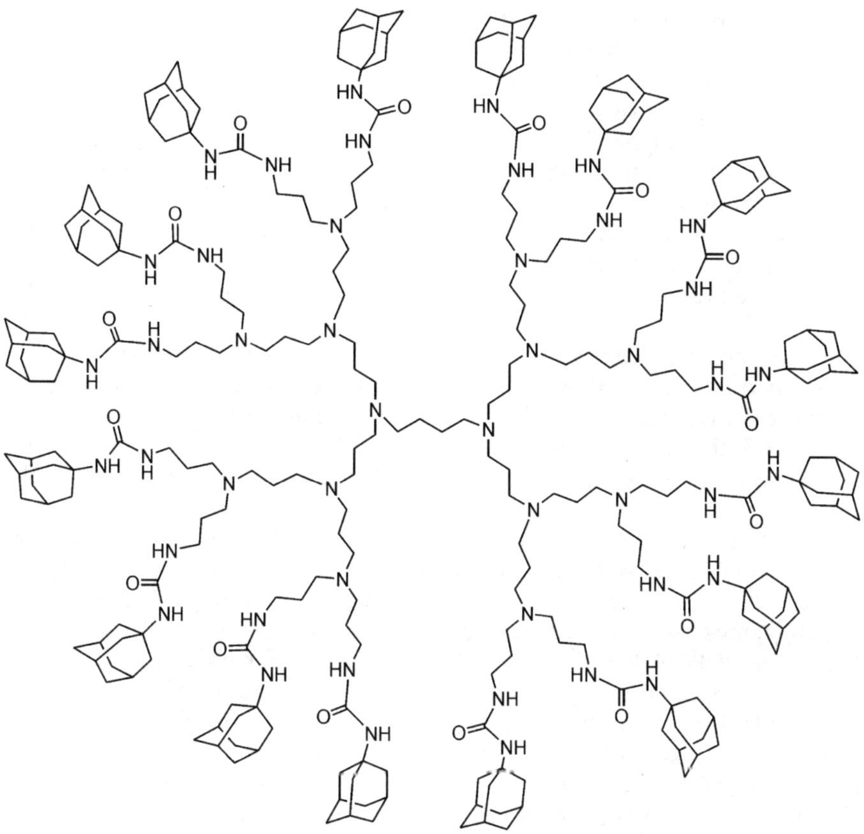

36

nonaphthalnene-1-sulfonate (ANS) is taken up into dendrimers of generations 2–5 because of electrostatic and acid–base interactions. Dye emission intensity is increased because dendrimers (particularly those of high generation) shield the probe from water, preventing quenching by OH groups.

In another example, 5-(dimethylamino)-1-naphthalenesulfonic acid (DNS) is encapsulated in a sixth generation amine-terminated poly(amido amine) (PAMAM) dendrimer in water solution [73]. Below pH 5.5, DNS is protonated and cannot be encapsulated by the host, while at pH>10, encapsulation efficiency is low because of the presence in the dendrimer of very few protonated primary amine units to which DNS can bind. Between pH 10 and 8, DNS emission intensity increases and its emission maximum is progressively blue-shifted. This observation demonstrates that the dendritic microenvironment polarity (always lower than that of bulk water) decreases and that the probe is sampling more densely packed binding sites with decreasing pH, because of structural rearrangement (backfolding of end groups) upon protonation of internal amine units.

Dendrimers containing a polar surface and an apolar interior can work as hosts of hydrophobic fluorescent probes in aqueous environment. In particular, much attention has been devoted to the interaction of dendrimers with pyrene in water solution, studying the polarity of the dendritic microenvironment, the hosting ability of the dendrimers as a function of pH and generation, and the aggregation between dendrimers and cationic or anionic surfactants. Indeed, pyrene is a highly fluorescent molecule, very sensitive to solvent polarity: the ratio of the first to the third band (I_1/I_3) of its vibrational structured emission spectrum increases as the environment polarity increases. Moreover, its emission can be quenched by an electron transfer mechanism involving tertiary amine units and an excimer emission is easily observed, thus demonstrating the presence of more than one pyrene molecule inside the same dendrimer host.

Particularly noteworthy are the investigations on the interaction of amine-terminated PAMAM [74] and POPAM dendrimers [75, 76] with pyrene. In the former case, the authors concluded that (i) the amount of solubilized pyrene increases with increasing dendrimer generation; (ii) concomitantly, the water content inside the dendritic environment decreases, as demonstrated by the decrease in the I_1/I_3 ratio; (iii) pyrene fluorescence is statically quenched by internal tertiary amine units of the dendrimer, thus demonstrating the formation of a host–guest system; (iv) excimer emission is observed even at very low concentration ([pyrene]/[dendrimer]$<10^{-3}$), where, according to a Poisson distribution, most of the dendrimers must contain only one fluorescent probe. In the case of POPAM dendrimers, similar results have been obtained [76] and the influence of pH on the recognition process has been investigated [75]. In particular, at pH 9 pyrene is encapsulated in the internal microcavities of the dendrimers since it cannot interact with the protonated amine shell. On the other hand, lowering the pH causes the release of pyrene into the water solution, as shown by an increase in fluorescence intensity (no quenching mechanism takes place in the bulk solution) and in the value of the I_1/I_3 ratio. This behavior is due to the protonation of all the amine units of the dendrimers, which leads to the formation of a polar interior. Therefore, this host–guest system can be controlled by pH stimuli. It should be noted that, even in the case of a fifth-generation dendrimer, the maximum host/guest ratio is 0.029, that is, one pyrene every 35 dendrimer molecules.

A recent example of photo-induced electron transfer involving a guest inside a dendrimer and an external quencher is offered by Zn(II)–mesoporphyrin coordinated to α-helical peptide-terminated PAMAM dendrimers in the presence of MV^{2+} or naphthalene sulfonate (NS$^-$) [77]. The fluorescence quenching of Zn(II)–mesoporphyrin by either MV^{2+} or NS$^-$, as well as the rate of MV^{2+} photoreduction [78], increases with dendrimer generation. However, the quenching mechanisms are different: static in the case of NS$^-$ and dynamic in the case of MV^{2+}, as demonstrated by the values of the excited state lifetimes. Indeed, for methylviologen, quencher association to the dendrimer is prevented by electrostatic repulsion between the arginine peptide positive shell and the positively charged MV^{2+}. This system has also been used as photosensitizer for photoinduced hydrogen evolution in the presence of triethanolamine as electron donor, MV^{2+} as electron carrier, and hydrogenase as hydrogen evolution catalyst [79]. The hydrogen evolution is more efficient in the presence of the den-

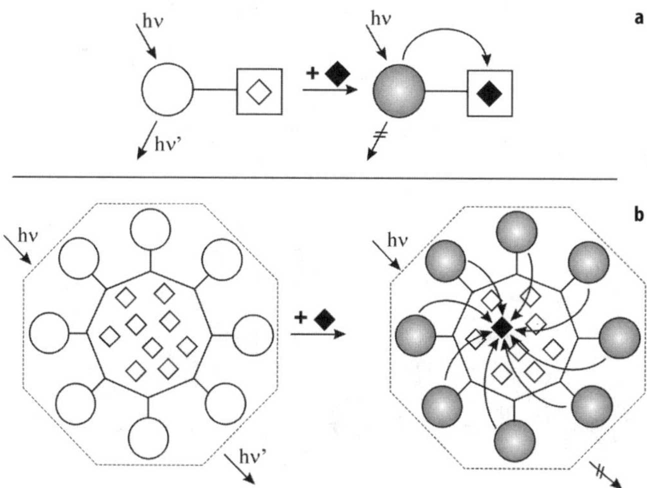

Fig. 3a, b. Schematic representation of (**a**) conventional fluorescent sensor and (**b**) fluorescent sensor with signal amplification. *Open rhombi* indicate coordination sites and *black rhombi* indicate metal ions. The *curved arrows* represent quenching processes. In the case of a dendrimer, the absorbed photon excites a single fluorophore component, which is quenched by the metal ion regardless of its position

dritic photosensitizer than in the case of an α-helix bundle structure and Zn(II)–mesoporphyrin alone.

It has been demonstrated that dendrimers can be used also as fluorescent sensors for metal ions. Poly(propylene amine) dendrimers functionalized with dansyl units at the periphery like **34** can coordinate metal ions by the aliphatic amine units contained in the interior of the dendrimer [80]. The advantage of a dendrimer for this kind of application is related to the fact that a single analyte can interact with a great number of fluorescent units, which results in signal amplification. For example, when a Co^{2+} ion enters dendrimer **34**, the fluorescence of all the 32 dansyl units is quenched with a 32-fold increase in sensitivity with respect to a normal dansyl sensor. This concept is illustrated in Fig. 3.

Dendrimer **37** is quite interesting since it contains 18 amide groups in the interior, which are known to strongly coordinate lanthanide ions, and 24 chromophoric dansyl units in the periphery, which, as mentioned above, show intense absorption bands in the near UV spectral region and an intense fluorescence band in the visible region [81, 82]. Addition of lanthanide ions to acetonitrile/dichloromethane (5:1 v/v) solutions of dendrimer **37** causes a quenching of the fluorescence of the dansyl units. At low metal ion concentration, each dendrimer can host not more than one metal ion and, when the encapsulated metal ion is Nd^{3+} or Eu^{3+}, the fluorescence of all the 24 dansyl units is quenched with unitary efficiency. Quenching by Nd^{3+} occurs by Förster-type energy transfer from the fluorescent excited state of the dansyl units to a manifold of Nd^{3+} energy levels and is accompanied by the sensitized emission in the near infrared region (λ_{max}=1,064 nm) of the lanthanide ion. Quenching by Eu^{3+} is not accompanied by any sensitized emission, since it occurs by electron transfer owing to

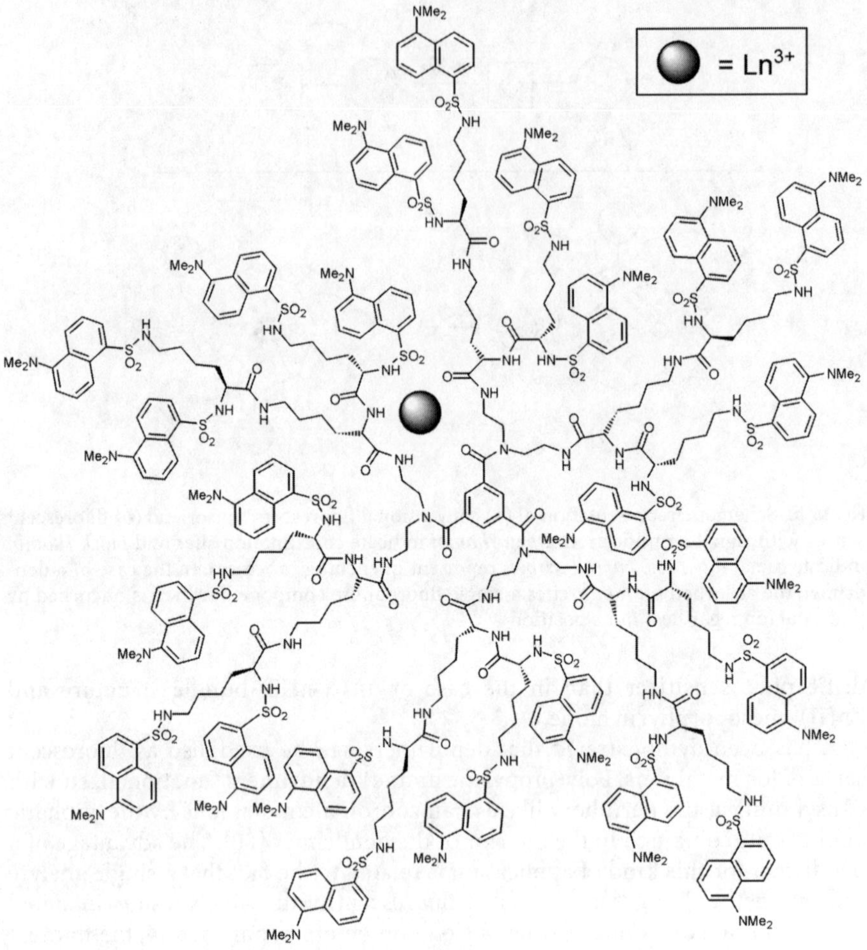

37

the low reduction potential of Eu^{3+}. In rigid matrix at 77 K, however, where electron transfer is disfavored, the quenching of the dansyl unit by Eu^{3+} takes place by energy transfer, as demonstrated by the presence of the sensitized Eu^{3+} emission [82].

Acknowledgements. This work has been supported by MIUR (Supramolecular Devices Project), University of Bologna (Funds for Selected Topics), and EC (HPRN-CT-2000–00029). C.S. acknowledges the Swiss National Science Foundation for financial support.

We would like to thank Professor F. Vögtle and his group for a long-lasting and most profitable collaboration in dendrimer chemistry.

8
References

1. Gilbert A, Baggot J (1991) Essentials of molecular photochemistry. Blackwell, Oxford, UK
2. Valeur B (2002) Molecular fluorescence. Wiley-VCH, Weinheim, Germany
3. Fabbrizzi L (2000) (ed) Special issue of Coord Chem Rev on Fluorescence sensing, vol 205
4. Newkome GR, Moorefield CN, Vögtle F (2001) Dendrimers and dendrons. Concepts, syntheses, applications. Wiley-VCH, Weinheim, Germany
5. Venturi M, Serroni S, Juris A, Campagna S, Balzani V (1998) Topics Curr Chem 197:193; Balzani V, Campagna S, Denti G, Juris A, Serroni S, Venturi M (1998) Acc Chem Res 31:26; Bosman AW, Janssen HM, Meijer EW (1999) Chem Rev 99:1665; Newkome GR, He E, Moorefield C (1999) Chem Rev 99:1689; Adronov A, Fréchet JMJ (2000) Chem Commun 1701
6. Balzani V, Ceroni P, Juris A, Venturi M, Campagna S, Puntoriero F, Serroni S (2001) Coord Chem Rev 219–221:545
7. Gilat SL, Adronov A, Fréchet JMJ (1999) Angew Chem Int Ed 38:1422; Hecht S, Fréchet JMJ (2001) Angew Chem Int Ed 40:75
8. Juris A (2001) In: Balzani V (ed) Electron transfer in chemistry. Biological and artificial supramolecular systems. Wiley-VCH, Weinheim, Germany, vol 3, p 655
9. Campagna S, Serroni S, Puntoriero F, Di Pietro C, Ricevuto V (2001) In: Balzani V (ed) Electron transfer in chemistry. Molecular-level electronics, imaging and information, energy and environment. Wiley-VCH, Weinheim, Germany, vol 5, p 186
10. Ceroni P, Juris A (2003) In: Abdel-Mottaleb MSA, Nalwa HS (eds) Handbook of photochemistry and photobiology. American Scientific Publishers (in press)
11. Zander Ch, Enderlein J, Keller RA (2002) (eds) Single molecule detection in solution. Methods and applications. Wiley-VCH, Weinheim, Germany
12. Feringa BL (ed) (2001) Molecular switches. Wiley-VCH, Weinheim, Germany
13. Balzani V, Scandola F (1991) Supramolecular photochemistry. Ellis Horwood, Chichester, England
14. Piotrowiak P (2001) (ed) In: Balzani V (ed) Electron transfer in chemistry. Principles, theories, methods, and techniques, vol 1, part 1. Wiley-VCH, Weinheim, Germany
15. Pleovets M, Vögtle F, De Cola L, Balzani V (1999) New J Chem 23:63
16. Vögtle F, Plevoets M, Nieger M, Azzellini GC, Credi A, De Cola L, De Marchis V, Venturi M, Balzani V (1999) J Am Chem Soc 121:6290
17. Zhou X, Tyson DS, Castellano FN (2000) Angew Chem Int Ed 39:4301
18. Kimura M, Shiba T, Muto T, Hanabusa K, Shirai H (2000) Tetrahedron Lett 41:6809
19. Kawa M, Fréchet JMJ (1998) Chem Mater 10:286
20. Sabbatini N, Dellonte S, Bonazzi A, Ciano M, Balzani V (1986) Inorg Chem 25:1738
21. Balzani V, Campagna S, Denti G, Juris A, Serroni S, Venturi M (1998) Acc Chem Res 31:26
22. Serroni S, Campagna S, Puntoriero F, Di Pietro C, McClenaghan ND, Loiseau F (2001) Chem Soc Rev 30:367
23. Baudin HB, Davidsson J, Serroni S, Juris A, Balzani V, Campagna S, Hammarstrom L (2002) J Phys Chem A 106:4312
24. Campagna S, Denti G, Serroni S, Juris A, Venturi M, Ricevuto V, Balzani V (1995) Chem Eur J 1:211
25. McClenaghan ND, Loiseau F, Puntoriero F, Serroni S, Campagna S (2001) Chem Commun 2634
26. Campagna S, Di Pietro C, Loiseau F, Maubert B, McClenaghan N, Passalacqua R, Puntotiero F, Ricevuto V, Serroni S (2002) Coord Chem Rev 229:67
27. Shortreed MR, Swallen SF, Shi Z-Y, Tan W, Xu Z, Devadoss C, Moore JS, Kopelman R (1997) J Phys Chem B 101:6318
28. Hofkens J, Latterini L, De Belder G, Gensch T, Maus M, Vosch T, Karni Y, Schweitzer G, De Schryver FC, Hermann A, Mullen K (1999) Chem Phys Lett 304:1; Karni Y, Jordens S, De Belder G, Hofkens J, Schweitzer G, De Schryver FC, Hermann A, Mullen K (1999) J Phys Chem B 103:9378; Hofkens J, Maus M, Gensch T, Vosch T, Cotlet M, Köhn F, Hermann A, Mullen K, De Schryver FC (2000) J Am Chem Soc 122:9278

29. Maus M, De R, Lor M, Weil T, Mitra S, Wiesler U-M, Hermann A, Hofkens J, Vosch T, Mullen K, De Schryver FC (2001) J Am Chem Soc 123:7668; Weil T, Reuther E, Mullen K (2002) Angew Chem Int Ed 41:1900

30. Maus M, Mitra S, Lor M, Hofkens J, Weil T, Hermann A, Mullen K, De Schryver FC (2001) J Phys Chem A 105:3961

31. Herrmann A, Weil T, Sinigersky V, Wiesler U-M, Vosch T, Hofkens J, De Schryver FC, Mullen K (2001) Chem Eur J 7:4844

32. Gronheid R, Hofkens J, Köhn F, Weil T, Reuther E, Müllen K, De Schryver FC (2002) J Am Chem Soc 124:2418

33. Gilat SL, Adronov A, Fréchet JMJ (1999) Angew Chem Int Ed 38:1422; Adronov A, Gilat SL, Fréchet JMJ, Ohta K, Neuwahl FVR, Fleming JR (2000) J Am Chem Soc 122:1175

34. Adronov A, Gilat SL, Fréchet JMJ, Ohta K, Neuwahl FVR, Fleming GR (2000) J Am Chem Soc 122:1175

35. Van Hutten PF, Krasnikov VV, Hadziioannou G (1999) Acc Chem Res 32:257

36. Meskers SJC, Bender M, Hübner J, Romanovskii YV, Oestreich M, Schenning APHJ, Mejier EW, Bässler H (2001) J Phys Chem A 105:10220

37. Pullerits T, Sundström V (1996) Acc Chem Res 29:381

38. Accorsi G, Armaroli N, Eckert J-F, Nierengarten JF (2002) Tetrahedron Lett 43:65

39. Cardona CM, Alvarez J, Kaifer AE, McCarley TD, Pandey S, Baker GA, Bonzagni NJ, Bright FV (2000) J Am Chem Soc 122:6139

40. Hu Q-S, Pugh V, Sabat M, Pu L (1999) J Org Chem 64:7528

41. Pugh VJ, Hu Q-S, Pu L (2000) Angew Chem Int Ed 39:3638

42. Gong L-Z, Hu Q-S, Pu L (2001) J Org Chem 66:2358

43. Pugh VJ, Hu Q-S, Zuo X, Lewis FD, Pu L (2001) J Org Chem 66:6136

44. Ma L, Lee SJ, Lin W (2002) Macromolecules 35:6178.

45. Kim S, Chang DW, Park SY (2002) Macromolecules 35:2753

46. Holten D, Bocian DF, Lindsey JS (2002) Acc Chem Res 35:57

47. Prodi A, Indelli MT, Kleverlaan CJ, Alessio E, Scandola F (2002) Coord Chem Rev 229:48

48. Jiang DL, Aida T (1998) J Am Chem Soc 120:10895

49. Choi M-S, Aida T, Yamazaki T, Yamazaki I (2001) Angew Chem Int Ed 40:3194; Choi M-S, Aida T, Yamazaki T, Yamazaki I (2002) Chem Eur J 8:2668

50. Yeow EKL, Ghiggino KP, Reek JNH, Crossley MJ, Bosman AW, Schenning APHJ, Meijer EW (2000) J Phys Chem B 104:2596

51. Kimura M, Shiba T, Muto T, Anabusa K, Shirai H (1999) Macromolecules 32:8237

52. Kimura M, Nakada K, Yamaguchi Y, Hanabusa K, Shirai H, Kobayashi N (1997) Chem Commun 1215

53. Tomioka N, Takasu D, Takahashi T, Aida T (1998) Angew Chem Int Ed 37:1531

54. Maruo N, Uchiyama M, Kato T, Arai T, Akisada H, Nishino N (1999) Chem Commun 2057

55. Kato T, Uchiyama M, Maruo N, Arai T, Nishino N (2000) Chem Lett 144

56. Mongin O, Hoyler N, Gossauer A (2000) Eur J Org Chem 1193

57. Vinogradov SA, Lo L-W, Wilson DF (1999) Chem Eur J 5:1338

58. Vinogradov SA, Wilson DF (2000) Chem Eur J 6:2456

59. Rozhkov A, Wilson D, Vinogradov SA (2002) Macromolecules 35:1991

60. Ng ACH, Li X-y, Ng DKP (1999) Macromolecules 32:5292

61. Capitosti GJ, Cramer SJ, Rajesh CS, Modarelli DA (2001) Org Lett 3:1645

62. Matos MS, Hofkens J, Verheijen W, De Schryver FC, Hecht S, Pollak KW, Fréchet JMJ, Forier B, Dehaen W (2000) Macromolecules 33:2967

63. Harth EM, Hecht S, Helms B, Malmstrom EE, Frèchet JMJ, Hawker CJ (2002) J Am Chem Soc 124:3926

64. Fukuzumi S, Guldi DM (2001) In: Balzani V (ed) Electron transfer in chemistry. Organic, organometallic, and inorganic molecules. Wiley-VCH, Weinheim, Germany, vol 2, p 270

65. Nierengarten JF (2000) Chem Eur J 6:3667; see also the extensive reference list in [66]

66. Rio Y, Accorsi G, Nierengarten H, Rehspringer J-L, Hönerlage B, Kopitkovas G, Chugreev A, Dorsselaer AV, Armaroli N, Nierengarten JF (2002) New J Chem 26:1146

67. Kunieda R, Fujitsuka M, Ito O, Ito M, Murata Y, Komatsu K (2002) J Phys Chem B 106:7193

68. Baars MWPL, Meijer EW (2001) Top Curr Chem 210:131
69. Balzani V, Ceroni P, Gestermann S, Gorka M, Kauffmann C, Maestri M, Vögtle F (2000) ChemPhysChem 224; Balzani V, Ceroni P, Gestermann S, Gorka M, Kauffmann C, Vögtle F (2002) Tetrahedron 58:629
70. Hahn U, Gorka M, Vögtle F, Vicinelli V, Ceroni P, Maestri M, Balzani V (2002) Angew Chem Int Ed 41:3595
71. Schenning APHJ, Peeters E, Mejier EW (2000) J Am Chem Soc 122:4489
72. Michels JJ, Baars MWPL, Meijer EW, Huskens J, Reinhoudt DN (2000) J Chem Soc Perkin Trans 2:1914
73. Chen W, Tomalia DA, Thomas JL (2000) Macromolecules 33:9169
74. Pistolis G, Malliaris A, Paleos CM, Tsiourvas D (1997) Langmuir 13:5870
75. Pistolis G, Malliaris A, Tsiourvas D, Paleos CM (1999) Chem Eur J 5:1440
76. Pistolis G, Malliaris A (2002) Langmuir 18:246.
77. Sakamoto M, Ueno A, Mihara H (2000) Chem Commun 1741
78. Sakamoto M, Ueno A, Mihara H (2001) Chem Eur J 7:2449
79. Sakamoto M, Kamachi T, Okura I, Ueno A, Mihara H (2001) Biopolymers 59:103
80. Balzani V, Ceroni P, Gestermann S, Kauffmann C, Gorka M, Vögtle F (2000) Chem Commun 853; Vögtle F, Gestermann S, Kauffmann C, Ceroni P, Vicinelli V, Balzani V (2000) J Am Chem Soc 122:10398
81. Vögtle F, Gorka M, Vicinelli V, Ceroni P, Maestri M, Balzani V (2001) ChemPhysChem 769
82. Vicinelli V, Ceroni P, Maestri M, Balzani V, Gorka M, Vögtle F (2002) J Am Chem Soc 124:6461

Top Curr Chem (2003) 228: 193–204
DOI 10.1007/b11011

Antenna Effects of Aromatic Dendrons and Their Luminescence Applications

Manabu Kawa

MCC-Group Science and Technology Research Center, Mitsubishi Chemical Corporation,
1000 Kamoshida-cho, Aoba-ku, Yokohama 227–8502, Japan
E-mail: 1508532@cc.m-kagaku.co.jp

Remarkable concentration of the photon energy absorbed by aromatic dendritic scaffolds toward the focal point (i.e., antenna effects) is one of the most distinct features of the dendritic macromolecules. Poly(benzyl ether) dendron is used as effective antennas in various luminescence systems. For example, organic photosensitive moieties (azo linkage, porphyrin), poly(phenyleneethynylene), metal cations (Tb^{3+}, Eu^{3+}, Ru^{2+}), and semiconductor nanocrystals (CdSe) are surrounded by the dendrons. "Energy funnels" are designed using poly(phenylacetylene) dendrimer and a multiporphyrin array arranged in a dendritic scaffold to demonstrate effective photon energy harvesting toward the focal point. The importance of symmetric structure for the antenna effect has been revealed both in the whole shape of the macromolecule and the dendritic repeating unit. Luminescence applications are discussed (e.g., temperature-responding gel, fluoroimmunoassay reagent).

Keywords. Dendrimer, Antenna, Photon energy, Luminescence, Metal cation

Abbreviations

PBE   poly(benzyl ether)
DMSO   dimethyl sulfoxide
FIA   fluoroimmunoassay
IR   infrared
PPhE   poly(phenyleneethynylene)
THF   tetrahydrofuran
TOPO   trioctylphosphine oxide

1
Introduction

Remarkable concentration of the photon energy absorbed by aromatic dendritic scaffold toward the focal point (herein called "antenna effects") is one of the most distinct features of the dendritic macromolecules. Poly(benzyl ether) (PBE) dendron [1] has been studied most extensively as an efficient antenna dendron. Careful studies have revealed the crucial role of the symmetric dendritic structure for the antenna effect.

The wedge-like skeletons of dendrons have inspired synthetic chemists to construct energy funnels which collect and concentrate photon energy effectively. This intuition partly came from the idea of mimicking of the natural light-harvesting systems. Another driving idea is the "shell" or "insulator" role of dendrons (i.e., site-isolation effect). The luminescence species located at the focal point can be engulfed within dendrimers; this isolates the luminescence species from the outside environment, which may otherwise be harmful for the luminescence, for example, by energy quenching *via* molecular collision or/and small coordinating molecules (e.g., water). The idea of site-isolation itself is a topological matter, having no relation with the antenna effect (an energy-related matter); however, the discovery of antenna effects followed the former. A simple but important benefit of the "shell" dendron should be stressed here: the dissolution of the luminescence species in desired media (e.g., solvents or resins) is indispensable to the development of practical applications.

2
Antenna Effects

2.1
Antenna Effect Towards an Organic Photosensitive Moiety at the Focal Point

Organic luminescence systems (macromolecules) consisting of covalent bonds have been reported. The two dendron subunits connected by an azo linkage illustrated in Fig. 1 showed unusual IR sensitivity [2, 3]. The focal azo linkage undergoes *cis–trans* isomerization at particular wavelengths of IR (1597 cm^{-1}= 6,262 nm, corresponding to the aromatic absorption) and ultraviolet irradiation. Only the dendron subunits larger than 3rd-generation allowed the IR-induced reaction. IR radiation induces relatively low-frequency molecular vibra-

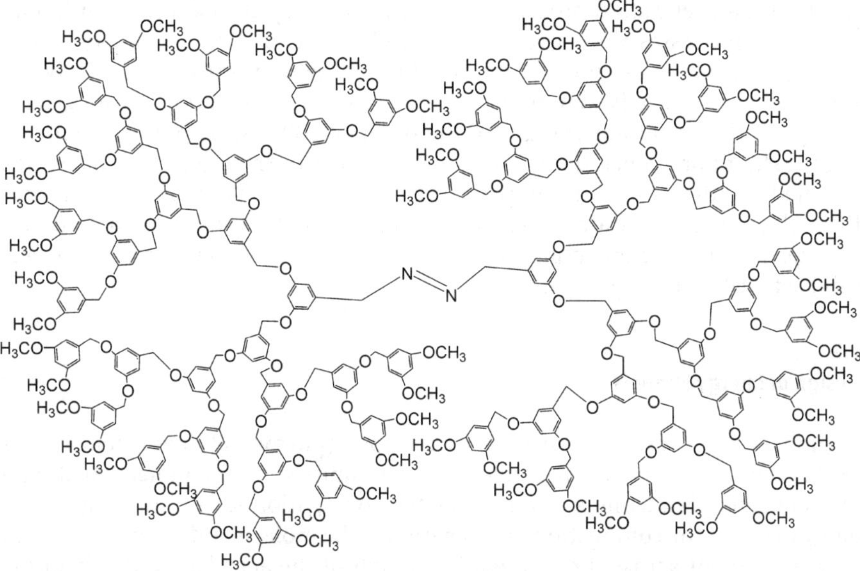

Fig. 1. Two poly(benzyl ether) dendron subunits connected by an azo linkage

tions. However, with the exception of hydrogen bond reorganization, the vibration energy is usually consumed rapidly through molecular collisions rather than inducing photochemical changes. Therefore the large dendrons are thought to insulate against molecular collisions, allowing the absorbed IR energy to induce the azo isomerization. Multiphoton intramolecular energy transfer was proposed as the mechanism of the unusual photoreaction. Precise measurement of the IR absorption behavior estimated the number of IR photons absorbed by a 4th-generation dendron subunit as only 10–3 (photons s^{-1}) [4]. The multiphoton energy was thought to be absorbed not simultaneously but sequentially, and long-term intramolecular energy storage was anticipated [5].

The antenna effect of the poly(benzyl ether) dendrimers remarkably depends on the whole macromolecular shape (i.e., the symmetry of the dendron subunits). A series of porphyrin-cored dendrimers were synthesized with different numbers of 4th-generation dendron subunits at the *meso*-positions of the porphyrin core [6]. The symmetrical morphology bearing four dendron subunits (a spherical morphology) allowed higher energy transfer quantum yield (80.3%) than the asymmetrical morphology bearing three dendron subunits (10.1–31.6%). Fluorescence depolarization studies indicated that the excitation energy migrates very efficiently in the symmetrical morphology through the dendritic array of 3,5-dioxybenzyl building units within the lifetime of the excited state.

The morphological dependence of the antenna effect was quantitatively studied by the comparison of the three focal isomers of benzene-cored 4th-generation dendrimers [7]. The energy of the photon absorbed by the dendron subunits must be dissipated either through radiative (luminescence) or nonradiative (thermal) relaxation processes. After excitation by UV light (244 nm), both relaxation processes were strictly measured for the *o*-, *m*-, and *p*-isomers in CH_2Cl_2 solution, using fluorescent and thermal lens spectroscopies. Only the *p*-isomer (i.e., the only symmetrical isomer) showed exceptionally long-term storage of the excitation energy (of the order of 100 ms) within the dendritic entity. The fluorescence decay times of the three isomers were 1.7 ns, which revealed that the anomalous properties of dendrimers did not originate in long-lived electronic excitation states, but in long-term storage of internal energy. To explain this phenomenon, a nonlinear conjugated oscillator model of Fermi–Pasta–Ulam theory was invoked.

2.2
Design of Energy Funnels

A pioneering study was the synthesis of poly(phenylacetylene) dendrimers [8–11]. The researchers constructed "energy funnels" by the molecular design of conjugated dendrimers; the shorter the conjugation length, the higher the energy level of the conjugation unit. Therefore, the molecular design imposed an energy gradient across the conjugation length of the acetylene units from the periphery (the highest energy) to the focal point (the lowest energy) of the dendrimer molecule (Fig. 2). Spectral [12] and computational [13] studies were also conducted to examine the energy gradient molecular design.

A large multiporphyrin array arranged in a dendritic scaffold (Fig. 3) was designed to demonstrate an energy funnel as a mimic of the bacterial light-harvesting antenna complex [14]. The macromolecule consists of four dendritic wedges of a zinc porphyrin heptamer (energy-donating units) anchored by a

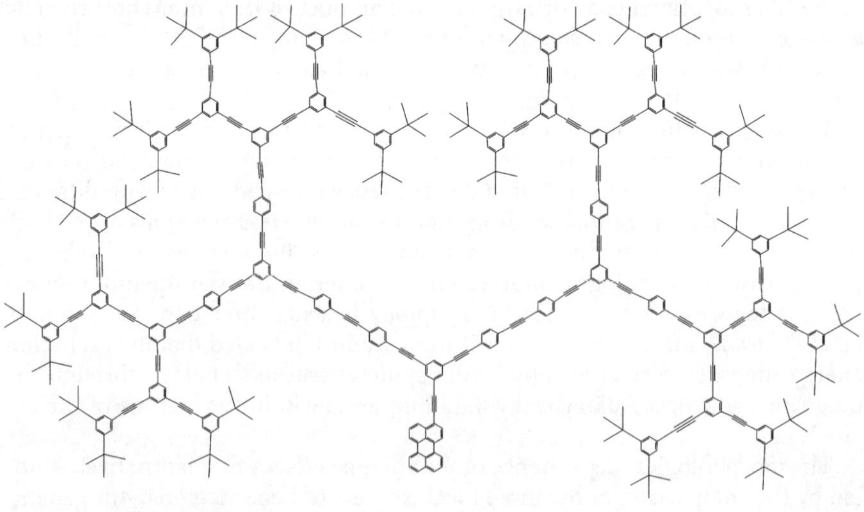

Fig. 2. A poly(phenylacetylene) dendrimer with an energy gradient across the conjugation length of the acetylene units from the periphery to the focal point

Fig. 3. A large multiporphyrin array arranged in a dendritic scaffold

focal free-base porphyrin (energy receptor). Second-generation PBE dendrons were attached to make the macromolecule soluble in common organic solvents. Fluorescence studies of the macromolecule led to an important conclusion, that is, an efficient migration of excitation energy takes place through the dendritic zinc porphyrin array before transfer to the focal free-base porphyrin. The zinc porphyrin units are thought to mutually cooperate to facilitate long-range energy migration and transfer. Each zinc porphyrin unit is separated by a benzyl ether dendron moiety, where π-electronic conjugation is discontinued by the CH_2–O– linkage. Thus, the dendritic location of the zinc porphyrin units within the macromolecule should be crucial for the energy migration and transfer, even in the absence of π-electronic conjugation. The importance of the symmetry of dendritic structure again appeared in this different chemical system.

2.3
Dendron-Grafted Conjugated Linear Polymers

Another remarkable aspect of the antenna effect of PBE dendrons has been demonstrated in the study of blue-luminescent poly(phenyleneethynylene) (PPhE), bearing the dendrons as the repeating side groups (Fig. 4) [15–17]. A series of the PPhE dendrimers were successfully synthesized by a transition-metal complex-catalyzed polymerization of a bis-ethynyl monomer bearing two dendrons and 1,4-diiodobenzene. The rod-like polymer grafted with the 3rd-generation PBE dendron emitted blue luminescence from the main chain when the PBE dendron was excited by UV light, indicating the energy transfer from the PBE dendron side chain to the PPhE main chain. The emission quantum yield appeared almost quantitatively (0.97). An important observation is that the grafting PBE dendrons smaller than 2nd-generation did not give the blue luminescence from the PPhE main chain, suggesting the contribution of the site isolation of the larger dendrons. Spin-coated films of the PPhE dendrimers were studied as well [18].

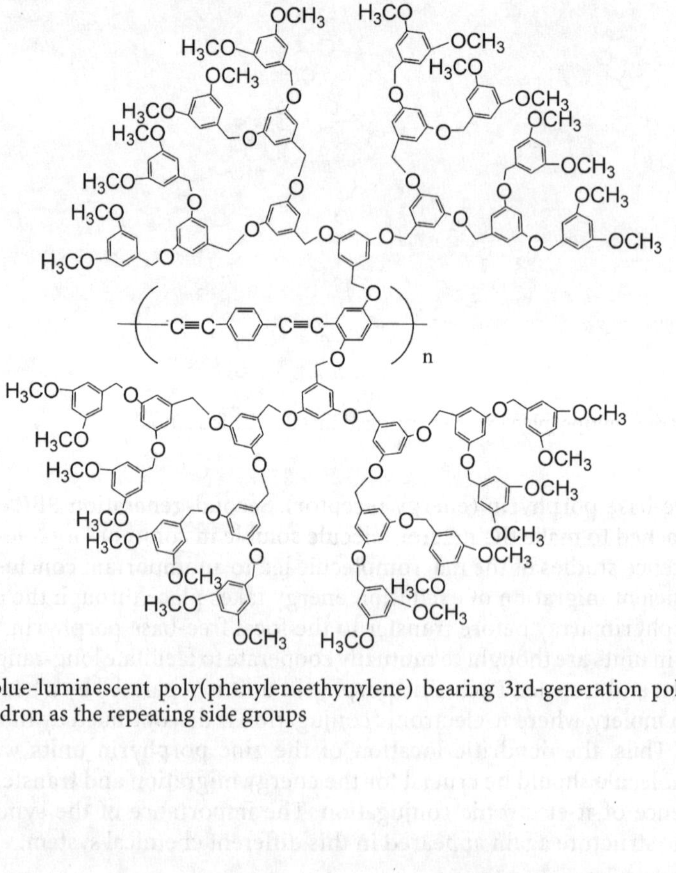

Fig. 4. A blue-luminescent poly(phenyleneethynylene) bearing 3rd-generation poly(benzyl ether) dendron as the repeating side groups

2.4
Dendrimer-Metal Cation Complexes

Self-assembly of aromatic dendron subunits has been tried by the design of coordination to multivalent metal cations (i.e., metal-cored dendrimer complexes). Several metal-cored dendrimer complexes have successfully exhibited luminescence by antenna effects.

Trivalent lanthanide cations (Tb^{3+} and Eu^{3+}) were coordinated by the PBE dendrons bearing focal carboxylate group, exhibiting remarkable luminescence from the lanthanide cation core by UV irradiation [19]. The luminescence of the lanthanide-cored dendrimer complexes was attributed to the transfer of the photon energy absorbed by the dendron subunits. The mechanism was supported by the coincidence between the excitation spectrum of the whole complex and the absorption spectrum of the dendron subunits. A remarkable observation both in solution and in the bulk state was that the intensity of the luminescence increased with increasing dendron generation. The maximum luminescence occurred by the excitation around 290 nm, corresponding to the absorption of monosubstituted benzene ring. Therefore, the absorption of photons by the phenyl terminals located at the farthest position in the dendron subunits worked as the major origin of the luminescence at the metal core, even through the largest 4th-generation dendron. The contribution of site isolation by the larger dendron subunits for the enhanced luminescence of the lanthanide core was also indicated by the bulk-state experiment [19].

The energy level of the excitation states between the lanthanide core and the dendron subunits should match for effective luminescence. In fact Tb^{3+} is the best illuminator to date matching with the carboxylate PBE dendrons, whereas Eu^{3+} illuminates less and other lanthanides (e.g., Dy^{3+}, Tm^{3+}) do not at all. Isomerism at the focal aromatic ring has been revealed to influence the intensity of the lanthanide core as well. The 3,4-dioxybenzoate moiety turned out to be the best focal building block for the Tb^{3+} core [20, 21], whereas the 2,4-dioxybenzoate moiety was the best for the Eu^{3+} core [22]. Indeed, the 3rd-generation Tb^{3+}-cored dendrimer complex bearing a focal 3,4-dioxybenzoate moiety illuminated stronger than the 4th-generation one bearing a focal 3,5-dioxybenzoate moiety.

In the context of the isomerism within the dendron structure, the 3,5-dioxybenzyl repeating unit of the PBE dendron proved to be the best isomeric structure [23]. Other isomeric repeating units (3,4- and 2,5-dioxybenzyl) apparently decreased the antenna effect. Thus, the remarkable antenna effects of the PBE dendrons in various systems are most probably attributed to the fortunate symmetrical 3,5-dioxybenzyl units spread amongst the dendritic skeleton. One more important aspect is that the electronic character of the individual aromatic building units (variable by the isomerism) seems to influence the energy transfer behavior of the whole PBE dendron, even though all the aromatic units are separated from each other by the nonconjugating $-CH_2-O-$linkages.

Poly(ether ketone) dendrons [24] have been observed to show an antenna effect toward lanthanide cations, especially for Eu^{3+}. Two focal coordination forms were examined, namely carboxylate [25] and 1,3-diketonate [26]. The use of the poly(ether ketone) dendrons instead of the PBE dendrons is practically advantageous for the following two reasons: a common UV source (365-nm mercury lamp) can be used as the excitation light and improved miscibility in the commercial transparent resins such like poly(methyl methacrylate) is observed. The two advantages come from the ketone moiety, which makes the π-conjugation longer (absorbing longer wavelength) and increases the dipole moment of the repeating unit.

PBE dendrimers with a cyclic polyamine core at the focal point have been synthesized to form transition-metal complexes [27]. Tb^{3+} complexes exhibited luminescence by the excitation of the dendrons.

PBE dendrons bearing a focal bipyridine moiety have been demonstrated to coordinate to Ru^{2+} cations, exhibiting luminescence from the metal cation core by the excitation of the dendron subunits [28–30]. The terminal peripheral unit was examined (e.g., phenyl, naphthyl, 4-*t*-butylphenyl) to control the luminescence. The Ru^{2+}-cored dendrimer complexes are thought to be photo/redox-active, and photophysical properties, electrochemical behavior, and excited-state electron-transfer reactions are reported.

A supramolecular assembly of macromolecules bearing antenna dendron has been reported. Pyrazole-anchored PBE dendrons were synthesized to examine the coordination behavior to transition-metal cations (Cu, Au, Ag) [31]. Self-assembled metallacycles were found. The Cu-metallacycle further formed luminescent fibers about 1 μm in diameter. The luminescence (605 nm) occurred by the excitation of the dendron (280 nm) and the excitation spectrum was coincident with the absorption spectrum of the dendron, suggesting the antenna effect. Interestingly, the luminescence of the Cu-metallacycle fiber disappeared when the fiber was dissociated into the individual metallacycles in C_2H_2.

2.5
Dendrimer Coordinating to Nanocrystals at the Focal Point

PBE dendrons coordinate to the surface of II–VI semiconductor nanocrystals (e.g., CdSe [33] and CdSe/ZnS core/shell structure [34, 35]) to modulate the photoluminescence of the nanocrystals [32]. Trioctylphosphine oxide (TOPO)-capped II–VI semiconductor nanocrystals of several-nanometers diameter have been synthesized by a pyrolysis reaction of organometallics in TOPO [33–35]. The capping ligand (TOPO) can be replaced by stronger ligands such as thiol compounds [36], suggesting that dendrons bearing sulfur atom(s) at the focal point replace TOPO as well.

A 1st-generation dendrimer having a thiocyanuric acid moiety at the focal point was synthesized as a directly replacing ligand. A simple mixing treatment of the TOPO-capped CdSe nanocrystal with the thiocyanuric acid-cored dendrimer in toluene gave a change in excitation spectra of the CdSe luminescence. The excitation spectra of the CdSe nanocrystal before and after the mixing treatment, measured at the same concentration in toluene (emissions at 552 nm), were compared. Coincidence in the longer wavelength region (>390 nm) between the two spectra indicated the preserved electronic states (i.e., band structure) of the CdSe nanocrystal, whereas an enhancement in the shorter wavelength region (280–390 nm) was obvious after the mixing treatment. The shorter wavelength region involves the absorption band of the dendrimer, suggesting energy transfer from the dendrimer toward the CdSe nanocrystal.

3
Applications

Two practical advantages of luminescence species engulfed in antenna dendrimer scaffolds are apparent, namely their miscibility with organic media (solvents or/and resins) and their ability to form thin films. For example the lanthanide-cored dendrimer complexes described in this chapter can be regarded as organic-soluble inorganic luminescers.

The PBE dendron has a glass transition at about 40 °C and is soluble in various organic solvents (e.g., THF, acetone, toluene). It is therefore a moldable, thermoplastic, film-forming material. This practical feature is maintained for the lanthanide-cored dendrimer complexes. The complexes are partially miscible with poly(methyl methacrylate), affording transparent luminescence compositions by mixing in solvent.

The PPhE bearing the PBE dendron as the repeating side chains is also soluble in THF, whereas the rigid main chain itself does not dissolve in any solvent. The blue-luminescence dendron-grafted rigid polymer forms thin films by spin coating [18].

Vinylphenyl-terminated PBE dendrons were prepared as polymerizable dendrons from 4-vinylbenzyl chloride [37]. The vinylphenyl-terminated PBE dendrons are useful to make the lanthanide-cored dendrimer complexes polymerizable. The 1st-generation Tb^{3+}-cored dendrimer complex bearing the vinylphenyl terminal on the dendron subunits (Fig. 5) was copolymerized with N-isopropylacrylamide in the presence of methylene bis-acrylamide (as crosslinker) in DMSO to give a green-luminescence transparent gel. The DMSO gel was con-

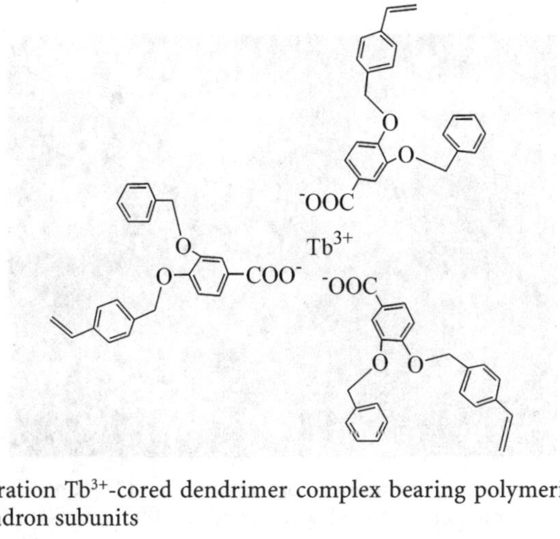

Fig. 5. A 1st-generation Tb^{3+}-cored dendrimer complex bearing polymerizable vinylphenyl groups on the dendron subunits

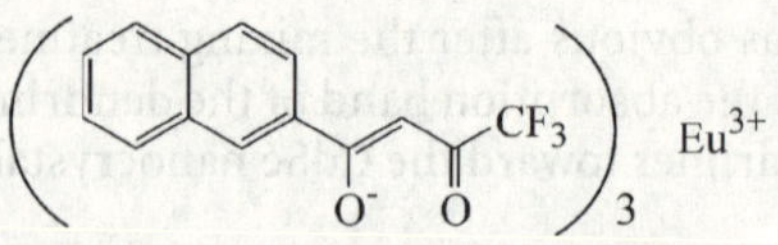

Fig. 6. A europium complex used in practical fluoroimmunoassay

verted into hydrogel by solvent exchange with excess water. The hydrogel still
exhibited a bright green luminescence when irradiated by UV. The brightness of
the luminescence was remarkable even with a little incorporation of the Tb^{3+}-
cored complex (less than 1 mol%) in the gel. The DMSO molecule might prevent
the well-known quenching of the lanthanide's fluorescence by hydration. One
more interesting feature of the Tb^{3+}-incorporated hydrogel is the phase transition around 33 °C, which is characteristic for the *N*-isopropylacrylamide hydrogel. When the Tb^{3+}-incorporated hydrogel was heated over 33 °C, the gel became
opaque. The opacity was reversible and disappeared on cooling. The gel phase
transition occurs in milliseconds; thus, the Tb^{3+}-incorporated hydrogel can be
considered as a quick switching luminescer modulated by temperature.

The Tb^{3+}-cored dendrimer complex bearing the vinylphenyl terminal on the
dendron subunits was copolymerized with styrene or/and methyl methacrylate
to give green-luminescence latex particles, useful as fluoroimmunoassay (FIA)
latex reagent [38]. FIA latex particle is about 500 nm in diameter, bearing antibodies on the surface, which binds selectively with cooperative antigens. A conventional luminescence species used for FIA is, for example, a europium complex bearing naphthyl groups (Fig. 6).

The red-luminescence (612 nm) europium complex is an excellent luminescer
in commercial use; however, the green-luminescence Tb^{3+}-cored dendrimer
complex enables a simultaneous assay at another wavelength (545 nm). The latex
formation was carried out by mini-emulsion radical polymerization of the
monomers dissolving the Tb^{3+}-cored dendrimer complexes. The polymeriza-

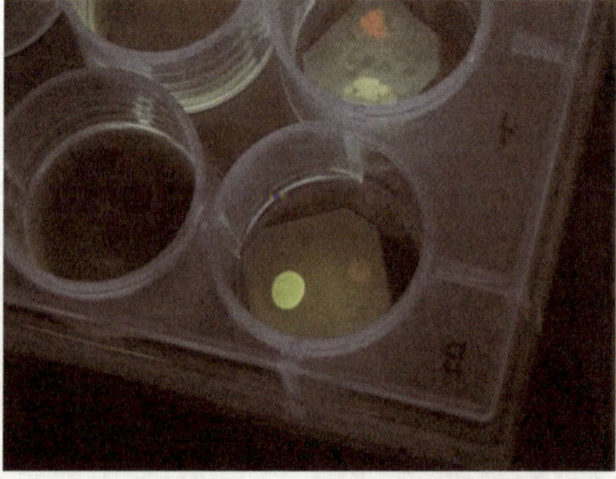

Fig. 7. A demonstration of fluoroimmunoassay. *Green circles* (Tb-cored dendrimer complex)
and *red circles* (conventional Eu complex) corresponds to different assays

tion afforded stable emulsions, milky-white under visible light but illuminating in green under UV irradiation. The 2nd-generation Tb^{3+}-cored dendrimer complex with the 3,4-dioxybenzoate focal moiety gave a UV-luminescent latex emulsion sufficiently bright for practical use. The Tb^{3+}-incorporated latex can attach desired antibody on the surface to allow practical FIA (Fig. 7). The unique aqueous latex emulsions might find various bulk applications such as coatings. The stability of the luminescence should be improved for tougher applications.

4
Conclusion

The antenna effect, the remarkably large photosensitization, of aromatic dendrons is one of the distinct features of strictly symmetrical dendritic hyperbranching. Not only the electronic character of dendrons for the efficient absorption and transfer of photon energy, but also the site-isolation (topological) ability of the dendritic hyperbranching is important for the whole antenna effects. The excellent solubility of aromatic dendrons (e.g., PBE dendron) is a practical advantage of the antenna systems, enabling unique polymeric composition materials. Further improvement in the stability of the luminescence under various practical conditions (e.g., long-time irradiation, heat, water, oxygen) is a key to enlarge their field of application.

5
References

1. Hawker C, Fréchet JMJ (1990) J Am Chem Soc 112:7638
2. Jiang DL, Aida T (1997) Nature 388:454
3. Aida T, Jiang DL, Yashima E, Okamoto Y (1998) Thin Solid Films 331:254
4. Wakabayashi Y, Tokeshi M, Jiang DL, Aida T, Kitamori T (1999) J Lumin 83:313
5. Wakabayashi Y, Tokeshi M, Hibara A, Jiang DL, Aida T, Kitamori T (2000) Anal Sci 16:1323
6. Jiang DL, Aida T (1998) J Am Chem Soc 120:10895
7. Wakabayashi Y, Tokeshi M, Hibara A, Jiang DL, Aida T, Kitamori T (2001) J Phys Chem B 105:4441
8. Zhang J, Moore JS, Xu Z, Aguirre RA (1992) J Am Chem Soc 113:2273
9. Xu Z, Moore JS (1993) Angew Chem Int Ed Engl 32:246
10. Xu Z, Moore JS (1994) Acta Polym 45:83
11. Xu Z, Kahr M, Walker KL, Wilkins CL, Moore JS (1994) J Am Chem Soc 116:4537
12. Kopelman R, Shortreed M, Shi ZY, Tan W, Xu Z, Moore JS, Bar-Haim A, Klafter J (1997) Phys Rev Lett 78:1239
13. Minami T, Tretiak S, Chernyak V, Mukamel S (2000) J Lumin 87–89:115
14. Choi MS, Aida T, Yamazaki T, Yamazaki I (2001) Angew Chem Int Ed Engl 40:3194
15. Sato T, Jiang DL, Aida T (1998) J Am Chem Soc 121:10658
16. Jiang DL, Sato T, Aida T (2001) Chinese J Polym Sci 19:161
17. Aida T, Jiang DL, Sato T, Kawa M (2000) JP Patent (Jpn Kokai Tokkyo Koho) 2000:239360A2
18. Masuo S, Yoshikawa H, Asahi T, Masuhara H, Sato T, Jiang DL, Aida T (2001) J Phys Chem B 105:2885
19. Kawa M, Fréchet JMJ (1998) Chem Mater 10:286
20. Kawa M, Motoda K (2000) Kobunshi Ronbunshu 57:855
21. Motoda K, Kawa M (2000) JP Patent (Jpn Kokai Tokkyo Koho) 2000:281618A2
22. Motoda K, Kawa M (2000) JP Patent (Jpn Kokai Tokkyo Koho) 2000:344712A2
23. Kawa M, Joyce L (unpublished result)

24. Morikawa A, Ono K (1999) Macromolecules 32:1062
25. Morikawa A, Kawa M (2002) JP Patent (Jpn Kokai Tokkyo Koho) 2002:121169A2
26. Morikawa A, Kawa M (2002) JP Patent (Jpn Kokai Tokkyo Koho) 2002:241339A2
27. Aida T, Jiang DL, Sase M, Kawa (2000) JP Patent (Jpn Kokai Tokkyo Koho) 2000:239373A2
28. Issberner J, Voegtle F, De Cola L, Balzani V (1997) Chem Eur J 3:706
29. Plevoets M, Voegtle F, De Cola L, Balzani V (1999) New J Chem 63
30. Voegtle F, Plevoets M, Nieger M, Azzellini GC, Credi A, De Cola L, De Marchis V, Venturi M, Balzani V (1999) J Am Chem Soc 121:6290
31. Enomoto M, Kishimura A, Aida T (2001) J Am Chem Soc 123:5608
32. Kawa M, Saita S, Saito Y (2001) Polym Mater Sci Eng 84:736
33. Murray CB, Norris DJ, Bawendi MG (1993) J Am Chem Soc 115:8706
34. Peng X, Schlamp MC, Kadavanich AV, Alivisatos AP (1997) J Am Chem Soc 119:7019
35. Dabbousi BO, Rodriguez-Viejo J, Mikulec FV, Heine JR, Mattoussi H, Ober R, Jensen KF, Bawendi MG (1997) J Phys Chem B 101:9463
36. Peng X, Wilson TE, Alivisatos AP, Schultz PG (1997) Angew Chem Int Ed Engl 36:145
37. Kawa M (2002) JP Patent (Jpn Kokai Tokkyo Koho) 2002:080535A2
38. Kawa M, Motoda K, Sorin T, Isomura T (2000) JP Patent (Jpn Kokai Tokkyo Koho) 2000: 345052A2

Top Curr Chem (2003) 228: 205–226
DOI 10.1007/b11012

Dendrimers for Optoelectronic Applications

Shiyoshi Yokoyama · Akira Otomo · Tatsuo Nakahama · Yoshishige Okuno
Shinro Mashiko

Communications Research Laboratory and PRESTO, Japan Science and Technology
Corporation (JST), 588-2 Iwaoka, Nishi-ku, Kobe 61-2492, Japan
E-mail: syoko@crl.go.jp

This manuscript describes the dendritic macromolecules for optical and optoelectronic applications, particularly stimulated emission, laser emission, and nonlinear optics. Dendrimers have been designed and synthesized for these applications based on simple concepts. A core-shell structure, through the encapsulation of active units by dendritic branches, or a cone-shaped structure, through the step-by-step reactions of active units, can provide particular benefits for the optical high-gain media and nonlinear optical materials. It also described experimental results that support the methods presented for designing and fabricating functionalized dendrimers for optoelectronic applications, and theoretical results that reveal the intermolecular electronic effect of the dendritic structure.

Keywords. Core-shell structure, Cone structure, Laser emission, Nonlinear optical property

1
Introduction

The widespread use of fluorescent dye in optical and optoelectronic applications has generated a renewed interest in the excitation and decay processes of various dyes. Emission from a dye-containing media is a classic process that can be altered significantly by the application under intensive optical excitation, resulting in stimulated emission. The stimulated emission yields a coherent optical gain. Amplified spontaneous emission and laser emission from organic high gain media are of particular interest because they can be applied to many applications.

From basic science, such as physics and spectroscopy, to medicine and industry, the dye laser has been shown to be an extremely flexible and useful tool. Indeed, it has been estimated that some 10,000 scientific articles have been published that mention dye lasers in their titles or abstract. Demonstration of light amplification from organic gain media rapidly followed the introduction of *mirrorless lasers* to fabricate miniature light sources in optical devices [1]. This growing interest can be attributed to the fact that fluorescent materials, including dye-doped polymer and conducting polymer, can become high-gain media to feedback the light in the simple forms such as films and fibers [2–9]. The performance of mirrorless lasers, which use a potentially inexpensive and easily interchangeable solid gain medium and do not require the use of a solvent or complex optical setup, has generated significant interest in their technology for a number of military and medical applications.

Polymers and supermolecules modified using electron push-pull chromophores are also of particular interest for nonlinear optics (NLO) [10–15]. NLO material has attracted much interest over the past 20 years and has been widely applied in various field (telecommunications, optical data storage, information processing, microfabrication, etc.). Chemists have developed ways to introduce NLO chromophores into many type of polymers, such as linear polymers, cross-linked polymers, and branched polymers, and have demonstrated their performance in NLO applications.

Dendritic macromolecules, called *dendrimers*, are a new category of hyperstructured material and have recently been introduced into optical and optoelectric applications [16–22]. Their long branching chains and the high degree of control over their molecular weight make it possible to create three-dimensional structures that are roughly spherical or globular. In optical and optoelectronic applications, functional chromophores can be placed at branches, cores, and the ends of the dendrimers to control their optical properties. For instance, the passivation of an active core with dendritic branches achieves the site-isolation effect, while the intermolecular coupling between branches enhances the performance of nonlinear optical activities.

Dendrimers, a relatively new class of macromolecules, differ from traditional linear, cross-linked, and branched polymers. The conventional way of introducing an active moiety into polymers is to link it chemically into the polymeric backbone or a polymer branch. This synthetic approach results in a topologically complex material. Therefore, a significant effort has to be devoted to improve the structural complexities and functions of the polymers.

The recent development of structurally controlled dendrimers has led to the development of a wide range of new functional macromolecules. These dendrimers were first applied in the fields of chemistry, including catalysis, pharmacology, and materials science [23–26]. More recently there have been several reports of dendrimers having electroactive, photoactive, and recognition elements [27–34]. Important applications in photonics have recently been exploited, though the number of reports is still limited.

We have found that dendrimers can be used to encapsulate active moieties, thereby preventing them from interacting. This passivation effect limits intermolecular interactions such as self-aggregation and molecular clustering. We also found that dendrimers can be made dipolar. This asymmetry in molecular orientation enables dendrimers to be used in NLO. In this chapter we describe our application of dendrimers to lasers and NLO.

2
Core-Shell Structured Dendrimers

In terms of quantum efficiency, organic laser-dye typically shows a large fluorescent yield, ranging from about 0.6 to near the optimum 1.0. In spite of this large yield, the dye concentration in both solutions and solids must be kept low to achieve highly efficient spontaneous emission. At higher concentrations, energy transfers in clustered molecules almost completely suppress the fluorescence. This is in contrast with the general tendency of photonic-active molecules having a π-electron-conjugated structure, which often self-aggregates to form complex structures. In these structures, fluorescence is often suppressed due to the intermolecular energy transfer. Therefore, a dye concentration of less than 10^{-3} mol/l is generally used in optical applications such as organic laser devices [35].

Optical excitation can be used to alter significantly the radiative action from a high-gain medium containing small particles, such as dendrimers. However, two problems remain to be solved: self-aggregation and self-quenching, both of which can occur at higher concentrations of dye in solutions and solids. To overcome them, we prepared the dendrimer illustrated in Fig. 1. We used a rhodamine B (RdB) chromophore as the core unit. The coupling reaction of the dendritic aryl ether blocks with the chromophore unit resulted in an RdB-core dendrimer. It has a mono-dispersed molecular weight, which we characterized by matrix-assisted laser-desorption-ionization time-of-flight (MALDI-TOF) mass spectroscopy and size exclusion chromatography (SEC). The MALDI-TOF mass spectrum showed a signal peak at m/z=9266, which is in good agreement with the expected molecular weight of 9264. The SEC measurement revealed a narrow peak, from which the purity was determined to be 99.0%.

The absorption and fluorescence spectra of a neat film made of RdB-dendrimer are shown in Fig. 2. The absorption spectrum in visible-wavelength region was similar to that obtained from a solution of RdB with a concentration less than 0.1 mmol/l. Interpretation of the fluorescence in terms of the Frank-Condon mechanism indicated that the core RdB chromophore behaved with a site-isolation effect and had little interaction with the neighboring chro-

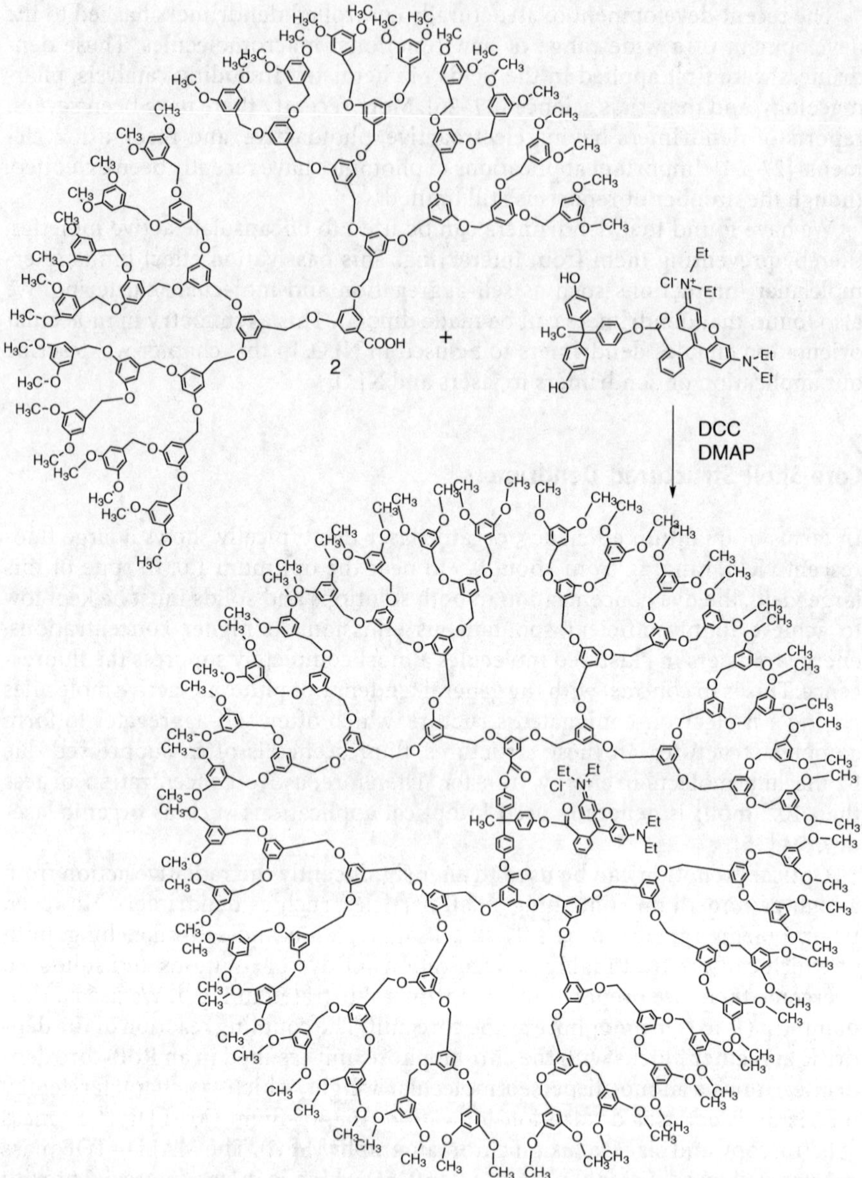

Fig. 1. Synthesis of rhodamine cored dendrimer (Rd/dendrimer)

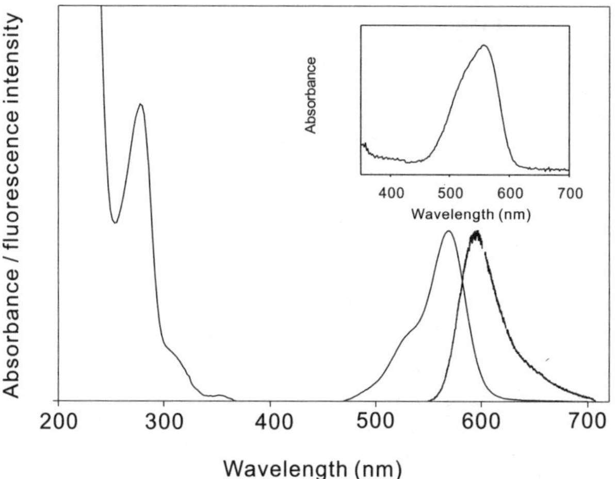

Fig. 2. Absorption (*solid line*) and fluorescence (*bold line*) spectra of Rd/dendrimer film. *Inset* is absorption spectra of the bared rhodamine B film

mophores. The absorption and fluorescence properties of a cast film of the RdB differed from those of the RdB-dendrimer. Its broad absorption spectrum (inset of Fig. 2) was due to molecular aggregation. Aggregation can also occur in solutions with higher concentrated level of the RdB [36, 37]; it results from weak intermolecular bonding involving dipole-dipole and other forces. As a result, the fluorescence in dimers or higher aggregates is nearly completely suppressed. This unwelcome alteration in dye dynamics is connected to the reduced lifetime of organic dyes with an aggregated form. Using time-resolved fluorescence decay measurement, we found that the RdB-dendrimer films exhibit an excited-state lifetime of 2.8 ns. This decay constant is identical to the lifetime of RdB chromophores in low concentration solutions [38]. Therefore, using RdB-dendrimers in optical gain media is advantageous.

In principle, other laser-dye molecules can replace the RdB unit depending on the desired application wavelength range, and the dendrimer shell can be modified to control solubility in the target host materials. However, a practical limitation on the use of dendrimers is their time-consuming synthesis; significant effort has gone into improving their structural complexities and functionality through repetitive step-by-step reactions. We have been investigating the quick preparation of a dye-doped dendrimer (guest/host) system, and have used it to demonstrate amplified spontaneous emission (ASE) and laser emission. The dye-doping of dendrimers is possible due to the unique property of dendrimers that enables them to be used in molecular encapsulation [39, 40]. Several kinds of dendrimers and hyper-branched polymers can now be purchased as chemical reagents. Starburst PAMAM dendrimers are constructed of repeating *tert*-amine and amide branching units, and their molecular weights ranging from 517 to 934,720. DAB-Am, another kind of dendritic macromolecule, is constructed of branching propyleneimine units. These dendrimers have been demonstrated to encapsulate smaller molecules. To enable the use of dendrimers as a

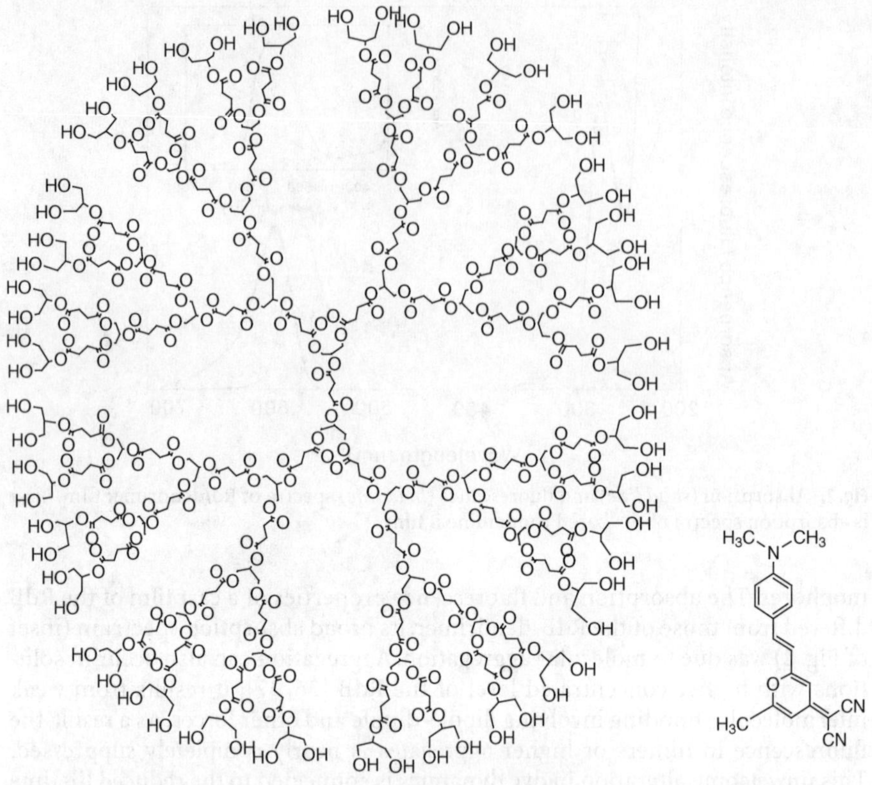

Fig. 3. Molecular structure of polyester dendrimer and DCM

host for laser-dyes, we have been investigating the development of electronically inert, i.e., neutral, dendrimers lacking the amine-base associated with the two commercial families of amine-containing dendrimers.

We chose the polyester dendrimer, illustrated in Fig. 3, because it can be synthesized using a previously reported method [41]. We slightly modified the reported method in order to produce rapidly high-molecular weight dendrimers of high purity with no means of purification other than solvent extraction and precipitation. Using MALDI-TOF mass spectroscopy, we confirmed its purity and determined its exact molecular weight (10,686). For the laser-dye we used 4-(dicyanomethylene)-2-methyl-6-(4-dimethylaminostyryl)-4H-pyrane (DCM). DCM has been widely used in dye-laser devices, and its conversion efficiency is as high as that of Rd B derivatives. DCM-doped dendrimers can be obtained by simply mixing DCM and dendrimers in a methanol solution. The DCM concentration was varied between 1.0 and 9.0 mmol/l in a 20 wt% dendrimer methanol solution. These concentrations are much higher than the highest concentration of pure DCM in methanol, about 1.0 mmol/l.

As shown in Fig. 4, the fluorescent intensity increased with the DCM concentration. This means that the fluorescence efficiency remains constant at high

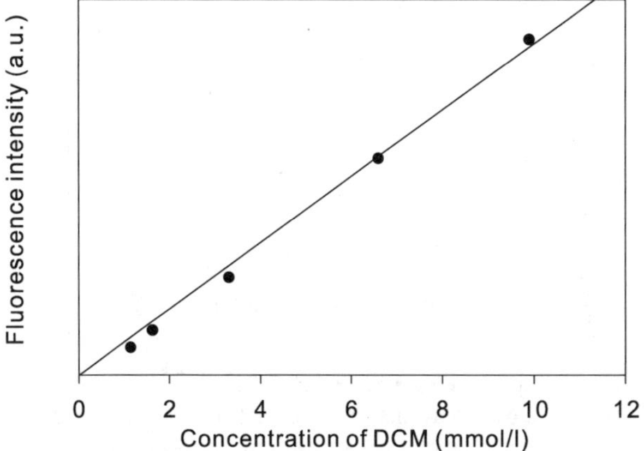

Fig. 4. Fluorescence intensity as a function of the concentration of DCM-dendrimer mixture

concentrations. Although the concentration was much higher than that of pure DCM in methanol at the saturated level, the absorption spectra were similar to the absorption spectrum of pure DCM. However, the fluorescence intensity of a pure DCM solution saturates at higher concentrations. The dendrimer is thus a good host for DCM, limiting cluster formation and intermolecular energy transfer, and promising high optical gain.

3
Stimulated Emission and Laser Emission from Dendrimer Media

Optically active media can take one of three forms; gas, liquid, or solid. In each form, it is the density of emitters in the excited state that determines the optical gain. Under optical excitation, dye molecules undergo transition from the ground state to the excited state, and then emit the light. There are two competing processes, which are spontaneous emission and stimulated emission. During spontaneous emission, the dye molecules emit incoherent photons, while, stimulated emission is the process which results in coherent optical gain. The rate of stimulated emission is proportional to three factors: the number density in the excited state, the propagation time of photons in a cavity, and the emission cross-section. Therefore, use of high-concentrated dye-doped media increases the population density and gain.

In this section, we describe a simple laser setup using a high-gain medium consisting of DCM-encapsulated dendrimers in a methanol solution. The results can be applied to a solid-state laser medium [42], described in the next section. In that case, we used RdB-dendrimer in the waveguide gain medium [43].

Figure 5 shows the dependence of the total emission intensity on the excitation intensity and its spectral width obtained from DCM-encapsulated dendrimers. A nitrogen laser (wavelength of 337 nm, pulse duration of 4 ns, and repetition rate of 10 Hz) was used as the excitation source. A cylindrical lens focused the excitation beam onto a stripe 200 µm wide on a quartz cuvette

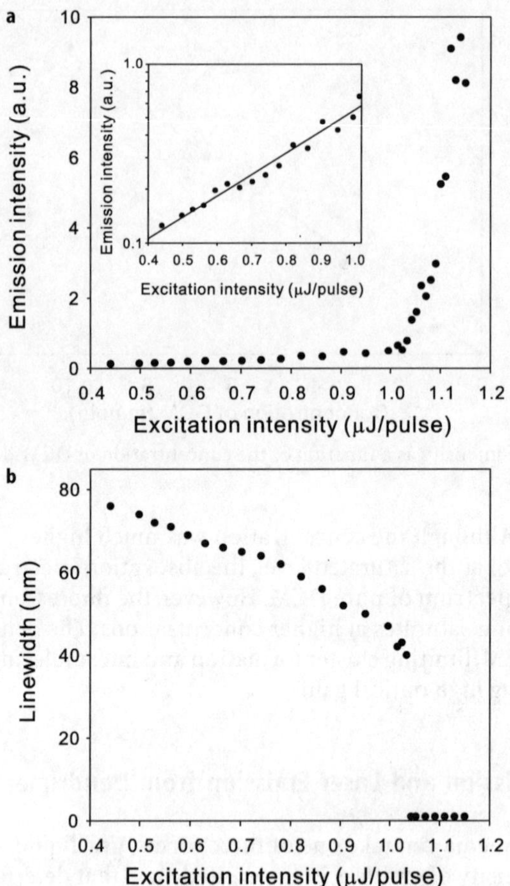

Fig. 5. a Total emission intensity. **b** Linewidth, both as functions of excitation intensity for DCM/dendrimer solution in cuvette. DCM concentration was 4.0 mmol/l. *Inset* in a shows plot in logarithmic scale at moderate excitation intensity

(2.0 mm optical length) containing a DCM-dendrimer solution (DCM concentration of 4.0 mmol/l). This optical setup, as shown in Fig. 6, could function as a very simple laser, in which both ASE and laser emission are induced along with gain guiding. We collected the emissions guided along the excitation from the side of the cuvette by using a round lens and spectrally analyzed them using a spectrometer and a charge-coupled device. From relatively low excitation intensities up to 1.0 μJ/pulse, the emission intensity grew exponentially with the excitation intensity (inset in Fig. 5a), and the emission spectra gradually narrowed from about 76 to 40 nm (Fig. 5b). These responses are attributed to the ASE model along with the gain guiding, which is described by $I_{se}=\beta(e^{(\gamma-\alpha)L}-1)$, where I_{se} is emission intensity, β is a constant that depend on the excitation geometry, L is the excitation stripe length, and γ and α are the optical gain and loss coefficients, respectively [44]. Since γ is linearly related to excitation intensity (I) in the sim-

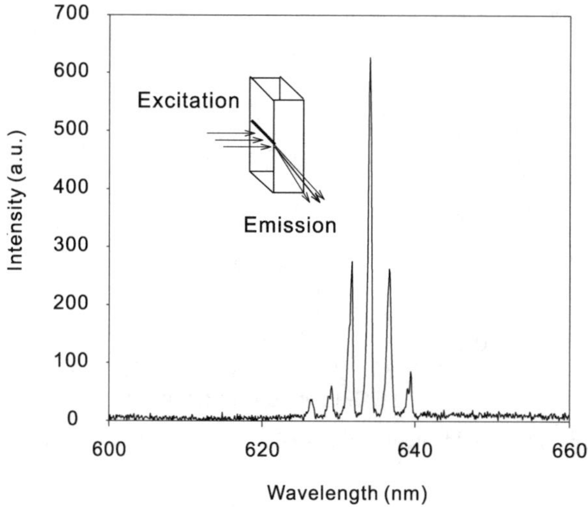

Fig. 6. Laser emission spectrum from DCM/dendrimer solution in cuvette. *Inset* schematically illustrates experimental setup

ple approximation, $\ln(I_{se})$ I for the ASE model is in agreement with the fitted experimental results (Fig. 5a insert).

When the excitation intensity was increased, the emission spectrum collapsed into multiple narrow lines as shown in Fig. 6. A clear threshold behavior in the emission-vs-excitation intensity plot and a second decrease in the linewidth at a higher excitation intensity indicated the onset of laser action (Fig. 5a,b). The strongly modulated laser spectrum with numerous evenly spaced peaks clearly indicated resonant modes. The laser beam was highly polarized in the longitudinal direction with a large polarization ratio (>150). Above the threshold, the total emission intensity increased much more rapidly with the excitation intensity. Similar lasing action was observed for the DCM-dendrimer solutions with DCM concentrations higher that 2.5 mmol/l. More importantly, the excitation energy required to reach the lasing threshold was reduced by increasing the concentration of DCM, as shown in Fig. 7. However, laser emission did not appear in a pure DCM solution with the same optical setup. With a pure DCM solution, a higher gain with a longer optical length, i.e., a larger cuvette, and higher excitation energy was required to achieve lasing action. By encapsulating DCM, dendrimers are thus a good host for producing a homogeneous high-gain media; using them increased the media concentration up to 9.0 mmol/l. A continuous decrease in the threshold intensity indicates that the gain for lasing should be much larger that the loss. The loss effect, for example that caused by scattering and self-quenching, was limited in a homogeneous DCM-dendrimer solution.

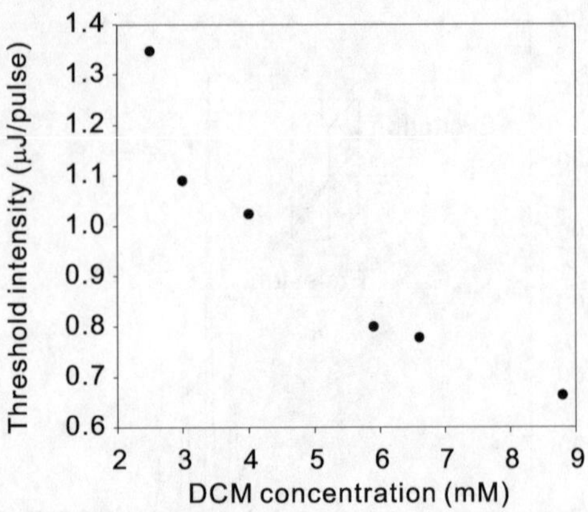

Fig. 7. Dependence of threshold intensity required for lasing on DCM-dendrimer concentration

4
Solid-State Application of Dendrimer Laser

Lasing action can be simply described as the result of optical feedback created when the gain under excitation exceeds the loss in an optical cavity. The gain of the light propagation in the cavity is proportional to the propagation length, as are the scattering losses causes by light escaping from the cavity. Recently there have been reports that pointed out that strong optical scattering could trap the light waves within a gain medium long enough to cause light feedback [45–47]. Such a self-formed cavity would create a random laser, which has been described as a laser not requiring a regular cavity structure.

The random laser is a simple optical system in which the strong optical scattering in the random medium forms an optical recurrent path. Recent reports on random lasers have described the emission of laser light by metal-oxide polycrystalline and micrometer-sized particles [46]. Because of its structural simplicity and small size, the single random laser is a promising miniature light source for optical devices, such as waveguides and optical switches.

For optical application of small particles, such as dendrimers, we have been investigating the lasing properties of the optical waveguides containing dendrimers and have observed supernarrowed laser emission without resonator mirrors. In these studies we postulated that the aggregated dendrimers in the host waveguide are responsible for the light scattering and lasing actions.

The waveguide material we used was prepared by mixing polymethylmethacrylate (PMMA) and RdB-dendrimers (Fig. 1) in cyclohexanone at various ratios (1–20 wt%). The solution was then spun onto a quartz substrate, which had been cut into 5-mm-wide rectangular shapes. The film thickness was about 0.7 μm so as to support a single waveguide mode at the emission wavelength of the dendrimer. A frequency-doubled Nd:YAG pulsed laser (wavelength of 532 nm, pulse duration of 3 ns, and repetition rate of 10 Hz) was used as an excitation source. The excitation beam was focused into a 0.1-mm-wide and 2.4-mm-long stripe

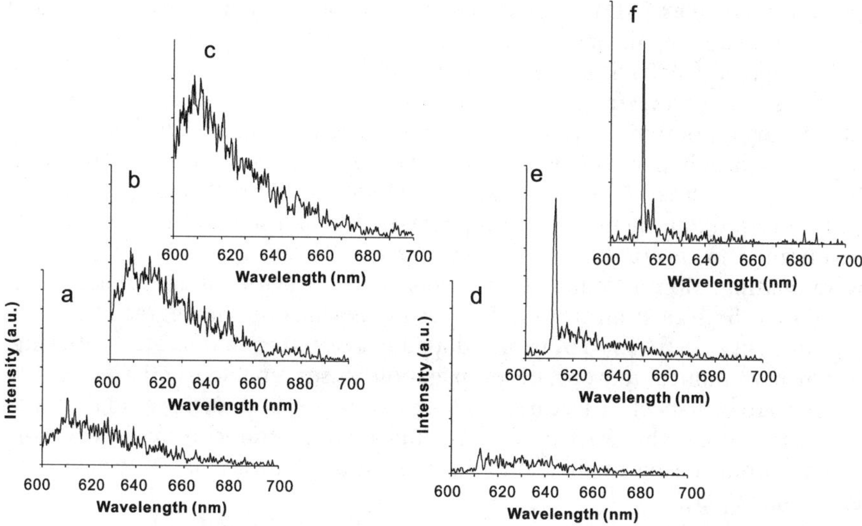

Fig. 8a–f. Evolution of emission spectra from waveguides with increasing excitation intensity for films with: **a–c** 2 wt%; **d–f** 10 wt% Rd/dendrimer content. The intensity increased for each spectrum from the bottom as follows: 1.7, 2.1, and 4.0 mJ/cm^2

through a cylindrical lens. The amplified emission guided along the stripe was collected with an objective lens and used to analyze the spectra.

Figure 8 shows the measured evolution of emission spectra from waveguides containing RdB-dendrimer with 2 or 10 wt% concentrations. At low excitation intensities, the emission spectra had a single broad peak for both concentrations. With the 10 wt% concentration, as the excitation intensity was increased, a narrow emission peak suddenly emerged and grew rapidly in intensity. The line width of the emission was less than 0.7 nm. Such a supernarrowing in emission spectra cannot be explained in terms of ordinary ASE, in which the spectrum gradually narrows as the excitation intensity in increased. In contrast, with the 2 wt% concentration, the waveguide exhibited normal ASE during excitation. Therefore, the observed lasing action apparently depended on the concentration of RdB-dendrimers. Moreover, the RdB-doped PMMA waveguide, which did not contain dendrimers, showed ASE-type spectral narrowing. Dendrimers evidently play an important role in optical feedback increasing the laser gain.

To clarify the optical feedback mechanism, we assumed that the multiple scattering effect might cause recurrent optical loops inside the waveguide, which then become self-formed cavities. Multiple scattering by dendrimers was verified by measuring the propagation loss of a waveguide, where the optical scattering loss increased 3 dB/cm for the waveguide with a 10 wt% dendrimer concentration. This scattering loss is presumably due to the formation of dendrimer aggregates in the PMMA matrix at high concentration. At higher concentration of dendrimer than 10wt% in the waveguide, the laser emission intensity was smaller than that for a waveguide with a 10 wt% concentration. This indicates that the dendrimer/PMMA system achieves a suitable balance between gain and loss for fine turning the laser emission.

The optical feedback process in the random laser is different from that in mirrorless micrometer-size lasers having well-defined resonators such as fibers, rings, photonic crystals, and meso-structured materials [48, 49]. For organic and polymeric materials, there has been controversy as to whether it is an ASE or a laser action process [50, 51]. Previously reported spectral narrowing in random media containing a laser-dye solution and micro-particles was limited to several nanometers of wavelength, but feedback lasing modes were missing. More recently reported results for organic random media indicated that the transition from ASE to lasing depends on the degree of scattering intensity [52]. In dendrimer-doped waveguides, moderate scattering, which may induce appropriate propagation losses, enabled emission with a supernarrowed spectral feature. Since the emission is directional and monochromatic along the waveguide axis, high-power monochromatic emission can be produced by further amplification in a mirrorless laser system with dendrimers [53].

5
Cone-Shaped Dipolar Dendrimers for Nonlinear Optical Application

The synthesis of dendrimers by using step-by-step reactions can be described as an "organic chemist's approach" to hyper-structured materials. It provides greater structural control than the ordinary approach using linear polymer synthesis. The ability to place precisely functional groups throughout the structure to modify selectively the focal point, branches, and chain end promises novel functions with wide applications. As a result of the organic chemist's challenge, various potential applications in molecular electronics and photonics have recently been reported for dendrimers. Nevertheless, there are still very few reports on their potential application to nonlinear optics.

In contrast to linear-shaped polymers, a widely branching chain of dendrimer results in a three-dimensional structure that is roughly spherical or globular. However, there have been reports describing a dendritic molecule having an electronically dipolar structure, which is essential for NLO activity such as second harmonic generation (SHG). In this section, we describe our experimental approach to the design and synthesis of dipolar dendrimers for NLO application and their molecular structure. The dendrimers we synthesized consisted of three distinct chemical units: the main NLO functional branch unit, the aliphatic functionality at the end, and the focal unit (Fig. 9) [54, 55]. The NLO unit is a chromophore with high molecular hyperpolarizability, H-azobenzene or $nitro$-azobenzene, having a π-electronic structure coupled with electron donor and acceptor groups. Since the electron-withdrawing effect of the carboxyl group is weak, a nitro group was introduced into the azobenzene at an $ortho$ position to optimize the second order susceptibility of the chromophore. Theoretically estimated structures and electronic properties supported some views. First, H- and nitro-azobenzenes have a rod-shaped structure with dominant dipole moments parallel to the molecular axis (Fig. 10). Second, the calcu-

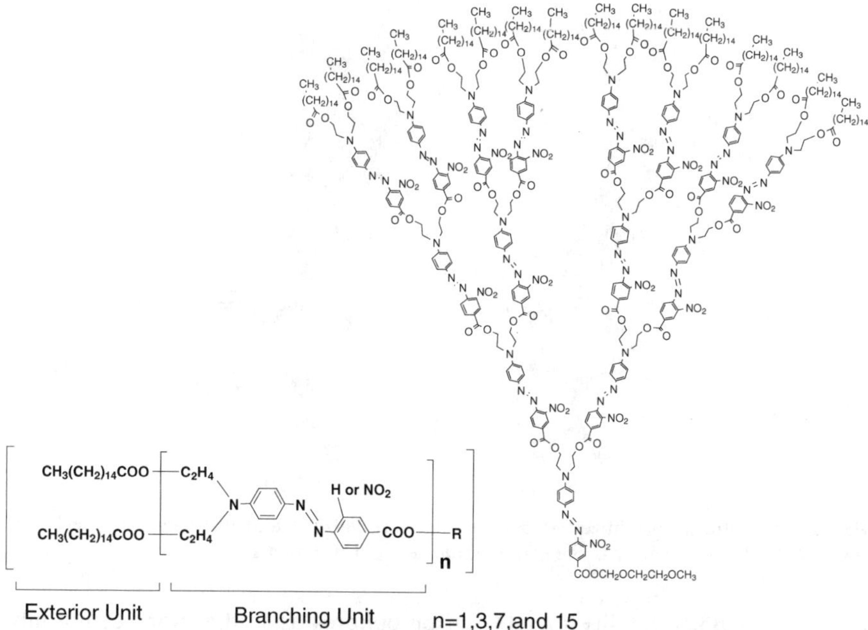

Fig. 9. Repeating structure of azobenzene dendrimers. G1 (n=1), G2 (n=3), G3 (n=7), and G5 (n=15). *Right structure* is G4 nitro-azobenzene dendrimer

lated molecular hyperpolarizability of the *nitro*-azobenzene is higher that that of *H*-azobenzene (Table 1).

An important goal of this study is to characterize the noncentrosymmetric arrangement of the azobenzene branches in dendrimers and the effect of these dendrimers on macroscopic second-order susceptibility. The NLO activity of the dendrimers can be characterized by their molecular hyperpolarizability. The hyper-Rayleigh scattering (HRS) method is a reliable way of measuring the mol-

Table 1. Dipole moments, charge transfers, and molecular hyperpolarizabilities for azobenzene chromophores, calculated using CNDO/S method[a]

	μ/D			Total	$\Delta\mu^{st}/D^{b}$ (OS)			$\beta_{0,vec}{}^{c}$
	x	y	z					
H azobenzene	8.28	2.07	1.66	8.70	4.80^2 (0.903)	9.74^4 (0.028)	3.48^6 (0.045)	15.0
NO_2 azobenzene	7.00	0.55	−4.61	8.41	7.79^2 (0.674)	12.07^3 (0.268)	10.63^5 (0.123)	24.7

[a] Molecular structures for CNDO/S calculations were optimized using HP/3–21G shown in Fig. 1.
[b] Dominant charge transfers at some selected states and their oscillator strength.
[c] β(E–30 esu).

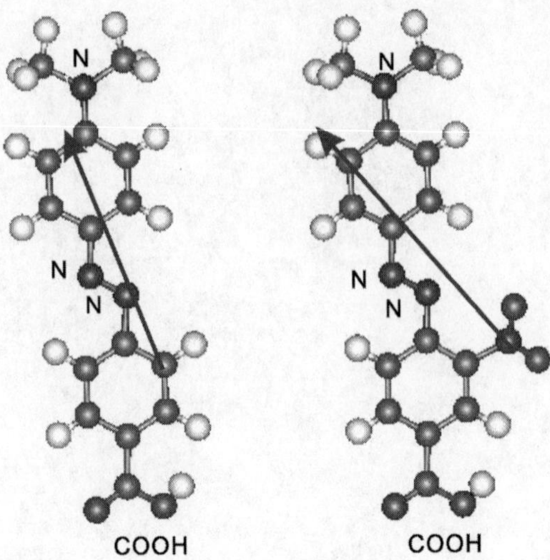

Fig. 10. Theoretically optimized structures of *H*-azobenzene and nitro-azobenzene by ab initio HP/3–21G calculations. The *arrows* indicate dipole moments

ecular hyperpolarizability of NLO chromophores as is the more commonly used electric field-induced second-harmonic generation method [56–60]. The main advantage of the HRS method over other NLO measurement methods is that it does not need an external electric field, and is carried in a solution. Therefore, the HRS can be used to characterize NLO polymers, particularly their molecular arrangement and geometric properties at the macroscopic level [61, 62].

The UV-visible spectra of the *H*- and *nitro*-azobenzene dendrimers in chloroform solution showed strong absorption bands within the visible region due to the π-π^* transitions of azobenzene chromophores (Table 2). Because of the stronger delocalization of π-electrons in nitro-azobenzene, the maximum absorption band is at a longer wavelength compared with that for *H*-azobenzene. There was little spectral shift of the absorption maximum for dendrimers with different numbers of azobenzene units, indicating that dendrimers did not form any special intermolecular aggregates.

Table 2. Linear and nonlinear optical properties of azobenzene dendrimers

Compound	n_c	NO$_2$ azobenzene			H azobenzene		
		λ^a_{max}	$\beta_0{}^a$	n_cF	λ^a_{max}	$\beta_0{}^a$	n_cF
G1	1	455	150	(1.0)	423	63	(1.0)
G2	3	458	500	3.33	427	280	4.49
G3	7	460	1430	9.53	428	740	11.7
G4	15	460	3010	20.1	428	1740	27.7

[a] Absorption spectra and HRS were measured in chloroform solution.

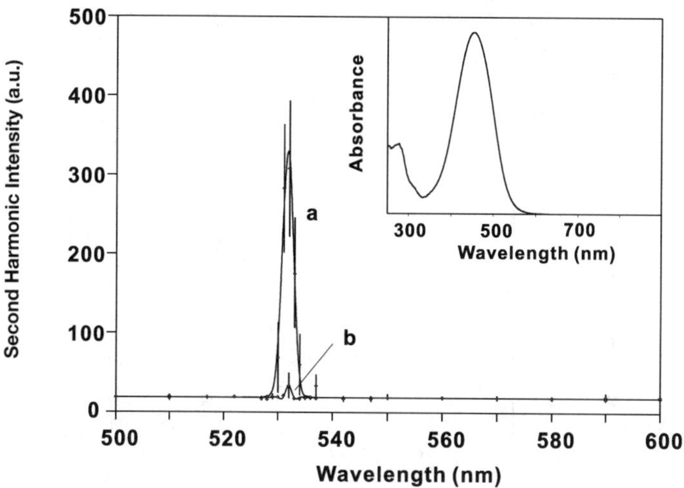

Fig. 11. Nonlinear scattering spectra for chloroform solution of nitro-azobenzene dendrimer (curve a) and neat chloroform solution (curve b). Inset is absorption spectrum of sample

In our HRS measurements, two incident photons at the fundamental frequency, ω, with a wavelength of 1064 nm generated a photon at the second harmonic frequency, 2ω, with a wavelength of 532 nm. The HRS spectrum from a sample solution showed an intense HRS signal at 532 nm (Fig. 11). The signal intensity level at other wavelengths was identical to that measured for the neat chloroform solution. Therefore, only the second-harmonic scattering light caused the detected signal, and there was little contribution from the two-photon induced fluorescence.

In the HRS measurements, the theoretically expected quadratic dependence of the HRS signal on the incident intensity, $I_{2\omega}=G<B^2>I^2_\omega$, was always observed, as shown in Fig. 12, where G is the scattering geometry and instrumental factor and $<B>$ is the average macroscopic second-order susceptibility. Depolarization measurement with the analyzer perpendicular or parallel to the laser polarization gave the ratio of $I_{x,2\omega}$ to $I_{y,2\omega}$. The HRS intensity ratio of $I_{x,2\omega}/I_{y,2\omega}$ is identical to the tensor component ratio of $\beta^2_{zzz}/\beta^2_{xzz}$ [63]. Depolarized measurement resulted in a ratio of $<\beta^2_{zzz}>/<\beta^2_{xzz}>=4.5$–$4.9$ for all dendrimers. The order of $<\beta^2_{zzz}>/<\beta^2_{xzz}>$ on a macroscopic scale is theoretically 5.0 for a linear conjugated molecule with one dominant β component, while it is smaller for macromolecules depending on the correlation length. The experimentally obtained $<\beta^2_{zzz}>/<\beta^2_{xzz}>$ was that expected for chromophores with a linear conjugated backbone and, more importantly, for dendrimers that can be modeled as a thin rod, where each of the chromophore units was arranged noncentrosymmetrically to be dipolar rather than have a spread molecular arrangement. This structural peculiarity is in contrast with the general tendency of dendritic macromolecules to be spherical or globular with spreading branches [62]. Other studies have shown that $<\beta^2_{zzz}>/<\beta^2_{xzz}>$ is small for molecules having high symmetry or spherically fictionalized dendrimers.

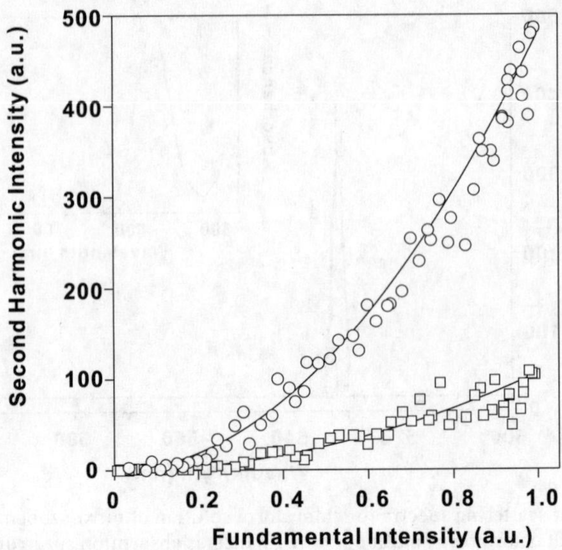

Fig. 12. Polarized HRS signal intensities as a function of fundamental laser intensities for a chloroform solution of nitro-azobenzene dendrimer. Detected signals were parallel (*circles*) and perpendicular (*squares*) to polarization of fundamental laser

6
Large Enhancement of Molecular Hyperpolarizabilities in Dipolar Dendrimers

To estimate the molecular hyperpolarizability of dendrimers, we used the molecular cluster, or assembly model, in which the effective molecular hyperpolarizability of the molecular system is summed over the contributions of each chromophore. The molecular hyperpolarizabilities of the dendrimers and chromophores in the HRS measurement are expressed as $\beta_{dendrimer}=n_c f \beta_{chromophore}$, where n_c is the number of chromophore units, and f is the local field factor due to the screening electric field generated by neighboring molecules. Using the known value of $\beta=-0.49\times10^{-30}$ esu for chloroform [64], the molecular hyperpolarizabilities of dendrimers were calculated as summarized in Table 2. In our experiments, the static hyperpolarizabilities, β_0, estimated using a two-level model expression [65], were compared to remove the effects of the electronic resonance from the hyperpolarizability. The values of β_0 for dendrimers having nitro-azobenzene units were higher that those for dendrimers having H-azobenzene units. Both β_0 and enhancement factor $n_c F$ increased with the number of chromophore units in the dendrimers. This increase was due to the macroscopic structural properties of the dendrimers, as discussed in the measurement of $<\beta^2_{zzz}>/<\beta^2_{xzz}>$: they had a noncentrosymmetric chromophore orientation. In such a molecular structure, each chromophore unit coherently contributes to the SHG. An enhancement of β_0 for each chromophore was clearly evident for the highly branched dendrimers. On passing from dendrimer G1 to

G2, to G3, and to G4, the effective enhancement was 10%, 36%, and 35% larger than the value estimated by the simple addition of monomeric β_0 values. The enhancement included the local field effect due to the screening electric field generated by neighboring molecules. Assuming the chromophore-solvent effect on the second-order susceptibility is independent of the number of chromophore units in the dendrimers, β enhancement can be attributed to the intermolecular dipole-dipole interaction of the chromophore units. Hence, such an intermolecular coupling for the β enhancement should be more effective with the dendrimers composed of the NLO chromophore, whose dipole moment and the charge transfer are unidirectional parallel to the molecular axis.

The HRS measurement indicated that dendrimers having H-azobenzene branches also had large β enhancements. The enhancement was more effective than that observed for dendrimers having nitro-azobenzene units. The effective enhancements were found to be 50%, 67%, and 85% for dendrimers G2, G3, and G4, respectively. These large enhancements are attributed to the electronic structure of H-azobenzene: it has a dipole moment parallel to the molecular axis and only for a unidirectional charge transfer. Nitro-azobenzene also had a large β_0, while it had a tilting dipole moment relative to the molecular axis and some nonzero components of charge transfer. These electronic properties of the azobenzenes clearly explain the experimental finding that β enhancement is more effective with the dendrimers composed of H-azobenzene units than those composed of nitro-azobenzene unit.

7
Theoretical Description of Intermolecular Electronics in Dendrimers

Large molecular hyperpolarizabilities for azobenzene dendrimers were experimentally shown to be due to the cone-shaped arrangements in which all the chromophore units are organized noncentrosymmetrically. Such arrangements might be possible after molecular dynamics calculation, as shown in Fig. 13. To clarify the origin of the electrical interactions between azobenzene branches, we carried out ab initio molecular orbital calculations using four model species (Fig. 14): H-azobenzene monomer (M, as shown in Fig. 10); H-azobenzene dimer in which a pair of monomers are linked in a head-to-tail configuration (BD); two monomers in which each monomer has a linear orientation (LD); and two monomers in which each monomer has a parallel orientation (PD) [66, 67]. All the calculations discussed here were performed at the Hartree-Fock level of the theory with a 6–31G* basis set, using the GAUSSIAN98 program package. The GAMESS program package was used to calculate analytically some tensors of molecular second hyperpolarizability. To stabilize the structure of the models, we first obtain the geometries of M, BD, and PD. The geometry of LD was optimized by artificial modulation of the optimized geometry of BD.

The geometrically optimized model of BD had a roughly linear conformation. This spontaneous ordering was unexpected given the general orientation of dipolar molecules. Azobenzenes that have permanent dipoles parallel to the molecular axis would intuitively be expected to tend to pair with their dipole oriented in the opposite direction. The linear geometry is probably due to the

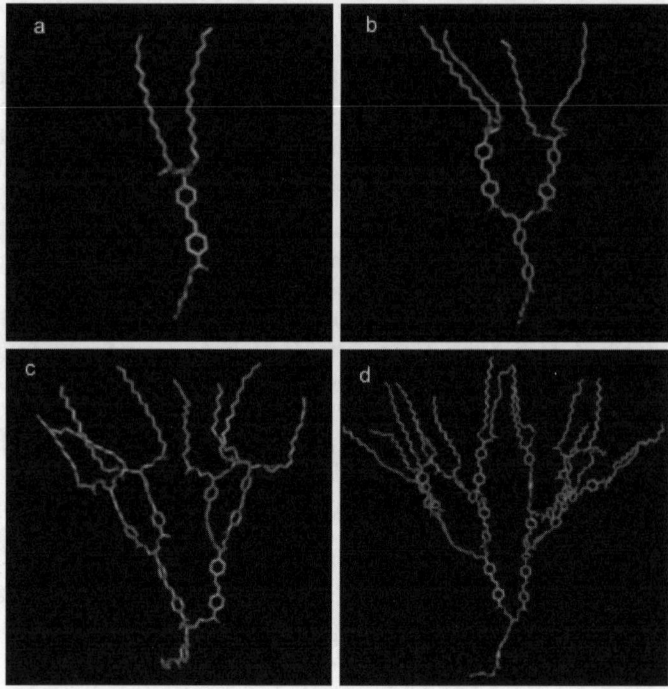

Fig. 13a–d. Energy-minimized conformation geometry of azobenzene dendrimers after molecular dynamics calculations: **a** G1; **b** G2; **c** G3; **d** G4

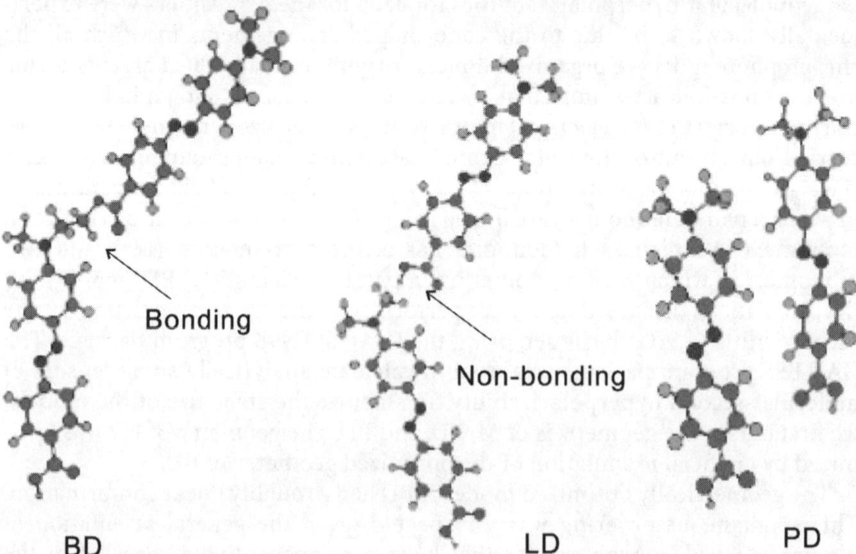

Fig. 14. Optimized structures of covalently bonded linear dimer (BD), non-bonded dimer (LD), and parallel dimer (PD)

Table 3. Calculated hyperpolarizabilities (10^{-30} esu) and dipole moments (D)

	β_{tot}	β_{zxx}	β_{zyy}	β_{zzz}	μ	β_{tot}^{SUM}
M	34.7	−2.1	−0.4	37.1	5.10	–
BD	75.5	12.6	−0.2	63.1	8.89	–
LD	75.1	12.8	−0.3	62.6	9.16	61.2
PD	57.6	−1.8	−2.2	61.6	9.41	69.2

steric hindrance effect caused by the interaction of one azobenzene unit with another azobenzene unit, resulting from the strain caused by bending the linear geometries of the covalently bonded dimers. We can thus attribute the noncentrosymmetric arrangement of the azobenzene dendrimers to the stable linear geometry of the azobenzene branching chains.

Table 3 summarizes the calculated molecular hyperpolarizabilities, their tensors, and dipole moments. The tensor sum of total molecular hyperpolarizability, β_{tot}, for the BD dimer was more than twice that for the monomer. This indicates that the hyperpolarizability is increased by interaction between monomer units. Based on the orbital energy calculations, the difference in orbital energies between the π and π^* levels for BD was in good agreement with the difference for the monomer unit. This agreement indicates that the non-conjugated linking of azobenzene units does not influence the π-π^* transition energy of the individual monomers. This is supported by the experimental results. The absorption maximum of the azobenzene dendrimers due to the π-π^* transition did not strongly depend on the difference in the number of azobenzene units between different-sized dendrimers. Therefore, both the calculated and experimentally measured results indicate that azobenzene dendrimers not only exhibit large second-order nonlinear optical susceptibility but also suppress the undesirable bathochromic shift in the optical absorption maximum.

The calculation of β_{tot} for the LD model also indicated an enhancement effect of molecular hyperpolarizability. The value of β_{tot} for the LD model was in good agreement with that for the BD one, indicating that the total molecular hyperpolarizability is barely affected by the covalent bonding between azobenzene monomers. We also found that the ab initio values of β_{zzz} for the LD model were larger than the ones obtained from the sum of the hyperpolarizability tensors of the individual monomers. In contrast, the ab initio value of β_{zxx} for the LD model nearly coincided with the β_{zxx} value obtained from the sum of the hyperpolarizability tensors of the individual monomers. These results indicate that the enhancement of the molecular hyperpolarizability by the head-to-tail arrangement of the monomers stems from the enhancement of its zzz component.

The PD model has two monomers in a parallel orientation, which are geometrically optimized by initially using a side-by-side arrangement. We found that a parallel dimer geometry, in which the two monomers are parallel and in a slightly staggered arrangement, is relatively stable. The stabilized energy of the parallel mode relative to that of the monomer was −2.2 kcal/mol, indicating that the interaction between monomers arranged in rows contributes to the stabilization of the parallel orientation despite the repulsive dipole-dipole interaction. Therefore, the interaction between the monomeric units, except for the dipole-dipole interaction, should contribute to the formation of a cone-shaped configuration of azobenzene dendrimers. The value of β_{tot} for the PD model was found to be larger than that for the monomer; however, there was no enhancement effect in this dimer orientation. The calculated β_{tot} value for the PD model was smaller than that obtained from the sum of the molecular hyperpolarizability, β_{tot}^{SUM}, of the monomers. This result indicates that the two monomers contribute to the total molecular hyperpolarizability coherently because of the parallel orientation; however, molecular interaction due to the electrostatic force reduces their hyperpolarizability.

8
Conclusion

From the viewpoints of organic and polymeric materials for optical and optoelectronic applications, many kinds of molecular systems have been demonstrated. It is apparent that constructing nanoscale building blocks, in which the chemistry and physics are well organized, is important for targeted applications. In this review, we described the structural and functional versatility exhibited by dendrimers. Optically active units or building blocks can be introduced into the chain end, branch, or core of a dendrimer to construct a precise molecular structure. Although great efforts are not necessary to obtained high-performance dendrimers in optical and nonlinear optical applications, the activities in these applications become much more effective than previous organics and polymers. The novel responses to be gained by using in dendrimers will open up new directions in optical and optoelectronic applications.

9
References

1. Wiersma D (2000) Nature 406:132
2. Qian S, Snow JB, Tzeng H, Chang RK (1986) Science 231:486
3. Lawandy NM, Balachandran RM, Gomes ASL, Sauvain E (1994) Nature 368:436
4. Frolov SV, Vardeny ZV, Yoshino K (1998) Phys Rev B 57:9141
5. Hide F, Díaz-García MA, Schwartz BJ, Andersson MR, Pei Q, Heeger AJ (1996) Science 273:1833
6. Fozlov VGK, Bulovic V, Burrows PE, Forrest SR (1997) Nature 389:362
7. Tessler N, Denton GJ, Frend H (1996) Nature 382:695
8. Frolov SV, Fujii A, Chinn D, Vardeny ZV, Yoshino K, Gregory RV (1998) Appl Phys Lett 72:2881
9. Marlow F, McGehee MD, Zhao D, Chmlka BF, Stucky GD (1999) Adv Mater 11:632
10. Chemla D, Zyss J (1987) Nonlinear optical properties of organic materials, vols 1 and 2. Academic Press, New York
11. Prasad P, Williams D (1991) Nonlinear optical effects in molecules and polymers. Wiley, New York
12. Zyss J (1993) Molecular nonlinear optics; materials, physics and devices. Academic Press, Boston MA
13. Hornal LA (1992) Polymers in lightwave and integrated optics, technology and applications. Marcel Dekker, New York

14. Wong CP (1993) Polymers for electronic and photonic applications. Academic Press, New York
15. Mitata S, Sasabe H (1997) Poled polymers and their applications to SHG and EO devices. Gordon and Breach Science Publishers, Amsterdam
16. Yokoyama S, Nakahama T, Otomo A, Mashiko S (2000) J Am Chem Soc 122:3174
17. Yokoyama S, Nakahama T, Otomo A, Mashiko S (1999) Thin Solid Films 331:248
18. Ma H, Jen AK (2001) Adv Mater 12:1201
19. Ma H, Chen Q, Sasa T, Dalton LR, Jen AK (2001) J Am Chem Soc 123:986
20. Adronov A, Fréchet JM, He GS, Kim KS, Chung SJ, Swiatkiewicz J, Prasad PN (2000) Chem Mater 12:2838
21. Chung SJ, Kim KS, Lin TC, He GS, Switatkiewicz J, Prasad PN (1999) J Phys Chem B 103:10,741
22. McDonagh AM, Humphrey MG, Samoc M, Luther-Davies B (1999) Organometallics 18:5195
23. Albrecht M, Gossage RA, Lutz M, Spek AL, van Koten G (2000) Chem Eur J 6:1431
24. Mager M, Becke S, Windisch H, Denninger U (2001) Angew Chem Int Ed 40:1898
25. Liu M, Fréchet MJ (1999) Pharm Sci Technol Today 2:393
26. Liu M, Kono K, Frechet MJ (2000) J Controlled Release 65:121
27. Balzani V, Campagna S, Denti G, Juris A, Serroni S, Venturi M (1998) Acc Chem Res 31:26
28. Wan PW, Liu YJ, Devadoss C, Bharahi P, Moore JS (1996) Adv Mater 8:237
29. Miller LL, Duan RG, Tully DC, Tomalia DA (1997) J Am Chem Soc 119:1005
30. Tanaka S, Iso T, Doke Y (1997) Chem Commun 2603
31. Zimmerman SC, Wang Y, Bharathi P, Moore JS (1998) J Am Chem Soc 120:2172
32. Newkome GR, He E, Godinez LA (1998) Macromolecules 31:4382
33. Gitsov I, Lambrych KR, Remnant VA, Practitto R (2000) J Polym Sci Part A Polym Chem 389:2711
34. Vetter S, Koch A, Schluter AD (2001) J Polym Sci Part A Polym Chem 39:1940
35. Duarte FJ, Hillman LW (1990) Dye laser principle, chap 7. Academic Press, San Diego
36. MacDonald RI (1990) J Biol Chem 265:13,533
37. Johansson LB, Niemi A (1987) J Phys Chem 91:3020
38. Nakashima K, Duhamel J, Winnik MA (1993) J Phys Chem 97:10,702
39. Jansen JFGA, Meijer EW (1995) J Am Chem Soc 117:4417
40. Jansen JFGA, de Brabander-van den Berg EMM, Meijer EW (1994) Science 266:1226
41. Carnahan MA, Grinstaff MW (2001) Macromolecules 34:7648
42. Yokoyama S, Otomo A, Mashiko S (2002) Appl Phys Lett 80:7
43. Otomo A, Yokoyama S, Nakahama T, Mashiko S (2000) Appl Phys Lett 77:3881
44. Siegman AE (1986) Laser. University Science Books, California
45. Wiersma DS, Bartolini P, Lagendijk A, Righini R (1997) Nature 390:671
46. Cao H, Xu JY, Zhang DZ, Chang SH, Ho ST, Seelig EW, Liu X, Chang RPH (2000) Phys Rev Lett 84:5584
47. Frolov SV, Vardeny ZV, YoshinoK (1999) Phys Rev B 59:5284
48. Painter O, Lee RK, Scherer A, Yariv A, O'Brien JD, Dapkus PD, Kim I (1999) Science 284:1819
49. Vietze U, Krauß O, Laeri F, Ihlein G, Abraham M (1998) Phys Rev Lett 81:4628
50. Lawandy NM, Balachandran RM, Gomes ASL, SauvianE (1994) Nature 368:436
51. Wiersma DS, van Albada MP, Lagendijk A (1995) Nature 373:203
52. Cao H, Xu JY, Chang SH, Ho ST (2000) Phys Rev E 61:1985
53. Otomo A, Otomo S, Yokoyama S, Mashiko S (2001) Thin Solid Films 393:278
54. Yokoyama S, Nakahama T, Otomo A, Mashiko S (1997) Chem Lett 1137
55. Yokoyama S, Nakahama T, Otomo A, Mashiko S (2000) J Am Chem Soc 122:3174
56. Terhune RW, Maker PD, Savage CM (1965) Phys Rev Lett 14:681
57. Clays K, Persoons A (1991) Phys Rev Lett 66:2980
58. Clays K, Persoons A, De Maeyer L (1991) Adv Chem Phys 66:2980
59. Singer KD, Garito AF (1981) J Chem Phys 75:3582
60. Kim DY, Tripathy SK, Li L, Kumar J (1995) Appl Phys Lett 66:1166

61. Kauranen M, Verbiest T, Boutton C, Teerenstra MN, Clays K, Schouten AJ, Nolte RJM, Persoons A (1995) Science 270:966
62. Put EJH, Clays K, Persoons A, Biemans HAM, Luijkx CPM, Meijer EW (1996) Chem Phys Lett 260:136
63. Heesink GJ, Ruiter AGT, van Hulst NF, Bölger B (1993) Phys Rev Lett 71:999
64. Kajzar F, Ledoux I, Zyss J (1987) Phys Rev A 6:2210
65. Chemla DS, Zyss J (1987) Nonlinear optical properties of organic molecules and crystals, vol. 1. Academic Press, New York
66. Okuno Y, Yokoyama S, Mashiko S (2001) J Phys Chem B 105:2163
67. Okuno Y, Yokoyama S, Mashiko S (2000) Nonlinear Opt 26:83

Top Curr Chem (2003) 228: 227–236
DOI 10.1007/b11013

Gene Transfer in Eukaryotic Cells Using Activated Dendrimers

Jörg Dennig

QIAGEN GmbH, Max-Volmer Strasse 4, 40724 Hilden, Germany
E-mail: joerg.dennig@qiagen.com

Gene transfer into eukaryotic cells plays an important role in cell biology. Over the last 30 years a number of transfection methods have been developed to mediate gene transfer into eukaryotic cells. Classical methods include co-precipitation of DNA with calcium phosphate, charge-dependent precipitation of DNA with DEAE-dextran, electroporation of nucleic acids, and formation of transfection complexes between DNA and cationic liposomes. Gene transfer technologies based on activated PAMAM-dendrimers provide another class of transfection reagents. PAMAM-dendrimers are highly branched, spherical molecules. Activation of newly synthesized dendrimers involves hydrolytic removal of some of the branches, and results in a molecule with a higher degree of flexibility. Activated dendrimers assemble DNA into compact structures via charge interactions. Activated dendrimer–DNA complexes bind to the cell membrane of eukaryotic cells, and are transported into the cell by non-specific endocytosis. A structural model of the activated dendrimer–DNA complex and a potential mechanism for its uptake into cells will be discussed.

Keywords. Polyamidoamine dendrimers, Activation of PAMAM dendrimers, Transfection, Gene transfer, DNA-dendrimer complex

List of Abbreviations

DEAE   Diethylaminoethyl
PAMAM Polyamidoamine

1
Introduction

Gene transfer into eukaryotic cells, which is also called transfection of cells, has become an important tool in modern molecular biology. Transfection of eukaryotic cells allows biochemical characterization and mutational analysis of specific genes, investigation of gene expression on cell growth, analysis of gene regulatory elements, and overexpression of specific proteins for purification.

Cells can be stably and transiently transfected. Stable transfection means the integration of the transfected gene into the genome of the transfected cell, whereas transient transfection means maintenance of the transfected nucleic acid for a restricted time. Stable integration of a transfected gene into the genome is a rare event, which occurs during transient transfection. The difference between stable and transient transfection is the selection of stably transfected cells in a large pool of transiently transfected cells. During the last 35 years a large number of different gene transfer technologies have been developed. Since 1993 polyamidoamine (PAMAM) dendrimers have also been used for transfection experiments [1]. This review gives an overview of the use of PAMAM dendrimers in gene transfer experiments.

2
Gene Transfer into Eukaryotic Cells

2.1
Overview

Different kinds of nucleic acids can be used for transfection experiments, including plasmid DNA, small DNA oligonucleotides consisting of only approximately 10–100 nucleotides, and RNA. In this review we want to focus on plasmid DNA that is most commonly used. Plasmids are DNA vectors which allow for propagation and amplification in bacteria like, e.g., *E. coli*. The sequence of plasmids can be easily modified. Most plasmids used in gene transfer experiments have several common features including sequences coding for the gene to be expressed, a eukaryotic promoter sequence, and one or more reporter genes. Reporter genes allow for monitoring of the efficiency of transfection and analysis of connected promoter sequences and can also encode eukaryotic antibiotic-resistance genes, allowing selection of stably transfected cells using the relevant antibiotic.

The size of plasmids used for transfection can vary considerably, but most plasmids are 4,000 to 10,000 base pairs in size. Despite their differences all plasmids face the same barriers when transfected. Transfected DNA has to cross the cell membrane or the endosomal membrane, it has to be transported into the nucleus, and it has to be protected against cellular nucleases and degradation inside the cell. Therefore, gene transfer into eukaryotic cells is difficult and requires specialized technologies. Early approaches to transfection of eukaryotic cells were based on the use of diethylaminoethyl (DEAE)-dextran and calcium phosphate. Later, methods were developed in which DNA was introduced directly into the cell by electroporation or microinjection. Later still, cationic lipids and viral vectors have been developed for gene transfer. These techniques are discussed below.

2.2
Classical Transfection Technologies

Classical gene transfer methods still in use today are diethylaminoethyl (DEAE)-dextran and calcium phosphate precipitation, electroporation, and microinjection. Introduced in 1965, DEAE-dextran transfection is one of the oldest gene transfer techniques [2]. It is based on the interaction of positive charges on the DEAE-dextran molecule with the negatively charged backbone of nucleic acids. The DNA-DEAE-dextran complexes appear to adsorb onto cell surfaces and be taken up by endocytosis.

The calcium phosphate method was first used in 1973 to introduce adenovirus DNA into mammalian cells [3]. DNA-Calcium-phosphate complexes are formed by mixing DNA in a phosphate buffer with calcium chloride. These complexes adhere to the cell membrane and enter the cytoplasm by endocytosis. Disadvantages of DEAE-dextran and calcium phosphate transfection are a certain level of cytotoxicity, a complicated transfection procedure, and the fact that not all cell types can be transfected using these methods.

Electroporation is based on the use of high voltage pulses to introduce DNA into cultured cells and was first established by Wong and Neumann using fibroblasts [4, 5]. Cells are subjected to a short high-voltage pulse that causes the membrane potential of the cells to break down. As a result, pores are formed through which macromolecules such as plasmid DNA can enter. The main drawback of electroporation is the high cell mortality. Microinjection, where DNA is injected into the cell nucleus using very fine needles, is useful for studying individual cultured cells. However, it is a highly skilled and laborious procedure, in which only a relative low number of cells can be transfected.

Although still in use, these techniques have been largely superceded by more modern methods that are presented below.

2.3
Viral Vectors and Liposomal Reagents

Viruses are infectious particles formed by nucleic acid, proteins, and in some cases lipids. As viruses (for example, retro- and adenoviruses) transfer viral genes into cells with high efficiency, modified forms are sometimes used as vectors for gene transfer. However, procedures using virus-based vectors are often significantly more complicated and time-consuming than other transfection methods. In addition, viral vectors are potentially hazardous, and biological safety issues need to be considered carefully. Therefore, techniques that combine some features of viral particles with the safety and simplicity of synthetic reagents have been developed. One such technique for gene transfer is the use of liposome technology.

The use of liposomes as transfection reagents was first described in 1987 by Felgner and coworkers [6]. Liposomes consist of cationic lipids organized into lipid bilayer structures. Complexing of DNA with lipid is based on the interaction of the positively charged head groups of the cationic lipids with the negatively charged phosphate groups of the nucleic acid. The assembled complexes bind to cell surfaces and are taken into the cell via endocytosis. In most cases liposomes give higher transfection efficiencies than the calcium phosphate or DEAE dextran precipitation methods. However, some liposomes show cytotoxic effects.

3
PAMAM Dendrimers

Transfection using PAMAM dendrimers (from the Greek: *dendron*, tree) offers an alternative to the methods described above. PAMAM dendrimers are highly branched spherical polymeric molecules that carry positively charged surface groups that allow them to bind to the negatively charged phosphate backbone of the DNA, forming a tight complex. This complex binds to the cell surface and is incorporated by unspecific endocytosis. It has been shown that gene transfer technologies based on PAMAM dendrimers allow efficient transfection of many different cell types and cell lines [1, 7, 8]. The focus of this review is the structure and synthesis of PAMAM dendrimers, and their use in gene transfer experiments. We will start with the structure of PAMAM dendrimers.

3.1
Structure of PAMAM Dendrimers and Activated PAMAM Dendrimers

Polyamidoamine dendrimers are highly branched spherical polymers. They consist of a multifunctional amine as core molecule from which branches radiate which again branch out and terminate at amino groups. The terminal amino groups provide the positive charge for the interaction with the DNA phosphate backbone. The dendrimer's branched structure is formed by alternating addition of methylacrylate and ethylendiamine to the growing branches. The resulting PAMAM dendrimers contain alternating amido and amine bonds and can be described as built up by layers of 'shells' called generations (Fig. 1). After generation four, steric factors cause PAMAM dendrimers to become spherical. The terminal amine groups give PAMAM dendrimers a net positive charge at physiological pH (pH 7–8), where both protonated and unprotonated amine groups are present. PAMAM dendrimers, which give good results in gene transfer experiments, typically are of generation six or seven. They have a diameter of 6–10 nm and a molecular mass of 30–50 kDa.

Newly synthesized PAMAM dendrimers have a defined size and shape and can be used in gene transfer experiments, but the efficiency of transfection can be greatly increased by a process called activation of the dendrimer. In activation, some of the tertiary amines are removed, resulting in a molecule with a higher

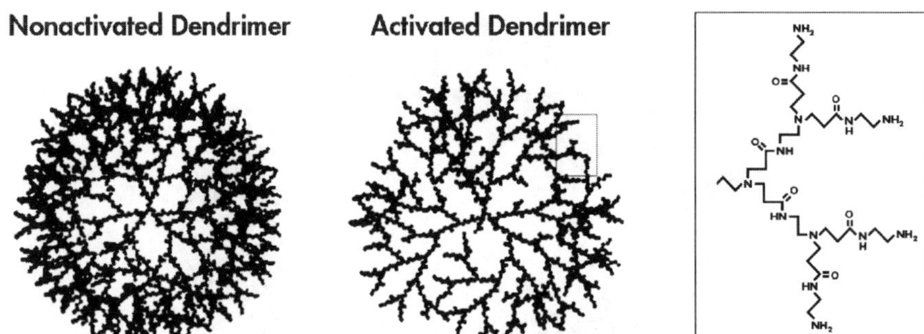

Fig. 1. Non-activated and activated PAMAM-dendrimers. Schematic diagram of a non-activated (*left*) and activated dendrimer (*middle*). The *right panel* shows a magnification of the dendrimer branches

degree of flexibility (Fig. 1). Activated dendrimers yield a transfection efficiency two to three orders of magnitude higher than non-activated dendrimers [9].

3.2
Synthesis and Activation of PAMAM Dendrimers

PAMAM dendrimers are synthesized in a multistep process. Starting from a multifunctional amine (for example ammonia, ethylenediamine, or tris(2-aminoethyl)amine) repeated Michael addition of methylacrylate and reaction of the product with ethylenediamine leads to dendrimers of different generation numbers [1, 9]. Two methylacrylate monomers are added to each bifunctional ethylenediamine generating a branch at each cycle. Unreacted ethylenediamine has to be completely removed at each step to prevent the initiation of additional dendrimers of lower generation number. Excess methylacrylate has also to be removed. Bridging between two branches of the same or of two different dendrimers by ethylenediamine can also be a problem, and has to be avoided by choosing appropriate reaction conditions.

PAMAM dendrimers are activated by solubilization in an appropriate solvolytic solvent and heating for a defined period of time. Some of the amide bonds from the inner part of the dendrimer molecule are randomly hydrolyzed, removing some of the branches (Fig. 1). Carboxyl groups are formed at the cleavage sites, and the molecular mass of the molecule is reduced by 20–25%. The resulting population of dendrimers has slight differences in molecular mass and structure, but the dendrimer's overall size and shape do not change following activation.

3.3
Activated Dendrimers as Transfection Reagents

PAMAM dendrimers have the following characteristics which are important for their use as transfection reagents. They bind and form complexes with nucleic acids, allow transfer of the DNA-dendrimer complex into the cytoplasm of the

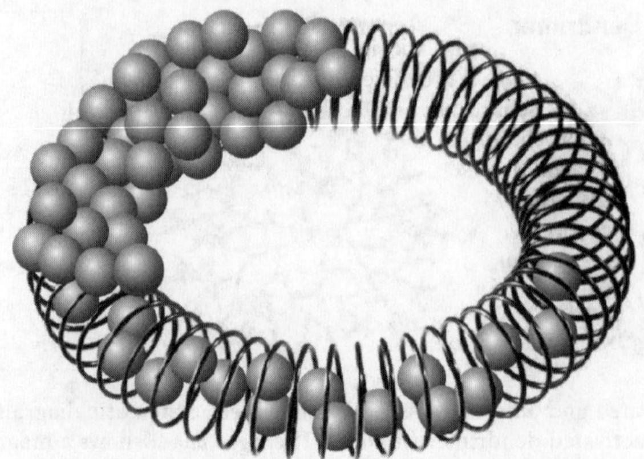

Fig. 2. Model of a toroid-like complex between plasmid DNA and dendrimers. Dendrimer molecules (spheres) are located both inside and outside the coiled DNA molecules, and more than one plasmid DNA molecule can be present in one toroid-like complex. The *left part* of the graphic shows the most probable situation; the dendrimers on the outside of the complex are not shown in the remainder of the graphic in order to give a better view of the inner part of the complex. The diameter of such a complex is approximately 50 to 100 nm

cell, and enable release of DNA from the DNA-dendrimer complex. PAMAM dendrimers therefore provide a vehicle for transport of DNA into the cell allowing, for example, for expression of genes encoded on a plasmid.

Both activated and non-activated PAMAM dendrimers assemble DNA into compact structures through the electrostatic interaction of negatively charged phosphate groups of nucleic acids and positively charged terminal amino groups of the dendrimers (Fig. 2). However, in cultured eukaryotic cells, the transfection efficiency of activated dendrimers is 2–3 orders of magnitudes higher than that of non-activated dendrimers [9]. This effect is probably caused by the higher flexibility of activated PAMAM dendrimers in comparison to the more rigid structure of non-activated PAMAM dendrimers. It is possible that this higher flexibility allows activated dendrimers to bind DNA more efficiently and may also be important for the release of DNA from the endosome inside the cell. A detailed theory of the mechanism of gene transfer using activated PAMAM dendrimers has been developed by Francis C. Szoka and coworkers [9] and is discussed below.

3.4
Theory of Gene Transfer with Activated PAMAM Dendrimers

Activated PAMAM dendrimers interact with DNA to form a DNA-dendrimer complex with a toroid-like structure (Fig. 2). Such DNA-dendrimer complexes have diameters of 50–100 nm [10], which means that the DNA molecules are highly condensed in these complexes. A 6-kb plasmid alone, for example, has an extended structure several hundred nanometers in diameter. In transfection experiments, typically an 8- to 12-fold excess of positive amino groups over negatively charged phosphate groups is used. This positive net charge allows interaction of the transfection complex with negatively charged molecules on the cell surface. After binding to the cell surface, DNA-dendrimer complexes are probably incorporated into the cell by unspecific endocytosis and transported to the endosomes.

The highly condensed complex protects DNA from degradation by endosomal nucleases. In the acidic environment of the endosomes, terminal amino groups of the dendrimers, which were unprotonated at neutral pH, can become protonated, buffering the endosomal environment and inhibiting pH-dependent endosomal nucleases [9]. In the model proposed by Tang et al., the buffering properties of the dendrimer also appear to be important for release of DNA from the endosome [9]. Activated dendrimers are proposed to have a fully extended conformation at neutral pH due to electrostatic repulsion between protonated primary amines at the branch termini. Charge neutralization of these terminal groups by DNA causes dendrimers to collapse into a compact form in the DNA-dendrimer complex. In the endosome, protonation of interior tertiary amines increases the positive charge of the dendrimer. Fewer dendrimer molecules are required to maintain charge neutralization of the DNA and excess dendrimers are released from the complex. Once the dendrimers are released from the complex, they convert into a fully hydrated form and increase in volume, causing the endosome to swell and lyse, and releasing DNA or DNA-dendrimer complexes into the cytoplasm. This model would also offer an additional explanation as to why activated dendrimers allow higher transfection efficiencies than non-activated dendrimers: the more rigid structure of non-activated dendrimers may prevent them from contracting in the DNA-dendrimer complex. In this case no swelling of the endosome would occur, and release of the DNA into the cytoplasm would be less efficient. The mechanism for transport of DNA into the nucleus is currently unclear.

3.5
Transfection Procedures

Despite the complexity of the mechanisms underlying gene transfer using PAMAM dendrimers, the practical procedure used to transfect eukaryotic cells with the dendrimers is simple. DNA and PAMAM dendrimer are mixed in a suitable ratio (depending on cell type and application) to obtain the desired charge ratio. Serum components, such as proteins, can inhibit efficient DNA-dendrimer complex formation. Complex formation is therefore performed under serum-free conditions: in water, a suitable buffer, or serum-free cell culture medium. The PAMAM dendrimer-DNA complexes are formed within a few minutes at room temperature and applied to the cells. Depending on the dendrimer and cell type complexes are either removed from the cells after a few hours incubation or are left on the cells until analysis of the transfected cells. Time point and method of analysis of the success and effects of the gene transfer depend on the experimental design. Alternatively, transfected cell populations can be selected for stably transfected cell clones.

3.6
Transfection of Eukaryotic Cells

In the last decade numerous cell types and cell lines have been successfully transfected using activated PAMAM dendrimers. Two activated dendrimer-based transfection reagents (SuperFect and PolyFect Transfection Reagent, introduced in 1997 and 2000, respectively) have been developed by the biotechnology company QIAGEN. These two activated PAMAM dendrimers differ in their generation number, their core molecule, and their activation conditions. Experience with PAMAM dendrimers has increased over the years. Activated PAMAM dendrimers have been used for the transfection of fibroblasts and epithelial cells [7], neuronal cell lines [11], T cell lines [8], bone marrow cells [12], primary aortic smooth muscle cells [13], and mouse embryo fibroblasts [14]. PAMAM dendrimers have also allowed gene transfer in *Plasmodium falciparum*, a parasite of humans [15]. In addition to transfer of plasmids, activated dendrimers have been used for successful transfection of DNA oligonucleotides [16], recombinant adenoviral vectors [17], and baculovirus in insect cells [18].

The different parameters that influence the transfection efficiency of a specific activated PAMAM dendrimer in a specific cell type must be optimized to achieve best results. Two important aspects of dendrimer structure are generation number and activation time. The generation number determines the overall size of the molecule, which may influence the nature of the DNA-dendrimer complex [1]. For the interaction of the dendrimer with DNA the flexibility of the dendrimer molecule plays a crucial role. As discussed above, flexibility is directly influenced by the activation step during dendrimer synthesis. If the reaction time is too short, the dendrimer will not be flexible enough for efficient transfection. Conversely, if the activation reaction proceeds for too long, the dendrimer will lack sufficient charge density to form a complex with DNA and/or lack sufficient mass to swell and rupture the endosome [9]. As yet, it is not possible to predict theoretically the transfection characteristics of an activated PAMAM dendrimer. This means optimal generation number as well as optimal activation time and conditions have to be determined experimentally. Dendrimers can also be synthesized using different core moieties. By varying all these different parameters, a range of activated PAMAM dendrimers with different chemical properties can be synthesized, enabling the development of activated PAMAM dendrimers suitable for a broad range of cell types and applications.

Once the optimal parameters of the dendrimer structure have been defined, the transfection procedure has to be optimized. This means maximizing the percentage of transfected cells and the level of gene expression in each transfected cell, and minimizing cytotoxic effects. All three factors can be influenced by the amount of activated dendrimer and DNA used for transfection and the ratio of dendrimer to DNA. By varying these parameters, gene transfer can be optimized in terms of efficiency and cytotoxicity. Figure 3 shows results obtained in HeLa and COS-7 cells with PolyFect Transfection Reagent compared to calcium phosphate precipitation, a method widely used for transfection in these cell types.

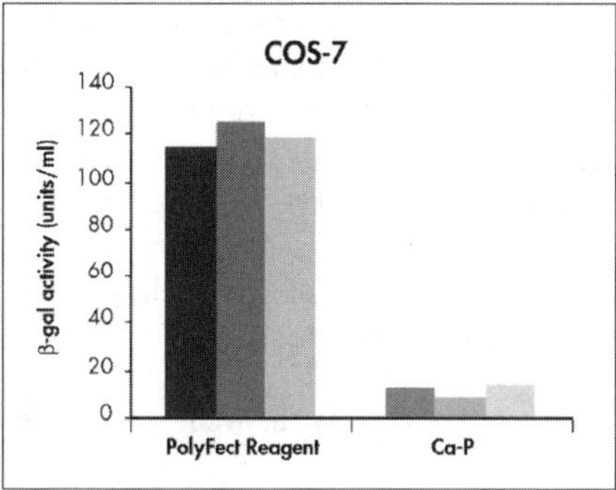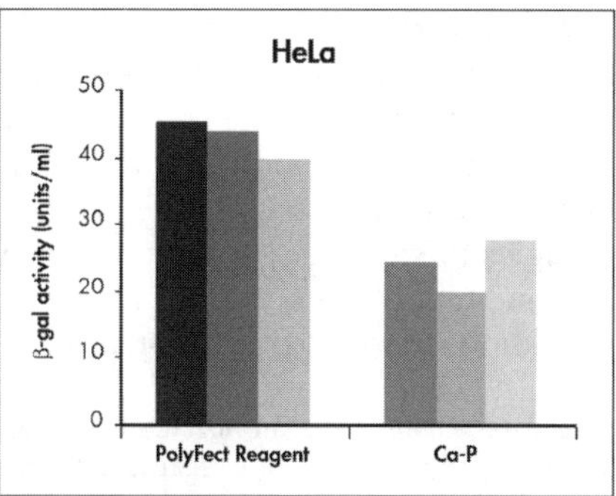

Fig. 3. Comparison of transfection efficiencies obtained using PolyFect Reagent, a dendrimer-based transfection reagent, and a calcium phosphate-mediated procedure. COS-7 and HeLa cells were transfected in six-well plates with a β-galactosidase expression plasmid using the appropriate protocol. For the calcium phosphate-mediated transfection, 6 μg of plasmid DNA was used and the medium was changed after 5 h incubation. Transfections were performed in triplicate, and transfection efficiency was measured by monitoring the β-galactosidase activity of extracts obtained from the transfected cells. The amount of β-galactosidase activity in the extracts correlates with the transfection efficiency. Cells were harvested 48 h post-transfection

The generation number, activation procedure, and core moiety of PolyFect Transfection Reagent have been developed specifically for transfection of certain cell lines, including HeLa and COS-7. Optimized amounts of DNA and dendrimer delivered transfection efficiencies in both cell lines significantly higher than those obtained using a standard transfection method.

4
Conclusion and Prospects

Activated PAMAM dendrimers have proven to be a versatile tool for gene transfer in many different applications. Gene transfer technologies will play an important role in many developing fields of cell biology where research is moving from widely used, well characterized cell lines, to primary cells, which are directly obtained from tissue, or even the in vivo situation. Primary cells are usually more difficult to transfect than continuous cell lines. This is also the case for some types of continuous cell-like cell lines derived from B or T cells or non-dividing cells. In vivo application of gene transfer vectors requires even more sophisticated methods and technologies. Potential toxicity plays an important role when choosing a transfection technology for use in gene transfer in multicellular organisms or gene therapy. The stability of the transfection complex and the influence of serum proteins and the extracellular matrix on this complex are two factors that have to be considered for efficient gene transfer in gene therapy. Targeted and specific transfection of distinct cell populations is often desired in gene therapy. Potentially, this can be achieved by DNA vectors that allow cell-specific expression of the gene of interest or by a transfection method developed for the transfection of specific cell types.

Much work is necessary to solve all the problems in modern applications of gene transfer. PAMAM dendrimers have already been used successfully in different fields of gene transfer and may provide a basis for the development of transfection technologies for the applications discussed above. Nevertheless, it may be necessary to combine activated PAMAM dendrimers with other technologies to succeed. The advantages of PAMAM dendrimers over other transfection technologies suggest that dendrimers will be one of the technologies providing solutions to current and future problems in transfection.

Acknowledgements. The author wishes to thank Drs Ute Krüger, Martin Weber, and Jason Smith for helpful comments regarding the manuscript.

5
References

1. Haensler J, Szoka FC (1993) Bioconjugate Chem 4:372
2. Vaheri A, Pagano JS (1965) Virology 27:434
3. Graham FL, Van der Eb AJ (1973) Virology 52:456
4. Wong TK, Neumann E (1982) Biochem Biophys Res 107:584
5. Neumann E, Schaefer-Ridder M, Wang Y, Hofschneider PH (1982) EMBO J 1:841
6. Felgner PL, Gadek TR, Holm M, Roman R, Chan HW, Wenz M, Northrop JP, Ringold GM, Danielsen M (1987) Proc Natl Acad Sci USA 84:7413
7. Albritton LM (1997) J NIH Res 9:52
8. Denisenko ON, Bomsztyk K (1997) Mol Cell Biol 17:4707
9. Tang MX, Redemann CT, Szoka FC (1996) Bioconjugate Chem 7:703
10. Tang MX, Szoka FC (1997) Gene Ther 4:823
11. Daniels RH, Hall PS, Bokoch GM (1998) EMBO J 17:754
12. Boden SD, Titus L, Hair G, Liu Y, Viggeswarapu M, Nanes MS, Baranowski C (1998) Spine 23:2486
13. Rafty LA, Khachigian LM (1998) J Biol Chem 273:5758
14. Kamijo T, Zindy F, Roussel MF, Quelle DE, Downing JR, Ashmun RA, Grosveld G, Sherr CJ (1997) Cell 91:649
15. Mamoun CB, Troung R, Gluzman I, Akopyants NS, Oksman A, Goldberg DE (1999) Mol Biochem Parasitol 103:107
16. Bolz SS, Fisslthaler B, Pieperhoff S, De Wit C, Fleming I, Busse R, Pohl U (2000) FASEB J 14:255
17. Tang DC, Jennelle RS, Shi Z, Garver RI Jr, Carbone DP, Loya F, Chang CH, Curiel DT (1997) Hum Gene Ther 8:2117
18. Zandi E, Chen Y, Karin M (1998) Science 281:1360

Top Curr Chem (2003) 228: 237–258
DOI 10.1007/b11014

Antibody Dendrimers

Hiroyasu Yamaguchi · Akira Harada

Department of Macromolecular Science, Graduate School of Science, Osaka University,
Toyonaka, Osaka 560-0043, Japan. *E-mail: harada@chem.sci.osaka-u.ac.jp*

Supramolecular formations of antibodies by their specific molecular recognition to antigens
are investigated. Linear and network supramolecular architectures have been constructed by
using immunoglobulin G (IgG) and divalent or trivalent antigens, respectively. An amplification method of the detection signals for the target molecule in the biosensors based on the
surface plasmon resonance (SPR) has been devised using the signal enhancement in the
supramolecular assembly of the antibody with multivalent antigens. Novel dendritic
supramolecular complexes are designed and prepared by using immunoglobulin M (IgM) or
protein A/G as a core and IgGs as branches. One of the "antibody dendrimers" is composed of
proteins with a molecular weight of about 2 million and constructed by non-covalent bonds.
The dendrimer can bind antigens strongly with high specificity. The biosensor technique
based on SPR shows that the antibody dendrimer has the advantage of amplification of detection signals for antigens.

Keywords. Antibodies, Biosensors, Non-covalent bond, Atomic force microscopy, Supramolecular chemistry

1
Introduction

In biological systems, life processes are led by the unique behavior of macromolecules such as proteins and DNAs. Molecular recognition by the macromolecules plays an especially important role, for example, in substrate specificities of enzyme and antigen–antibody reactions in human life. Selective recognition among macromolecules is achieved by non-covalent bonds with a large number of weak bonding interactions including hydrogen bonds, van der Waals, dipole and/or electrostatic interactions. We have focused our attention on the functions and structures of antibodies, because they can recognize larger and more complicated compounds with high specificity and efficiency than enzymes or synthetic host molecules. The immune system can selectively generate antibodies against virtually any molecule of interest. Antibodies are host proteins produced in response to the presence of foreign molecules in the body. Recently, with the advent of cell technology, it has become possible to prepare individual immunoglobulins called "monoclonal antibodies" in large amounts and in homogeneous form [1]. Monoclonal antibodies have become more and more important offering high possibilities as new chemical and tailor-made materials.

The basic structure of all antibody or immunoglobulin (Ig) molecules consists of four protein chains as shown in Fig. 1. The most common immunoglobulin class is IgG. Two identical heavy chains of approximately 50,000 Da and two identical light chains of about 25,000 Da are cross-linked each other by disulfide bonds. The molecule adopts a conformation that resembles the letter Y [2, 3]. There are two identical binding sites at the top of Fab fragments of IgG which are bound by flexible hinges with a single constant stem (Fc). The complementarity-determining regions in the antigen binding site differ in length and sequence between different antibodies and are mainly responsible for the specificity (recognition) and affinity (binding) of the antibodies to their target molecules. Fc does not bind the antigen, but it has other important biological activities, including the mediation of responses termed effector functions. Antibodies are divided into five classes: IgG, IgM, IgA, IgE, and IgD, on the basis of the number of Y-like units and the type of heavy-chain polypeptide they contain (Fig. 2). Each type has common characteristic sequences and variable sequences characteristic of antibodies in five types, respectively. IgM (Mw=960,000) has a pen-

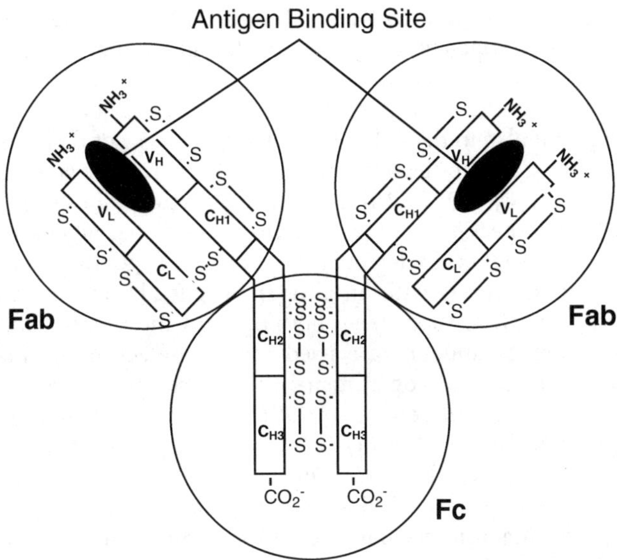

Fig. 1. The schematic representation of the structure of immunoglobulin G (IgG)

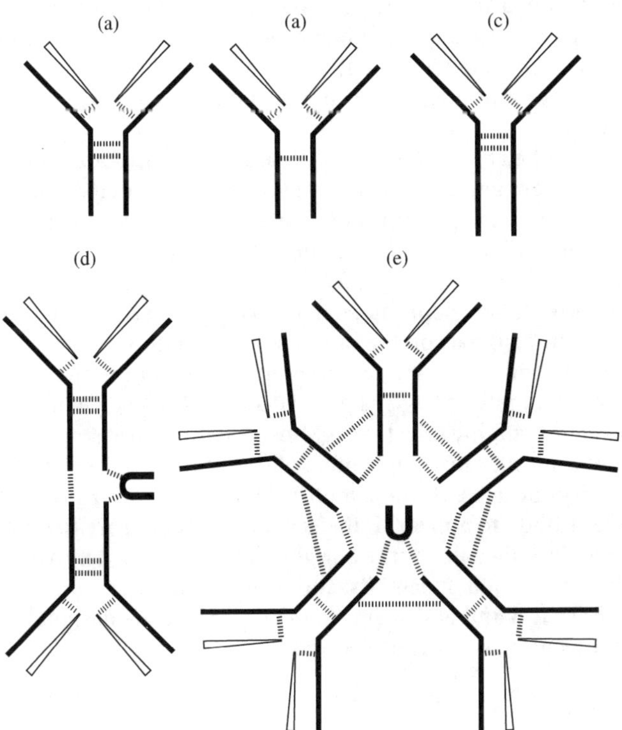

Fig. 2a–e. Structures of immunoglobulins. IgG (**a**), IgD (**b**), IgE (**c**), IgA (**d**), and IgM (**e**)

tameric structure of IgG and ten antigen binding sites in a single molecule [4]. The presence of ten antigen binding sites enables IgM to bind tightly to antigens containing multiple identical epitopes.

Recently, much attention has been directed toward antibodies not only in the field of biology but also in the field of chemistry because of their unique structures and functions. Antibodies, immunoglobulins, have been studied as sensors [5, 6], diagnostics [7, 8], DDS [9, 10], catalysts (catalytic antibodies) [11–13], and components for nanotechnology [14–16]. Antibodies especially have been widely used as efficient reagents to detect target molecules [17]. Based on the principle that an antibody reacts with an antigen specifically and by non-covalent bonds, several procedures have been developed in the immunosorbent assay. Labeled antibodies or antigens are used for the detection, localization, and quantification of biological constituents. More recently, an optical technique based on surface plasmon resonance (SPR) [18–22] or a microgravimetric quartz-crystal-microbalance (QCM) [23–25] technique has been found to be useful for measuring and characterizing macromolecular interactions in the increasingly expanding area of biosensor technology. SPR in particular has great potential for macromolecular interaction analysis in terms of sensitivity and signal translation. The use of biosensors based on SPR has made it possible to determine kinetic parameters in real time and without any labeling of biomacromolecules for detection. However, the SPR response reflects a change in mass concentration at the detector surface as molecules bind or dissociate; the specific sensing of substrates with low molecular weight is difficult. In such a case, functional molecules with a high molecular weight such as antibodies have a great potential for amplification of the response signals expressing molecular recognition event.

Here, we describe the design and preparation of antibody supramolecular complexes and their application to a highly sensitive detection method. The complex formation between antibodies (IgG) and multivalent antigens is investigated. When an antibody solution is mixed with divalent antigen, a linear or cyclic supramolecule forms [26–29]. With trivalent antigens, the antibody forms network structures. These supramolecular formations are utilized for the amplification of detection signals on the biosensor techniques.

Sensitivity and specificity are required and important in the construction of excellent detection methods. IgG is generated in a final stage of immunization, so it is matured and highly selective. IgM has a pentameric structure of IgG and ten antigen binding sites in a single molecule. It binds large multivalent antigens strongly (avidity) because of their multivalent structures. However, IgM is the first class of antibody to appear in the serum after exposure to an antigen; it is unmatured and less specific for the antigen than IgG. In order to design an antibody system with a high specificity and a high affinity, a combination of the functions of both IgG and IgM seems to be important. We design dendritic antibody supramolecules as artificial antibodies.

2
Experimental Section

2.1
Materials

2.1.1
Anti-Viologen Antibodies

Monoclonal antibodies have been elicited for the viologen derivative, 4,4'-bipyri-dinium 1-(carboxypentyl)-1'-methyl-dichloride (1) (Fig. 3a). The hapten 1 was coupled to keyhole limpet hemocyanin (KLH) or bovine serum albumin (BSA) using carbonyldiimidazole (CDI). KLH-1 and BSA-1 were purified by size exclusion chromatography using Sephadex G-150 and used as an antigen for the immunization to mice and immunoassay, respectively. Balb/c mice aged 8 weeks were immunized with KLH-1 emulsified in Freund's complete adjuvant 6 times at weekly intervals. Three days after the final injection (boost), spleen cells were removed and used for the fusion experiments. Spleen cells from a mouse were fused with the SP 2/0 mouse myeloma cells [30]. The hybridomas secreting antibodies for viologen were detected by enzyme-linked immunosorbent assay (ELISA). The tissue culture supernatants were added onto the ELISA plate coated with 0.3 mg mL^{-1} BSA-1 and incubated 90 min. The amount of antibody bound to the antigen was measured using goat anti-mouse immunoglobulins labeled with alkaline phosphatase. Hybridomas secreting anti-viologen antibodies were cloned two times by limiting dilution. The ascites fluid was harvested after 10 days. The monoclonal antibodies were purified by affinity chromatography using protein A (Amersham Ampure PA kit). The purity of antibodies was checked by SDS-PAGE electrophoresis (Phast System, Pharmacia LKB, Uppsala, Sweden). Antibody solutions were stored and treated in 0.1 M phosphate borate buffer (pH 9.0).

2.1.2
Anti-Porphyrin Antibodies

A monoclonal antibody (IgM) for a cationic porphyrin has been prepared using [5-(4-carboxyphenyl)-10,15,20-tris-(4-methylpyridyl)]porphine (3MPy1C) as a hapten [31]. IgG specific for anionic porphyrin, *meso*-tetrakis(4-carboxy-phenyl)porphine (TCPP), has been prepared [32, 33]. The cationic porphyrin, 3MPy1C has been attached to the IgG via activation of carboxylic acid in 3MPy1C using the condensation agent, carbonyldiimidazole. The IgG-cationic porphyrin conjugate was purified by column chromatography using Sephadex G-150 to remove the porphyrins that did not react with the antibody.

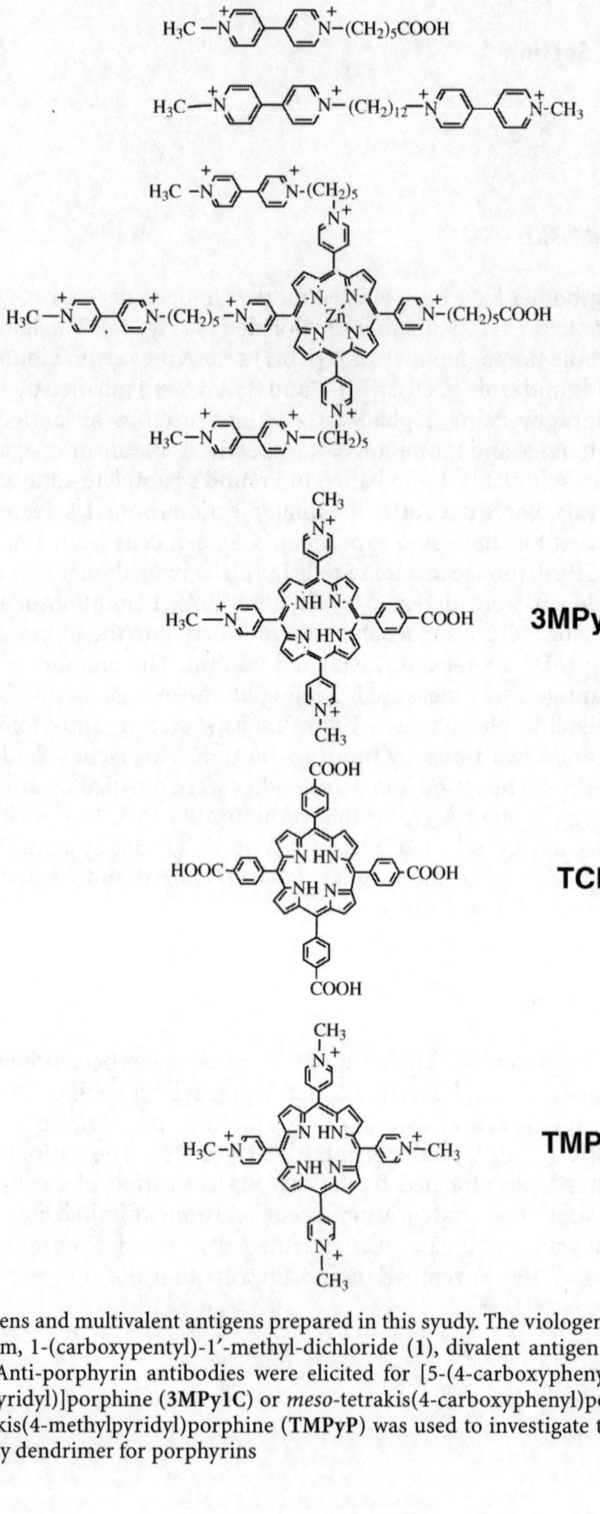

1

2

3

3MPy1C

TCPP

TMPyP

Fig. 3. Haptens and multivalent antigens prepared in this syudy. The viologen derivative, 4,4′-bipyridinium, 1-(carboxypentyl)-1′-methyl-dichloride (**1**), divalent antigen **2**, and trivalent antigen **3**. Anti-porphyrin antibodies were elicited for [5-(4-carboxyphenyl)-10,15,20-tris-(4-methylpyridyl)]porphine (**3MPy1C**) or *meso*-tetrakis(4-carboxyphenyl)porphine (TCPP). *meso*-Tetrakis(4-methylpyridyl)porphine (**TMPyP**) was used to investigate the specificity of the antibody dendrimer for porphyrins

2.2
Measurements

2.2.1
Competition ELISA

The antibody solution (1.6×10^{-8} M) and substrate solutions with various concentration from 10^{-9} M to 10^{-3} M were mixed on a BSA-coated plate. The mixed solution of antibodies and substrates was allowed to stand for 1 day at room temperature, and then transported to the ELISA plates pre-coated with BSA-hapten and BSA blocking buffer. Absorbance at 405 nm for the resulting enzymatic hydrolysis product (*p*-nitrophenolate) by alkalinephosphatase of the second antibody was recorded on an Immuno-Mini NJ-2300 to determine the amount of antibody bound to BSA–hapten.

2.2.2
Biosensor Technique

Biospecific interaction analysis (BIA) was performed by using a BIAcore X system (BIACORE, Uppsala, Sweden) based on the surface plasmon resonance. One of the anti-viologen antibodies was immobilized on the sensor chip by activating carboxymethyl groups on the surface matrix by a mixture of N-hydroxysuccinimide (NHS) and N-ethyl-N'-(dimethylaminopropyl)-carbodiimide (EDC). The response signal (RU) of the BIAcore apparatus is proportional to the change in the refractive index at the surface and is generally assumed to be proportional to the mass of substance bound to the chip. Kinetics of association to the immobilized antibody were studied by measuring the amount of bound substrate as a function of time when a solution containing viologen derivative or antibody passes over the chip surface. The dissociation kinetics were monitored by detecting the time-dependence of the mass decrease after the surface was subsequently washed with buffer.

2.2.3
AFM Measurement

A total of 2 µL of the antibodies or the antibody-supramolecules (0.5 mg mL^{-1} in 0.1 M phosphate borate buffer, pH 9.0) solutions was used for the experiment. These samples were air-dried onto the surface of freshly cleaved HOPG (highly oriented pyrolytic graphite, 1.2 cm×1.2 cm) substrate. The sample was allowed to stand for 1 day in a desiccator with $CaCl_2$ to dry the sample before it was transferred into the AFM. The sample surface was observed by contact AFM under mild conditions such that any damage caused by, for example, the preparation was minimized [34]. AFM measurements were carried out by using a Si_3N_4 tip with a radius of approximately 40 nm. All measurements were taken on a multimode Nanoscope IIIa (Digital Instruments, Santa Barbara, CA). The line scan speed was 2 Hz with 512 pixels per line. All scales were calibrated against a standard sample (5 µm×5 µm grid) and rechecked with graphite.

3
Supramolecular Formation of Antibodies with Multivalent Antigens

In this study, methyl viologen is selected as one of the target molecules to be detected. Although viologens are harmful, they are well-known functional molecules which act as herbicides and electron acceptors. Methyl viologen has been suggested as a potential etiologic factor in Parkinson's disease because of the structural similarity to the known dopaminergic neurotoxicant, 1-methyl-4-phenyl-1,2,3,6-tetrahydropyridine [35, 36]. Anti-viologen antibodies [37–40] may be expected to be useful not only as a highly sensitive reagent of viologens but also as a functional material to control the electron transfer from electron donors to viologens. It is important to use the anti-viologen antibodies for the sensitive and specific detection of viologens by the SPR biosensor because methyl viologen is a charged substrate with low molecular weight. However, the detection of viologens at low concentrations is difficult owing to the low sensitivity (small response) in a common SPR biosensor technique using corresponding antibodies. To improve the sensitivity, it is important to detect methyl viologen as a large response signal caused by the antibody bindings. A solution to this problem is thought to be the inhibition of the supramolecular formation between the anti-viologen antibody and viologen dimer by methyl viologen. We investigated the complex formation of one of the antibodies, 10D5, with methyl viologen or viologen dimer **2** and trivalent antigen **3** using the SPR biosensor.

3.1
Supramolecular Formation of Antibodies with Divalent Antigens

The specific binding of the antibody (divalent) and antigen dimer (divalent) produces the supramolecules such as linear or cyclic antigen–antibody oligomers as shown in Scheme 1 [26, 27]. Stepwise additions of antigen dimer and antibody to the sensor chip of the biosensor, whose surface is modified with the antibody, leads to the enhancement of the response signal intensity in SPR due to the formation of antigen–antibody supramolecules. When methyl viologen is added to the flow cell instead of viologen dimer, the binding sites of the

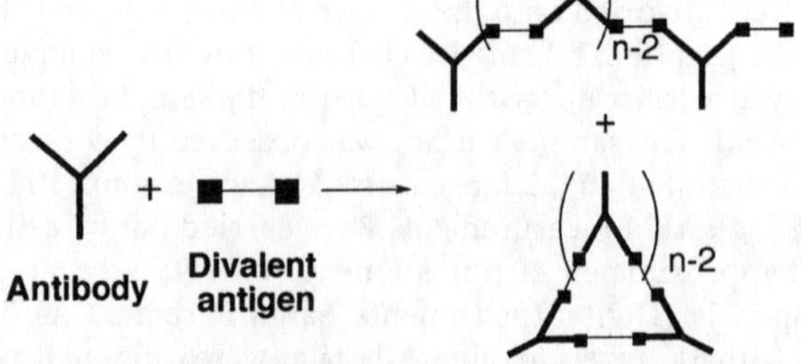

n-2

+

n-2

Antibody Divalent antigen

Scheme 1. Complex formation of antibodies (IgG) with divalent antigens

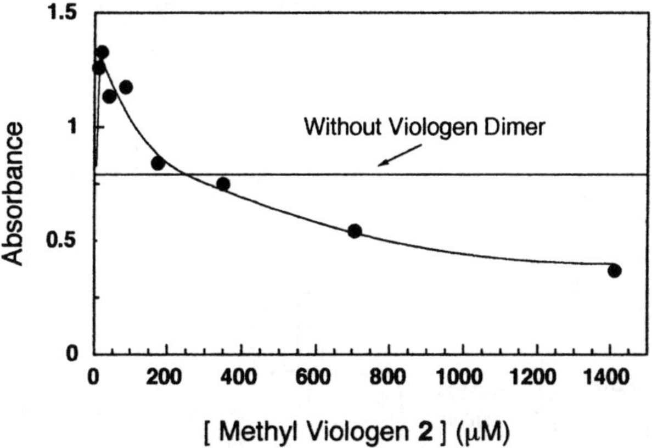

Fig. 4. Competition ELISA between antibody 10D5 and viologen dimer 2. A number of antibody molecules are immobilized on the ELISA plate at lower concentrations of viologen dimer 2

antibody at the end of the supramolecule are occupied by methyl viologen and cannot bind viologen dimer that acts as a connector of the antibodies in the supramolecular formation. The addition of methyl viologen may cause a reduction in the response signal enhancement. In this method, the binding of methyl viologen to the antibody affects the next growing step in the supramolecular formation between the antibody and viologen dimer. The amount of methyl viologen (Mw=257) is expressed as the amount of the antibody (Mw=150,000) that cannot form the supramolecules between the viologen dimer and the antibody.

The antibody 10D5 (IgG$_1$) binds hapten 1 with the dissociation constant K_d=2.0×10^{-7} M. The dissociation constant between antibody 10D5 and methyl viologen was found to be 2.0×10^{-7} M. The antibody 10D5 recognizes the bipyridinium moiety with high specificity. Figure 4 shows the result of competition ELISA between antibody 10D5 and viologen dimer 2. At low concentrations of viologen dimer, the absorbance of the product of enzymatic reactions is higher than that in the absence of viologen dimer. It is suggested that a number of antibody molecules are immobilized to the ELISA plate in the presence of viologen dimer. The signal intensity on the SPR biosensor also increased on the addition of an aqueous solution of antibody 10D5 to the sensor chip on which the viologen dimer–antibody complex was pre-coated. Figure 5 shows the sensorgram of the repeated injection of the aqueous viologen dimer and antibody solutions. The signal intensity enhanced by the binding of antibody to the viologen dimer–antibody complex. The viologen dimer molecule is considered to act as a connector between antibodies. It is suggested that the phenomena observed in the ELISA and SPR measurements are ascribable to the higher-order complex formation between antibodies and viologen dimer.

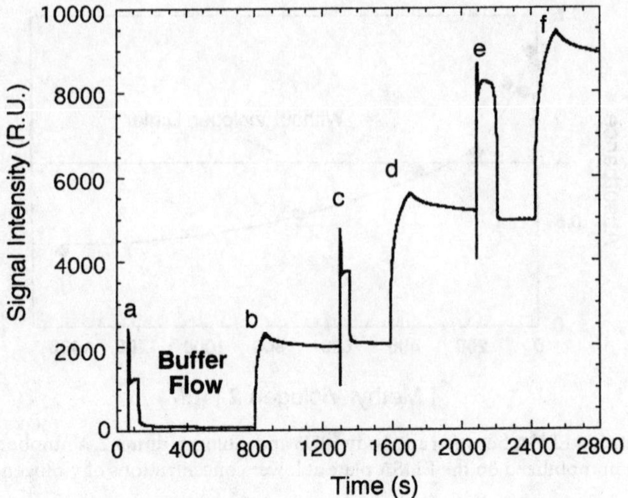

Fig. 5a–f. The sensorgram of the repeated injection of the aqueous viologen dimer 2 (**a, c, e**) and the antibody (**b, d, f**) solutions. [Viologen dimer 2]=2.0 µM and [antibody]=2.0 µM in phosphate borate buffer. Injection period 60 s for **a–c** and 120 s for **d–f**. A solution of viologen dimer 2 or the antibody passes over the surface of the sensor chip for 60 or 120 s at a constant flow rate of 20 µL min⁻¹. The surface of the sensor chip was subsequently washed with buffer

3.2
Applications for Highly-Sensitive Detection Method of Small Molecules by the Supramolecular Complexes between Antibodies and Multivalent Antigens

We found that the supramolecular formation of antibodies and viologen dimer 2 produces an SPR signal enhancement of the biosensor. The additional binding of the antibody to the viologen dimer–antibody complex gives a remarkable increase of signal intensities. On the other hand, the addition of methyl viologen (viologen monomer) instead of viologen dimer is expected to block the antigen binding sites and to inhibit the additional antibody binding. A small amount of methyl viologen can be detected as a decrease of signal enhancement due to the inhibition of the complex formation between viologen dimer and antibody by methyl viologen, compared with the signal intensity of complete supramolecular formation between the antibody and viologen dimer. Scheme 2 shows the strategy for the amplification of detection signals for methyl viologen based on the signal enhancement by the supramolecular formation between the antibody and viologen dimer on the biosensor technique. This system includes a two-step procedure as follows: (i) the injection of the aqueous solution of anti-viologen antibody with methyl viologen to the sensor chip whose surface is modified with the antibody–viologen dimer complex, and then (ii) the addition of antibody–viologen dimer (1:2) complex to the previous state. The changes of the signal intensities in the presence of methyl viologen are compared with that in the absence of methyl viologen.

The amount of antibody immobilized on the sensor chip decreased with increasing concentration of methyl viologen. To enlarge the difference in the sig-

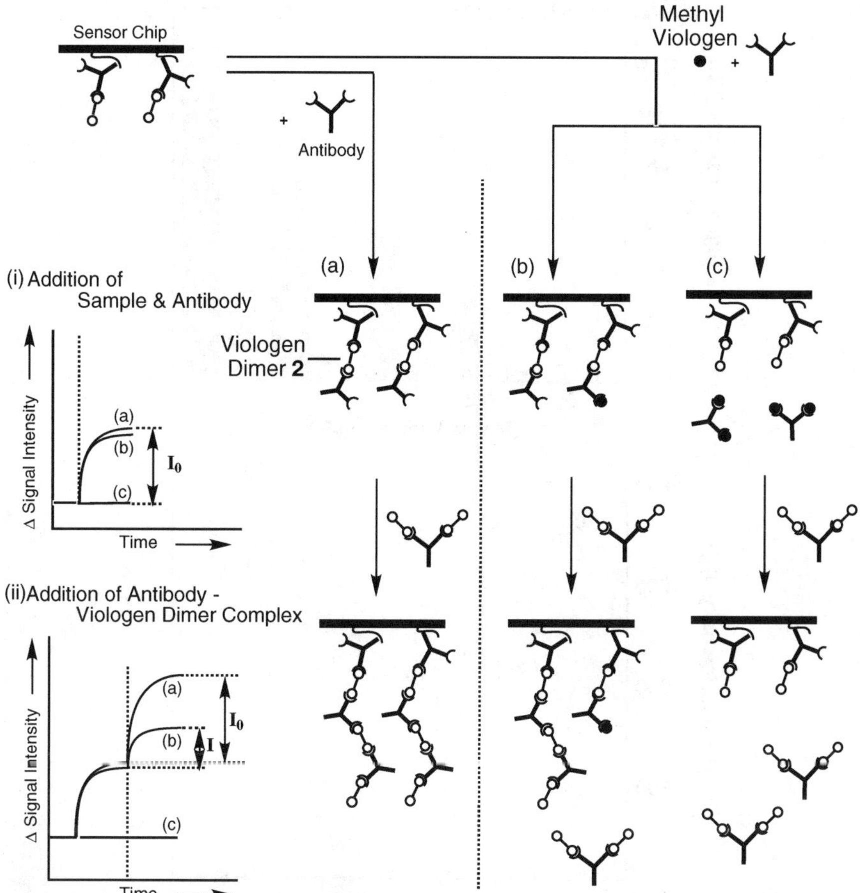

Scheme 2. Strategy for the highly sensitive detection of viologen based on the SPR biosensor technique. Inhibition of the complex formation of the antibody with viologen dimer by methyl viologen and the signal enhancement due to the supramolecular formation between the antibody and viologen dimer. Antibody 10D5 viologen dimer complex is immobilized onto the surface of the sensor chip. An aqueous solution of antibody 10D5 and a sample including methyl viologen or that without methyl viologen is injected into the flow cell (*i*) before the addition of the complex between antibody 10D5 and viologen dimer 2 (*ii*). The supramolecular formation between antibody 10D5 and viologen dimer 2 without methyl viologen (**a**), and that in the presence methyl viologen (**b**) and (**c**). [Methyl viologen] <[antibody combining site]: (**b**) and [methyl viologen] >>[antibody combining site]: (**c**)

nal intensities in the presence of a small amount of methyl viologen, a solution of the antibody–viologen dimer complex was added to the previous state (i). Figure 6a shows the differences in the response signal intensities (I_0–I) between the complete supramolecular system in the absence of methyl viologen (I_0) and that in the presence of various concentrations of methyl viologen (I) at each step. The differences in signal intensities due to the binding of the additional antibody (0.9 µM) in the presence of methyl viologen ranging the concentration

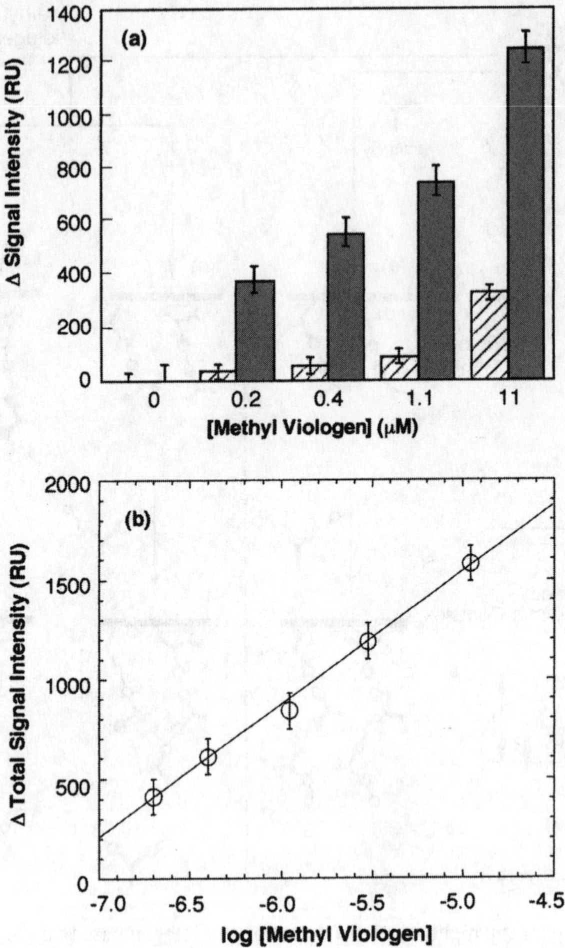

Fig. 6a,b. The differences in the response signal intensities between the complete supramolecular system in the absence of methyl viologen and that in the presence of various concentrations of methyl viologen at each step. **a** Changes in the signal intensities by the addition of an aqueous solution of antibody 10D5 and methyl viologen in the step (i) (*left side*) and those by the addition of the solution including the antibody–viologen dimer complex in the step (ii) (*right side*). **b** The relationship between total changes in the signal intensities and the concentration of methyl viologen (logarithm)

from 0.2 to 1.1 μM is slight in the step (i). However, further addition of viologen dimer **2** and antibody solutions in the step (ii) causes a clear difference in the response signal intensities in the same concentration range of methyl viologen. The total changes in the signal intensities were found to have a linear relationship with the logarithm of the concentration of methyl viologen as shown in Fig. 6b. The sensitivity in this system is 140-fold larger than that in the simple addition of methyl viologen to the antibody immobilized on the surface of the sensor chip. It is clear that this system can be utilized for the quantitative detec-

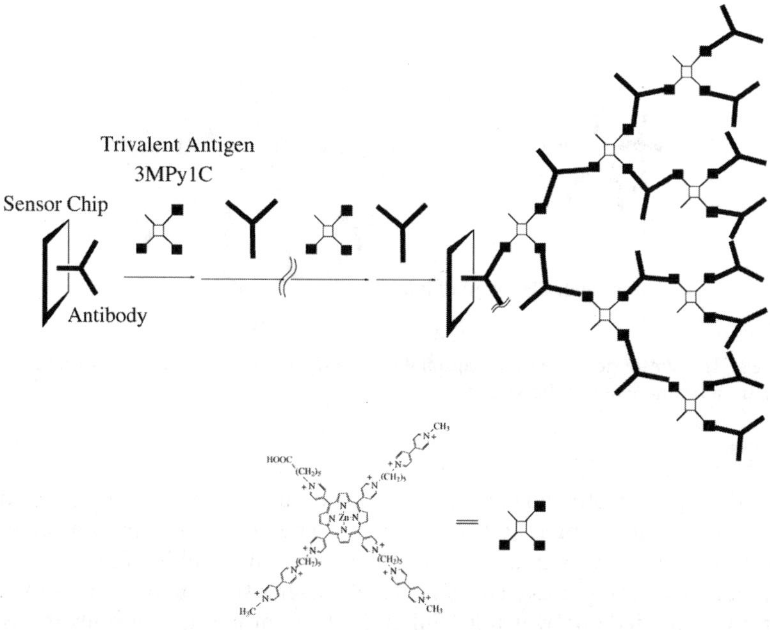

Fig. 7. A proposed structure of the complex of the antibody with the trivalent antigen immobilized on the surface of the sensor chip

tion of methyl viologen. Amplification of methyl viologen sensing processes is realized by the inhibition of complex formation between the antibody and viologen dimer–antibody complex and signal enhancement due to the supramolecular formation of the antibody and viologen dimer.

In the case of trivalent antigen **3**, the signal intensities on SPR biosensors increased by the addition of the antibody in the presence of the trivalent antigen were twice that in the presence of the divalent antigen. Figure 7 shows the schematic representation of the complex formation between the antibody and the trivalent antigen immobilized onto the surface of the sensor chip. AFM images of the complex obtained from the homogeneous solution of the mixture also suggested the presence of the network structures.

4
Dendrimers Constructed by IgM and Chemically Modified IgG

4.1
Preparation of Antibody Dendrimers and their Topological Structures

A novel antibody supramolecule is designed and prepared by using immunoglobulin M (IgM) as a core and chemically modified IgGs as branches as shown in Scheme 3. The characteristic binding ability and specificity of IgG were found to remain during the chemical modification of IgG with 3MPy1C. When IgM for

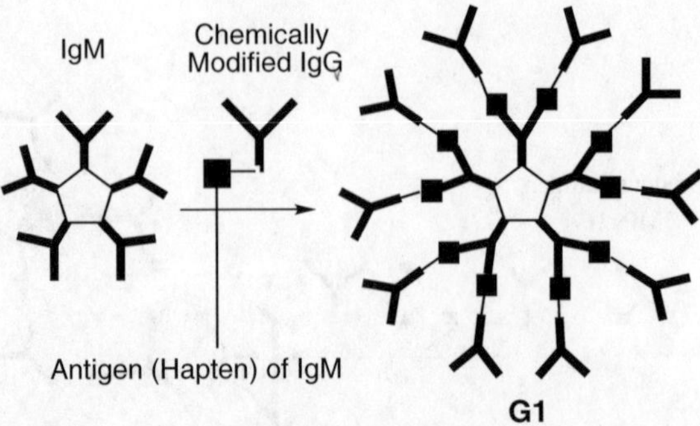

Scheme 3. The synthetic route of the complete antibody dendrimers. An ideal structure of the dendritic supramolecule is shown as **G1**

3MPy1C is treated with IgG covalently bound cationic porphyrin, IgM binds the cationic porphyrin attached on the IgG to give a dendritic antibody supramolecule "antibody dendrimer" (**G1** in Scheme 3). The dendritic supramolecules are composed of proteins with molecular weight of about 2 million and constructed from non-covalent bonds. The structural observation of the antibody dendrimer was carried out by using AFM. At first, an individual antibody molecule, IgG or IgM, was observed by AFM at room temperature depositing from a solution onto a freshly cleaved HOPG surface. Figure 8 shows AFM images of the IgG molecule whose molecular weight is 150,000. Characteristic T or Y shape molecules can be seen [41]. The overall lateral dimension is 40–50 nm as shown A, B, and C in Fig. 8, which is in good agreement with the expected values [42]. Although these molecules are monoclonal antibodies, each antibody molecular image is not identical. There are two fragments with the same height and one fragment with different height with the other two fragments in each molecular image. Fab fragments and an Fc fragment can be differentiated by comparison of the height among three fragments. The angles between the Fab fractions of each immunoglobulin are different. Each antibody molecules takes a somewhat

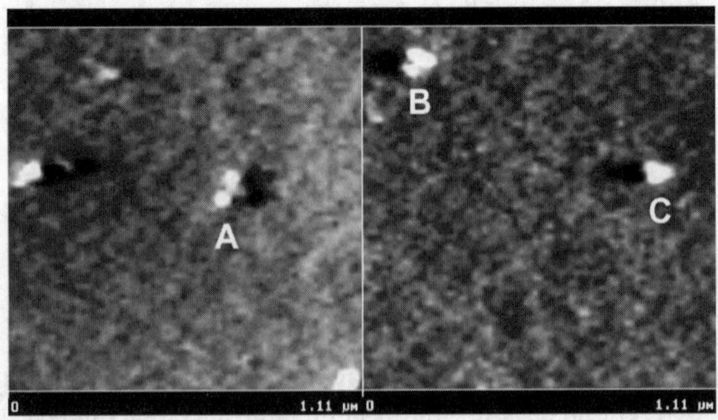

Fig. 8. AFM images of the individual antibody (IgG) molecules

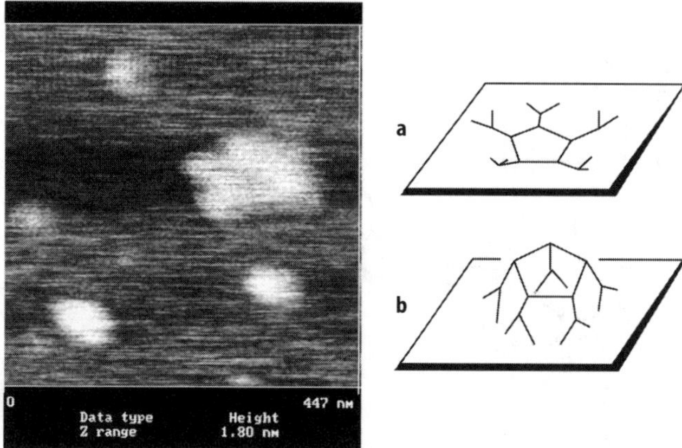

Fig. 9a,b. The molecular images of the monoclonal IgM on a cleaved highly orientated pyrolytic graphite (HOPG) surface and schematic representation of the images for IgM. A flat pentagram (**a**) and a smaller object with higher center (**b**)

different shape. This is probably due to the flexibility of its hinge region. Figure 9 shows IgM images. Two kinds of images are evident, although the object is a single species. The bigger one is a flat pentamer and the smaller one has a higher center. The results are consistent with those obtained by the cryo-AFM measurement [43, 44]. The sample surface was observed under conditions such that any damage caused by scanning the cantilever is minimized and that any nonspecific assembly among antibodies does not occur. Figure 10 shows an AFM image of the dendrimer. The image of the dendrimer was twice as large as that of starting IgM. Some branches (IgGs) can be seen outside of the IgM core. Such an assembled structure was not observed in a chemically modified IgG solution or an IgM solution alone.

The similar antibody dendrimers are prepared by the other route as shown in Scheme 4. The building block used as a branch was prepared by the specific

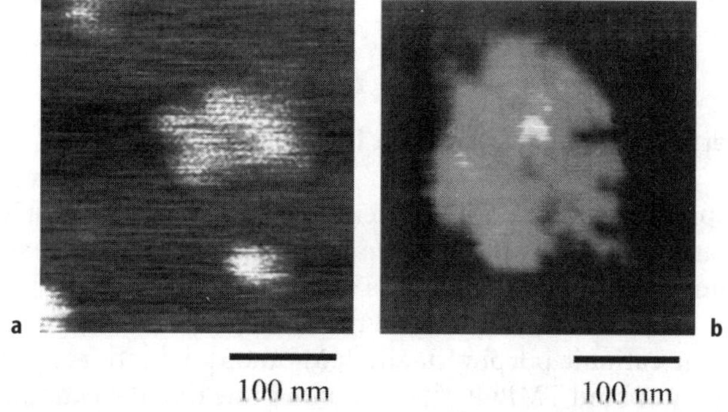

Fig. 10a,b. The AFM images of IgM (**a**) and the antibody dendrimer (**b**). A total of 2 µL of solutions of the antibodies (3.0×10⁻⁹ M) in 0.1 M phosphate borate buffer (pH 9.0) was deposited onto the surface of highly oriented pyrolytic graphite (HOPG) and air-dried

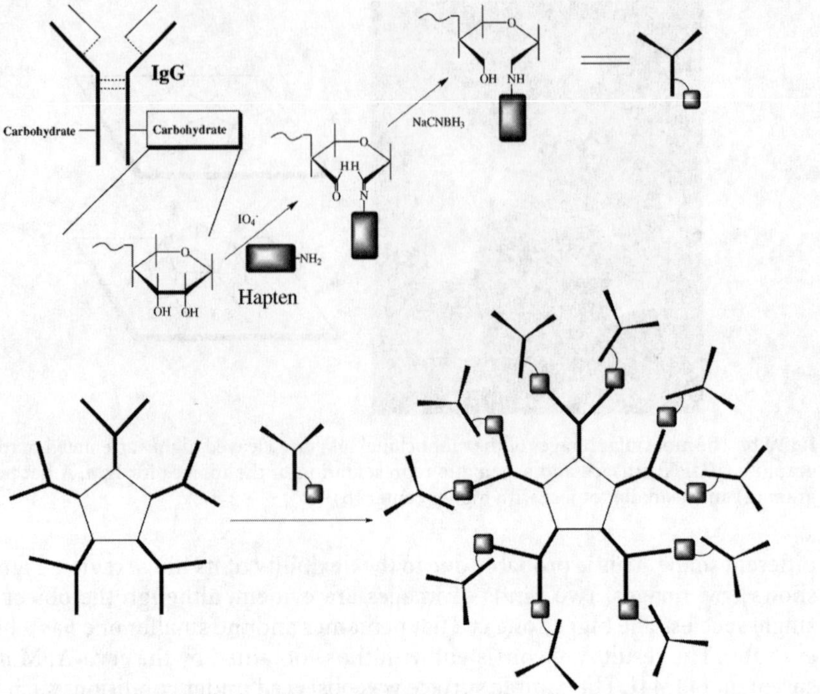

Scheme 4. Antibody dendrimers prepared by the combination of IgM with IgG attached hapten to the carbohydrate moiety in the Fc fragment

modification of a hapten to the carbohydrate moiety in the Fc fragment of the antibody molecule. Site-specific modification of antibody molecule makes it possible to prepare a homogeneous building block rather than amine coupling of the hapten to the antibody molecule. The combination of IgM and IgGs attached hapten to the carbohydrate part leads to the construction of a complete radial structure that possess antigen binding sites on the surface of the antibody dendrimer.

4.2
Binding Properties of Antibody Dendrimers for Antigens

The binding property of the antibody dendrimer (**G1**) with a cationic or anionic porphyrin was measured by the enzyme-linked immunosorbent assay (ELISA). Figure 11a shows the binding properties of the IgG, IgM, and **G1** with the cationic porphyrin, *meso*-tetrakis(4-methylpyridyl)porphine (TMPyP). Although IgG did not bind the cationic porphyrin and IgM bound the cationic porphyrin, the dendrimer did not bind TMPyP. These results show that the cationic porphyrin attached to IgG occupies the binding sites of IgM in the dendrimer, thus there are no free binding sites against TMPyP on IgM. Figure 11b shows the binding of IgG, IgM, and **G1** to TCPP. The IgM used in this study can bind both anionic

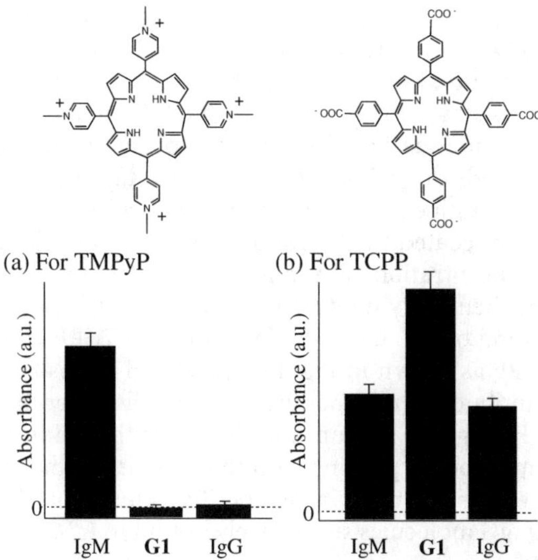

Fig. 11a,b. Binding affinities of IgG, IgM, and the antibody dendrimer (**G1**) with the cationic porphyrin (TMPyP) (**a**) and those with the anionic porphyrin (TCPP) (**b**) estimated by ELISA

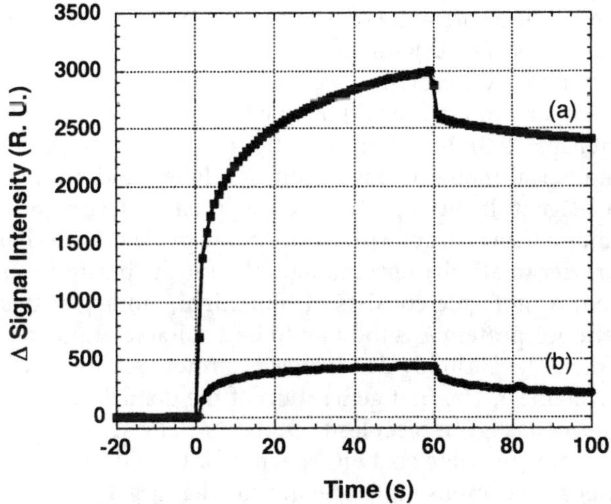

Fig. 12a,b. The sensorgrams for the binding of the antibody dendrimer (**a**) or IgG (**b**) to the anionic porphyrin immobilized onto the surface of the sensor chip. Phosphate borate buffer (0.1 M, pH 9.0) was used. TCPP was immobilized via hexamethylenediamine spacer onto the sensor chip and then a solution of IgG or the dendrimer was injected to the flow cell. After 60 s from the injection of the antibody solutions, flow cell was filled with buffer

and cationic porphyrins, due to the low specificity of IgM against porphyrins. Both IgM and IgG bound TCPP, while **G1** bound TCPP more efficiently than IgM or IgG. The increase in affinity of **G1** for the anionic porphyrin indicates that many IgG molecules attach to the surface of the IgM molecule.

The biosensor technique based on surface plasmon resonance (SPR) [19] shows that the antibody dendrimer allows an advantageous amplification of the detection signals for antigens. A solution of **G1** was added to the sensor chip on which TCPP was pre-coated by the coupling with hexamethylenediamine as a spacer. The total concentration of the antibody was fixed at 0.2 mg mL^{-1} (0.1 mg of IgM+0.1 mg of chemically modified IgG in 1 ml buffer for **G1**). The sensorgram for the binding of the antibody dendrimer to TCPP was compared with that of IgG to TCPP as shown in Fig. 12. The signal intensity increased by the injection of the antibody dendrimer was sufficiently larger than that of simple addition of IgG. Taking into account the change of the binding property of the antibody dendrimer for porphyrins with the increase in the amount of bound antibody to the anionic porphyrin on the SPR biosensor, the antibody dendrimer has many IgG molecules successively bound to IgM molecule.

5
Other Methods for the Construction of Antibody Dendrimers

Antibody dendrimers were prepared by using proteins with high affinity for the Fc fragment in IgG as a core and IgG as a branch. In this procedure, an IgM-like complex can be constructed without any modification of IgG. Staphylococcal protein A or streptococcal protein G were examined as a core component of the antibody dendrimer. Protein A or G plays an important role in molecular biology owing to its specific interaction with the Fc portion of immunoglobulins. Many immunological methods have been developed and refined using these proteins as a reagent, including immunoprecipitation techniques and double sandwich immunoassays. In addition, solid-phase protein A or G has been widely used as a carrier in affinity chromatography for the purification of antibodies. The protein A molecule consists of four highly homogeneous Fc-binding domains. Therefore, protein A is thought to be a suitable component of the core of the dendrimer. The antibody dendrimer of protein A with IgG was prepared according to Scheme 5. The first generation of the dendrimer was prepared by the addition of mouse IgG molecules to protein A. The second-generation dendrimer was prepared by using goat IgG specific for Fc of the mouse IgG antibody developed in goat and mouse IgG specific for the target molecule. Each dendrimer was applied for the detection of the antigens on SPR biosensors. The signal intensities of the binding of the antigen immobilized on the surface of the sensor chip increased in proportion to the number of the generation of the dendrimer as shown in Fig. 13.

The immobilization of the hyperbranched spherical structures onto physical transducers greatly increases the binding capacity of the surface and leads to enhanced sensitivity and extended linearity of biosensors. Nucleic acid dendrimers were prepared and their amplification properties for the detection of DNA were examined using mass-sensitive transducers [45, 46]. Antibodies

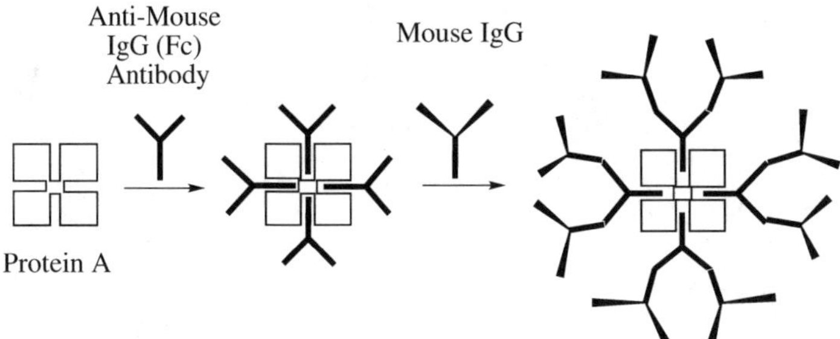

Scheme 5. Preparation of the dendritic supramolecules with protein A as a core and IgGs as branches

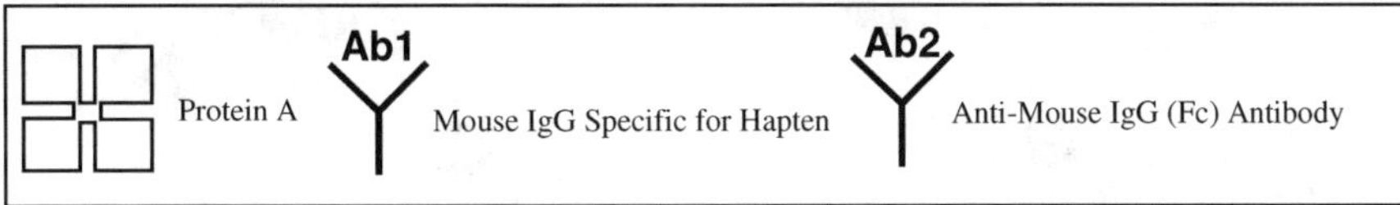

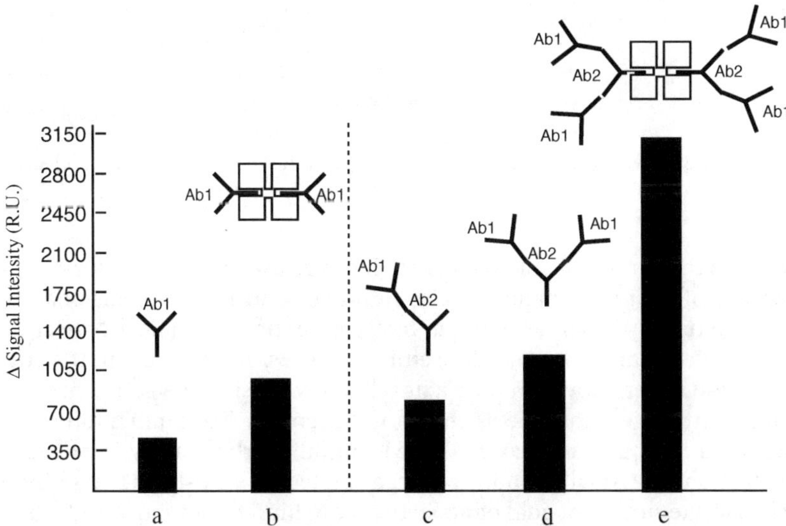

Fig. 13a–e. The increase of the signal intensities by the addition of the dendritic complexes composed of IgGs and protein A. The hapten was immobilized to the surface of the SPR sensor chip. The increase of the signal intensities on the complex formation of hapten with the antibodies were monitored. The addition of mouse IgG specific for hapten (**Ab1**) (**a**), the complex of the **Ab1** with protein A (**b**), one to one complex of **Ab1** with anti-mouse IgG (Fc) antibody (**Ab2**) (**c**), two to one complex of **Ab1** with **Ab2** (**d**), and two to one complex of **Ab1** with **Ab2** in the presence of protein A (**e**)

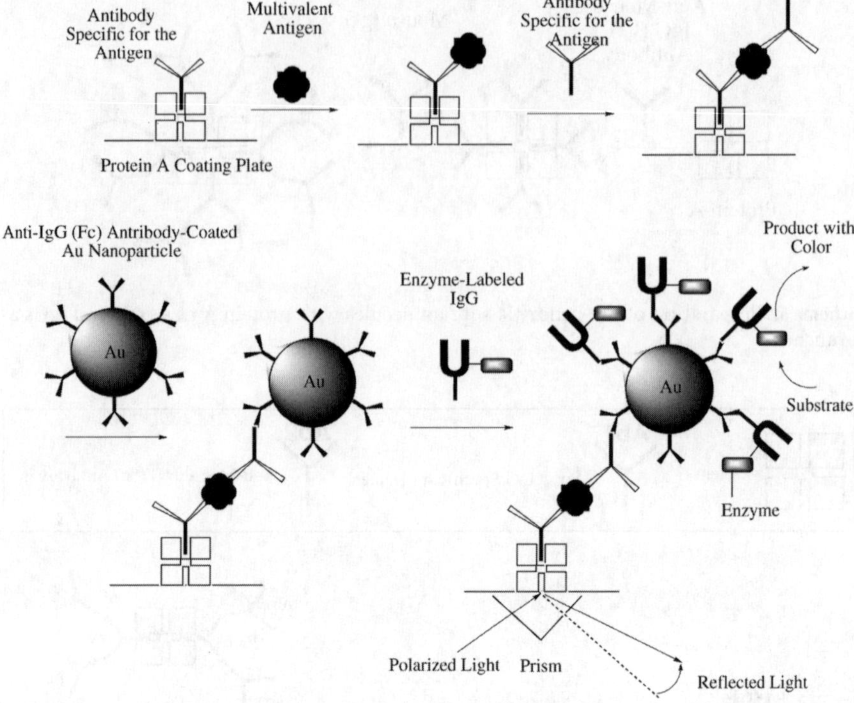

Scheme 6. A highly sensitive detection method using an antibody specific for IgG (Fc) and gold nanoparticle

attached to the surface of gold nanoparticles were used for the organization and patterning of inorganic nanoparticles into two- and three-dimensional functional structures [47]. These materials have a potential as chemical, optical, magnetic, and electronic devices with useful properties. Taking account the advantage of these materials, we propose a novel highly sensitive detection method of a multivalent target molecule as shown in Scheme 6. The combination of SPR biosensor technique and enzyme-linked immunosorbent assay is utilized for the detection of a small amount of target molecule. Anti-IgG (Fc) antibodies attached to the surface of gold nanoparticles can bind the antibody (IgG) bound to the target molecule immobilized onto the surface of the sensor chip and the antibodies labeled with enzymes. The detection signals are expressed as a color of the enzymatic reaction products and also as an enhancement of the SPR.

6
Conclusions

An enhancement of SPR signal intensity was observed by the addition of the antibody to the divalent antigen–antibody complex immobilized onto the surface of the sensor chip, indicating the formation of linear supramolecules. An amplification method of the detection signals for a target molecule has been devised by using the signal enhancement in the supramolecular assembly of anti-viologen antibodies and divalent antigens. Target substrate added to the flow cell of SPR can be detected quantitatively by monitoring the total amount of the antibody bound to the surface of the sensor chip. The sensitivity of this system was found to be two orders larger than that of the simple addition of target substrate to the antibody immobilized on the surface of the sensor chip. This method can be potentially applied to many compounds with a highly sensitivity and specificity by using corresponding antibodies and dimers of the target molecule (divalent antigen). A trivalent antigen formed a dendritic network structure and enhanced the response signal intensities on the SPR biosensor.

New antibody dendrimers were designed and prepared by the combination of IgG and IgM, that is, using IgM as a core and IgG as branches. Many binding sites of IgG were arranged radially on the surface of one object; the resulting artificial antibodies bound antigens more selectively than IgM and more strongly than IgG. The characteristic features of the antibody dendrimer are: (i) they are composed of proteins, (ii) have a large size with molecular weight of about 2 million, (iii) are constructed from non-covalent bonds, and (iv) bind antigen strongly with high specificity. The antibody dendrimer will be used as functionalized materials for sensitive detection of many kinds of chemicals, for diagnosis, and for drug delivery systems. The other antibody dendrimers are prepared by the specific modification of a hapten to the carbohydrate of antibody molecules. Antibody dendrimers consisting of protein A or G and IgG are also prepared. They are found to be useful as a functional material for the amplification of detection signals of target molecules in SPR sensors.

7
References

1. Kohler G, Milstein C (1975) Nature 256:495
2. Terry WD, Matthews BW, Davies DR (1968) Nature 220:239
3. Silverton EW, Navia MA, Davies DR (1977) Proc Natl Acad Sci USA 74:5140
4. Tomalia DA, Naylor AM, Goggard III WA (1990) Angew Chem Int Ed Engl 29:138
5. McDonnell JM (2001) Curr Opin Chem Biol 5:572
6. Scheller FW, Wollenberger U, Warsinke A, Lisdat F (2001) Curr Opin Biotechnol 12:35
7. Borrebaeck CAK (2000) Immunol Today 21:379
8. Stoll D, Templin MF, Schrenk M, Traub PC, Vöhringer CF, Joos TO (2002) Front Biosci 7:C13
9. Kaneda Y (2000) Adv Drug Deliv Rev 43:197
10. Garnett MC (2001) Adv Drug Deliv Rev 53:171
11. Golinelli-Pimpaneau B (2000) Curr Opin Struct Biol 10:697
12. Feiters MC, Rowan AE, Nolte RJM (2000) Chem Soc Rev 29:375
13. Lu Y, Berry SM, Pfister T D (2001) Chem Rev 101:3047
14. Shenton W, Davies SA, Mann S (1999) Adv Mater 11:449
15. Harada A, Yamaguchi H, Okamoto K, Fukushima H, Shiotsuki K, Kamachi M (1999) Photochem Photobiol 70:298
16. Yamaguchi H, Kamachi M, Harada A (2000) Angew Chem Int Ed 39:3829
17. Yamaguchi H, Harada A, Kamachi M (1998) Reactive Functional Polym 37:245
18. Löfas S, Johnsson B (1990) J Chem Soc Chem Commun 1526
19. Jönsson U, Fägerstam L, Ivarsson B, Johnsson B, Karlsson R, Lundh K, Löfas S, Persson B, Roos H, Ronnberg I, Sjölander S, Stenberg E, Stahlberg R, Urbaniczky C, Östlin H, Malmqvist M (1990) Biotechniques 11:620

20. Karlsson R, Michaelsson A, Mattsson L (1991) J Immunol Methods 145:229
21. Chaiken I, Rose S, Karlsson R (1992) Anal Chem 201:197
22. Malmqvist M (1993) Nature 361:186
23. Niikura K, Nagata K, Okahata Y (1996) Chem Lett 863
24. Bardea A, Dagan A, Ben-Dov I, Amit B, Willner I (1998) Chem Commun 839
25. Patolsky F, Lichtenstein A, Willner I (2000) J Am Chem Soc 122:418
26. Yamaguchi H, Oka F, Kamachi M, Harada A (1999) Kobunshi Ronbunsyu 56:660
27. Harada A, Yamaguchi H, Oka F, Kamachi M (1999) Proc SPIE Int Soc Opt Eng 3607:126
28. Yamaguchi H, Harada A (2002) Chem Lett 382
29. Yamaguchi H, Harada A (2002) Biomacromolecules 3:1163
30. Harada A, Okamoto K, Kamachi M, Honda T, Miwatani T (1990) Chem Lett 917
31. Harada A, Yamaguchi H, Tsubouchi K, Horita E (2003) Chem Lett 32:18
32. Harada A, Shiotsuki K, Fukushima H, Yamaguchi H, Kamachi M (1995) Inorg Chem 34:1070
33. Harada A, Fukushima H, Shiotsuki K, Yamaguchi H, Oka F, Kamachi M (1997) Inorg Chem 36:6099
34. Harada A, Yamaguchi H, Kamachi M (1997) Chem Lett 1141
35. Liou HH, Tsai MC, Chen CJ, Jeng JS, Chang YC, Chen SY, Chen RC (1997) Neurology 48: 1583
36. Brooks AI, Chadwick CA, Gelbard HA, Cory-Slechta DA, Federoff HJ (1999) Brain Res 823:1
37. Niewola Z, Hayward C, Symington BA, Robson RT (1985) Clinica Chimica Acta 148:149
38. Hogg PJ, Johnston SC, Bowles MR, Pond SM, Winzor DJ (1987) Molec Immun 24:797
39. Bowles MR, Pond SM (1990) Mol Immunol 27:847
40. Bowles MR, Hall DR, Pond SM, Winzor DJ (1997) Anal Biochem 244:133
41. Leatherbarrow RJ, Stedman M, Wells TNC (1991) J Mol Biol 221:361
42. Mazeran PE, Loubet JL, Martelet C, Theretz A (1995) Ultramicroscopy 60:33
43. Han W, Mou J, Sheng J, Yang J, Shao Z (1994) Biochemistry 34:8215
44. Zhang Y, Sheng S, Shao Z (1996) Biophys J 71:2168
45. Wang J, Jiang M, Nilsen TW, Getts RC (1998) J Am Chem Soc 120:8281
46. Esch MB, Baeumner AJ, Durst RA (2001) Anal Chem 73:3162
47. Shenton BW, Davis SA, Mann S (1999) Adv Mater 11:449

Author Index Volume 201–228

Hentze H-P, Co CC, McKelvey CA, Kaler EW (2003) Templating Vesicles, Microemulsions and Lyotropic Mesophases by Organic Polymerization Processes. *226*:197–223

Hergenrother PJ, Martin SF (2000) Phosphatidylcholine-Preferring Phospholipase C from *B. cereus*. Function, Structure, and Mechanism. *211*:131–167

Hermann C, see Kuhlmann J (2000) *211*:61–116

Heydt H (2003) The Fascinating Chemistry of Triphosphabenzenes and Valence Isomers. *223*:215–249

Hirsch A, Vostrowsky O (2001) Dendrimers with Carbon Rich-Cores. *217*:51–93

Hiyama T, Shirakawa E (2002) Organosilicon Compounds. *219*:61–85

Holmberg K, see Häger M (2003) *227*:53–74

Houseman BT, Mrksich M (2002) Model Systems for Studying Polyvalent Carbohydrate Binding Interactions. *218*:1–44

Hricovíniová Z, see Petruš L (2001) *215*:15–41

Idee J-M, Tichkowsky I, Port M, Petta M, Le Lem G, Le Greneur S, Meyer D, Corot C (2002) Iodiated Contrast Media: from Non-Specific to Blood-Pool Agents. *222*:151–171

Igau A, see Majoral J-P (2002) *220*:53–77

Iwaoka M, Tomoda S (2000) Nucleophilic Selenium. *208*:55–80

Iwasawa N, Narasaka K (2000) Transition Metal Promated Ring Expansion of Alkynyl- and Propadienylcyclopropanes. *207*:69–88

Imperiali B, McDonnell KA, Shogren-Knaak M (1999) Design and Construction of Novel Peptides and Proteins by Tailored Incorparation of Coenzyme Functionality. *202*:1–38

Ito S, see Yoshifuji M (2003) *223*:67–89

Jacques V, Desreux JF (2002) New Classes of MRI Contrast Agents. *221*:123–164

James TD, Shinkai S (2002) Artificial Receptors as Chemosensors for Carbohydrates. *218*:159–200

Jenne A, see Famulok M (1999) *202*:101–131

Junker T, see Trauger SA (2003) *225*:257–274

Kaler EW, see Hentze H-P (2003) *226*:197–223

Kamerling JP, see Haseley SR (2002) *218*:93–114

Kashemirov BA, see Mc Kenna CE (2002) *220*:201–238

Kato S, see Murai T (2000) *208*:177–199

Kawa M (2003) Antenna Effects of Aromatic Dendrons and Their Luminescene Applications. *228*:193–204

Kee TP, Nixon TD (2003) The Asymmetric Phospho-Aldol Reaction. Past, Present, and Future. *223*:45–65

Khlebnikov AF, see de Meijere A (2000) *207*:89–147

Kim K, see Lee JW (2003) *228*:111–140

Kirtman B (1999) Local Space Approximation Methods for Correlated Electronic Structure Calculations in Large Delocalized Systems that are Locally Perturbed. *203*:147–166

Kita Y, see Tohma H (2003) *224*:209–248

Kleij AW, see Kreiter R (2001) *217*:163–199

Klein Gebbink RJM, see Kreiter R (2001) *217*:163–199

Klibanov AL (2002) Ultrasound Contrast Agents: Development of the Field and Current Status. *222*:73–106

Klopper W, Kutzelnigg W, Müller H, Noga J, Vogtner S (1999) Extremal Electron Pairs – Application to Electron Correlation, Especially the R12 Method. *203*:21–42

Knochel P, see Betzemeier B (1999) *206*:61–78

Koser GF (2003) C-Heteroatom-Bond Forming Reactions. *224*:137–172

Koser GF (2003) Heteroatom-Heteroatom-Bond Forming Reactions. *224*:173–183

Kosugi M, see Fugami K (2002) *219*:87–130

Kozhushkov SI, see de Meijere A (1999) *201*:1–42

Kozhushkov SI, see de Meijere A (2000) *207*:89–147

Kozhushkov SI, see de Meijere A (2000) *207*:149–227

Krause W (2002) Liver-Specific X-Ray Contrast Agents. *222*:173–200

Ublacker GA, see Maul JJ (1999) *206*:79–105
Uemura S, see Nishibayashi Y (2000) *208*:201–233
Uemura S, see Nishibayashi Y (2000) *208*:235–255
Uggerud E (2003) Physical Organic Chemistry of the Gas Phase. Reactivity Trends for Organic Cations. *225*:1–34
Valdemoro C (1999) Electron Correlation and Reduced Density Matrices. *203*:187–200
Valério C, see Astruc D (2000) *210*:229–259
van Benthem RATM, see Muscat D (2001) *212*:41–80
van Koten G, see Kreiter R (2001) *217*:163–199
van Manen H-J, van Veggel FCJM, Reinhoudt DN (2001) Non-Covalent Synthesis of Metallo-dendrimers. *217*:121–162
van Veggel FCJM, see van Manen H-J (2001) *217*:121–162
Varvoglis A (2003) Preparation of Hypervalent Iodine Compounds. *224*:69–98
Verkade JG (2003) P(RNCH$_2$CH$_2$)$_3$N: Very Strong Non-ionic Bases Useful in Organic Synthesis. *223*:1–44
Vicinelli V, see Balzani V (2003) *228*:159–191
Vliegenthart JFG, see Haseley SR (2002) *218*:93–114
Vogler A, Kunkely H (2001) Luminescent Metal Complexes: Diversity of Excited States. *213*:143–182
Vogtner S, see Klopper W (1999) *203*:21–42
Vostrowsky O, see Hirsch A (2001) *217*:51–93
Waldmann H, Thutewohl M (2000) Ras-Farnesyltransferase-Inhibitors as Promising Anti-Tumor Drugs. *211*:117–130
Wang G-X, see Chow H-F (2001) *217*:1–50
Weil T, see Wiesler U-M (2001) *212*:1–40
Werschkun B, Thiem J (2001) Claisen Rearrangements in Carbohydrate Chemistry. *215*:293–325
Wiesler U-M, Weil T, Müllen K (2001) Nanosized Polyphenylene Dendrimers. *212*:1–40
Williams RM, Stocking EM, Sanz-Cervera JF (2000) Biosynthesis of Prenylated Alkaloids Derived from Tryptophan. *209*:97–173
Wirth T (2000) Introduction and General Aspects. *208*:1–5
Wirth T (2003) Introduction and General Aspects. *224*:1–4
Wirth T (2003) Oxidations and Rearrangements. *224*:185–208
Wrodnigg TM, Eder B (2001) The Amadori and Heyns Rearrangements: Landmarks in the History of Carbohydrate Chemistry or Unrecognized Synthetic Opportunities? *215*:115–175
Wyttenbach T, Bowers MT (2003) Gas-Phase Confirmations: The Ion Mobility/Ion Chromatography Method. *225*:201–226
Yamaguchi H, Harada A (2003) Antibody Dendrimers. *228*:237–258
Yersin H, Donges D (2001) Low-Lying Electronic States and Photophysical Properties of Organometallic Pd(II) and Pt(II) Compounds. Modern Research Trends Presented in Detailed Case Studies. *214*:81–186
Yeung LK, see Crooks RM (2001) *212*:81–135
Yokoyama S, Otomo A, Nakahama T, Okuno Y, Mashiko S (2003) Dendrimers for Optoelectronic Applications. *228*:205–226
Yoshifuji M, Ito S (2003) Chemistry of Phosphanylidene Carbenoids. *223*:67–89
Zablocka M, see Majoral J-P (2002) *220*:53–77
Zhang J, see Chow H-F (2001) *217*:1–50
Zhdankin VV (2003) C-C Bond Forming Reactions. *224*:99–136
Zhao M, see Crooks RM (2001) *212*:81–135
Zimmermann SC, Lawless LJ (2001) Supramolecular Chemistry of Dendrimers. *217*:95–120
Zwanenburg B, ten Holte P (2001) The Synthetic Potential of Three-Membered Ring Aza-Heterocycles. *216*:93–124

Subject Index

Printing: Saladruck Berlin
Binding: Stürtz AG, Würzburg